All the titles listed below can be obtained from good booksellers or from Cambridge University Press. For a complete series listing visit www.cambridge.org/mathematics.

119 M. Deza and M. Dutour Sikirić *Geometry of Chemical Graphs*
120 T. Nishiura *Absolute Measurable Spaces*
121 M. Prest *Purity, Spectra and Localisation*
122 S. Khrushchev *Orthogonal Polynomials and Continued Fractions*
123 H. Nagamochi and T. Ibaraki *Algorithmic Aspects of Graph Connectivity*
124 F. W. King *Hilbert Transforms I*
125 F. W. King *Hilbert Transforms II*
126 O. Calin and D.-C. Chang *Sub-Riemannian Geometry*
127 M. Grabisch *et al. Aggregation Functions*
128 L. W. Beineke and R. J. Wilson (eds.) with J. L. Gross and T. W. Tucker *Topics in Topological Graph Theory*
129 J. Berstel, D. Perrin and C. Reutenauer *Codes and Automata*
130 T. G. Faticoni *Modules over Endomorphism Rings*
131 H. Morimoto *Stochastic Control and Mathematical Modeling*
132 G. Schmidt *Relational Mathematics*
133 P. Kornerup and D. W. Matula *Finite Precision Number Systems and Arithmetic*
134 Y. Crama and P. L. Hammer (eds.) *Boolean Models and Methods in Mathematics, Computer Science, and Engineering*
135 V. Berthé and M. Rigo (eds.) *Combinatorics, Automata and Number Theory*
136 A. Kristály, V. D. Rădulescu and C. Varga *Variational Principles in Mathematical Physics, Geometry, and Economics*
137 J. Berstel and C. Reutenauer *Noncommutative Rational Series with Applications*
138 B. Courcelle and J. Engelfriet *Graph Structure and Monadic Second-Order Logic*
139 M. Fiedler *Matrices and Graphs in Geometry*
140 N. Vakil *Real Analysis through Modern Infinitesimals*
141 R. B. Paris *Hadamard Expansions and Hyperasymptotic Evaluation*
142 Y. Crama and P. L. Hammer *Boolean Functions*
143 A. Arapostathis, V. S. Borkar and M. K. Ghosh *Ergodic Control of Diffusion Processes*
144 N. Caspard, B. Leclerc and B. Monjardet *Finite Ordered Sets*
145 D. Z. Arov and H. Dym *Bitangential Direct and Inverse Problems for Systems of Integral and Differential Equations*
146 G. Dassios *Ellipsoidal Harmonics*
147 L. W. Beineke and R. J. Wilson (eds.) with O. R. Oellermann *Topics in Structural Graph Theory*
148 L. Berlyand, A. G. Kolpakov and A. Novikov *Introduction to the Network Approximation Method for Materials Modeling*
149 M. Baake and U. Grimm *Aperiodic Order I: A Mathematical Invitation*
150 J. Borwein *et al. Lattice Sums Then and Now*
151 R. Schneider *Convex Bodies: The Brunn–Minkowski Theory (Second Edition)*
152 G. Da Prato and J. Zabczyk *Stochastic Equations in Infinite Dimensions (Second Edition)*
153 D. Hofmann, G. J. Seal and W. Tholen (eds.) *Monoidal Topology*
154 M. Cabrera García and Á. Rodríguez Palacios *Non-Associative Normed Algebras I: The Vidav–Palmer and Gelfand–Naimark Theorems*
155 C. F. Dunkl and Y. Xu *Orthogonal Polynomials of Several Variables (Second Edition)*
156 L. W. Beineke and R. J. Wilson (eds.) with B. Toft *Topics in Chromatic Graph Theory*
157 T. Mora *Solving Polynomial Equation Systems III: Algebraic Solving*
158 T. Mora *Solving Polynomial Equation Systems IV: Buchberger Theory and Beyond*
159 V. Berthé and M. Rigo (eds.) *Combinatorics, Words and Symbolic Dynamics*
160 B. Rubin *Introduction to Radon Transforms: With Elements of Fractional Calculus and Harmonic Analysis*
161 M. Ghergu and S. D. Taliaferro *Isolated Singularities in Partial Differential Inequalities*
162 G. Molica Bisci, V. Radulescu and R. Servadei *Variational Methods for Nonlocal Fractional Problems*
163 S. Wagon *The Banach–Tarski Paradox (Second Edition)*
164 K. Broughan *Equivalents of the Riemann Hypothesis I: Arithmetic Equivalents*
165 K. Broughan *Equivalents of the Riemann Hypothesis II: Analytic Equivalents*
166 M. Baake and U. Grimm *Aperiodic Order II: Representation Theory and the Zelmanov Approach*

EQUIVALENTS OF THE RIEMANN HYPOTHESIS

Volume One: Arithmetic Equivalents

The Riemann hypothesis (RH) is perhaps the most important outstanding problem in mathematics. This two-volume text presents the main known equivalents to RH using analytic and computational methods. The books are gentle on the reader with definitions repeated, proofs split into logical sections, and graphical descriptions of the relations between different results. They also include extensive tables, supplementary computational tools, and open problems suitable for research. Accompanying software is free to download.

These books will interest mathematicians who wish to update their knowledge, graduate and senior undergraduate students seeking accessible research problems in number theory, and others who want to explore and extend results computationally. Each volume can be read independently.

Volume 1 presents classical and modern arithmetic equivalents to RH, with some analytic methods. Volume 2 covers equivalences with a strong analytic orientation, supported by an extensive set of appendices containing fully developed proofs.

Encyclopedia of Mathematics and Its Applications

This series is devoted to significant topics or themes that have wide application in mathematics or mathematical science and for which a detailed development of the abstract theory is less important than a thorough and concrete exploration of the implications and applications.

Books in the **Encyclopedia of Mathematics and Its Applications** cover their subjects comprehensively. Less important results may be summarized as exercises at the ends of chapters. For technicalities, readers can be referred to the bibliography, which is expected to be comprehensive. As a result, volumes are encyclopedic references or manageable guides to major subjects.

Encyclopedia of Mathematics and Its Applications

Equivalents of the Riemann Hypothesis

Volume One: Arithmetic Equivalents

KEVIN BROUGHAN
University of Waikato, New Zealand

CAMBRIDGE
UNIVERSITY PRESS

CAMBRIDGE
UNIVERSITY PRESS

University Printing House, Cambridge CB2 8BS, United Kingdom

One Liberty Plaza, 20th Floor, New York, NY 10006, USA

477 Williamstown Road, Port Melbourne, VIC 3207, Australia

314-321, 3rd Floor, Plot 3, Splendor Forum, Jasola District Centre, New Delhi-110025, India

79 Anson Road, #06-04/06, Singapore 079906

Cambridge University Press is part of the University of Cambridge.

It furthers the University's mission by disseminating knowledge in the pursuit of education, learning and research at the highest international levels of excellence.

www.cambridge.org
Information on this title: www.cambridge.org/9781107197046
DOI: 10.1017/9781108178228

First published 2017

A catalogue record for this publication is available from the British Library

Library of Congress Cataloging in Publication data
Names: Broughan, Kevin A. (Kevin Alfred), 1943– author.
Title: Equivalents of the Riemann hypothesis / Kevin Broughan, University of Waikato, New Zealand.
Description: Cambridge : Cambridge University Press, 2017– |
Series: Encyclopedia of mathematics and its applications ; 164 |
Includes bibliographical references and index. Contents: volume 1. Arithmetic equivalents
Identifiers: LCCN 2017034308 | ISBN 9781107197046 (hardback : alk. paper : v. 1)
Subjects: LCSH: Riemann hypothesis.
Classification: LCC QA246 .B745 2017 |
DDC 512.7/3–dc23 LC record available at https://lccn.loc.gov/2017034308

ISBN – 2 Volume Set 978-1-108-29078-4 Hardback
ISBN – Volume 1 978-1-107-19704-6 Hardback
ISBN – Volume 2 978-1-107-19712-1 Hardback

Dedicated to Jackie, Jude and Beck

RH is a precise statement, and in one sense what it means is clear, but what it is connected with, what it implies, where it comes from, can be very unobvious.

Martin Huxley

Contents for Volume One

Contents for Volume Two

Illustrations

Tables

Preface

Why have these two volumes on equivalences to the Riemann hypothesis been written? Many would say that the Riemann hypothesis (RH) is the most noteworthy problem in all of mathematics. This is not only because of its relationship to the distribution of prime numbers, the fundamental building blocks of arithmetic, but also because there exist a host of related conjectures that will be resolved if RH is proved to be true and which will be proved to be false if the converse is demonstrated. These are the RH equivalences. The lists of equivalent conjectures have continued to grow ever since the hypothesis was first enunciated, over 150 years ago.

The many attacks on RH that have been reported, the numerous failed attempts, and the efforts of the many whose work has remained obscure, have underlined the problem's singular nature. So too has its mythology. The great English number theorist, Godfrey Hardy, wrote a postcard to Harald Bohr while returning to Cambridge from Denmark in rough weather that read: "Have proof of RH. Postcard too short for proof." He didn't believe in a God, but was certain he would not be allowed to drown with his name associated with an infamous missing proof. David Hilbert, the renowned German mathematician, was once asked, "If you were to die and be revived after five hundred years, what would you then do?" Hilbert replied that he would ask "Has someone proved the Riemann hypothesis?" More recently, towards the end of the twentieth century, Enrico Bombieri, an Italian mathematician at the Institute for Advanced Study, Princeton, issued a joke email announcing the solution of RH by a young physicist, on 1 April of course!

There are several ways in which the truth of the hypothesis has been supported but not proved. These have included increasing the finite range of values $T > 0$ such that the imaginary part of all complex zeros of $\zeta(s)$ up to T all have real part $\frac{1}{2}$ [68], increasing the lower bound for the proportion of zeros that are on the critical line $\Re s = \frac{1}{2}$ [40], and increasing the size of the region in the complex plane where $\zeta(s)$ can be proved to be non-zero [63].

This volume includes a detailed account of some recent work that takes a different approach. It is based on inequalities involving some simple accessible arithmetic functions. Broadly outlined RH implies that an inequality is true for all integers, or all integers sufficiently large, or all integers of a particular type. If RH fails, there is an integer in the given range, or of the given type, for which the inequality fails, so the truth of the inequality is equivalent to the truth of RH. Progress under this approach is made whenever the nature of any counterexample is shown to be more restricted than previously demonstrated. The reader may also wish to consult the introductions to Chapters 3 to 7 for further details.

The relatively recent work depends critically on a range of explicit estimates for arithmetic functions. These have been derived using greater computing power than was available in the 1940s and 1960s when these sorts of estimates were first published. In many cases the details are included, along with simplified presentations.

Also included are a range of other equivalences to RH, some by now classical. The more recent work depends on these classical equivalences for both the results and techniques, so it is useful to set both out explicitly. It also shows how some equivalences are more fundamental than others. This is not to suggest new equivalences are easy consequences of older established ones, even though this may be true in some cases.

The aim of these volumes is to give graduate students and number theory researchers easy access to these methods and results in order that they might build on them. To this end, complete proofs have been included wherever possible, so readers might judge for themselves their depth and crucial steps. To provide context, a range of additional equivalences has been included in this volume, some of which are arithmetical and some more analytic. An intuitive background for some of the functions employed is also included in the form of graphical representations. Numerical calculations have been reworked, and values different from those found in the literature have often been arrived at.

To aid the reader, definitions are often repeated and major steps in proofs are numbered to give a clear indication of the main parts and allow for easy proof internal referencing. When possible, errors in the literature have been corrected. Where a proof has not been verified, either because this author was not able to fill gaps in the argument, or because it was incorrect, it has not been included. There is a website for errata and corrigenda, and readers are encouraged to communicate with the author in this regard at `kab@waikato.ac.nz`. The website is linked to the author's homepage: `www.math.waikato.ac.nz/~kab`.

Also linked to this website is a suite of *Mathematica*™ programs, called RHpack, related to the material in this volume, which is available for free download. Instructions on how to download the software are given in Appendix B.

The two volumes are distinct, with a small amount of overlap. This volume, Volume One, has an arithmetic orientation, with some analytic methods, especially those relying on the manipulation of inequalities. The equivalences found here are the Möbius mu estimate of Littlewood, the explicit $\psi(x)$ function estimate of Schoenfeld, the Liouville $\lambda(n)$ limit criterion of Landau, two Euler totient function criteria of Nicolas, the sum of divisors inequality of Ramanujan and Robin and its reformulation by Lagarias, the criterion of Caveney, Nicolas and Sondow based on so-called "extraordinary numbers", the criterion of Nazardonyavi and Yakubovich based on extremely abundant numbers, the estimate of Shapiro that uses the integral of $\psi(x)$, the Franel–Landau Farey fraction criterion, the divisibility matrix criterion of Redheffer, the Levinson–Montgomery criterion that uses counts of the zeros of the derivative $\zeta'(s)$, the inequality of Spira relating values of zeta at s and $1-s$, the self-adjoint operator criteria of Hilbert and Pólya, and the criterion of Lagarias and Garunkstis based on the real part of the logarithmic derivative of $\xi(s)$. In addition, Volume One has criteria based on the divisibility matrix of Redheffer and a closely related graph, the Dirichlet eta function, and an estimate for the size of the maximum order of an element of the symmetric group.

The Appendix includes tables of the numbers that appear in some of the equivalences and a mini-manual for RHpack.

Volume Two [32] contains equivalences with a strong analytic orientation. To support these, there is an extensive set of appendices containing fully developed proofs. The equivalences set out are named Amoroso, Hardy–Littlewood, Báez-Duarte, Beurling, Bombieri, Bombieri–Lagarias, de Bruijn–Newman, Cardon–Roberts, Hildebrand, Levinson, Li, Riesz, Sekatskii–Beltraminelli–Merlini, Salem, Sondow–Dumitrescu, Verjovsky, Weil, Yoshida and Zagier. For summary details, see the Preface for Volume Two and Chapter 1 of Volume Two. In addition, Bombieri's proof of Weil's explicit formula, a discussion of the Weil conjectures and a proof of the conjectures for elliptic curves are included.

In the case of the general Riemann hypothesis (GRH) for Dirichlet L-functions, in Volume Two the Titchmarsh criterion is given, as well as proofs of the Bombieri–Vinogradov and Gallagher theorems and a range of their applications. There is a small supporting *Mathematica* package, GRHpack.

The set of appendices for Volume Two gives comprehensive statements and proofs of the special results that are needed to derive the equivalences in that volume. In addition, there is a GRHpack mini-manual.

A note concerning the cover figure. This represents integral paths for the flow $\dot{s} = \zeta(s)$ in a small region of the upper complex plane, rotated and reflected in $\sigma = 0$. It was produced using an interactive program written by the author and Francis Kuo in Java®. It includes three critical zeros and three trivial zeros.

Many people have assisted with the development and production of these volumes. Without their help and support, the work would not have been possible, and certainly not completed in a reasonable period of time. They include Sir Michael Berry, Enrico Bombieri, Jude Broughan, George Csordas, Daniel Delbourgo, Tomás Garcia Ferrari, Pat Gallagher, Adolf Hildebrand, Geoff Holmes, Stephen Joe, Jeff Lagarias, Wayne Smith, Tim Trudgian, John Turner and Michael Wilson. The support of the University of Waikato and especially its Faculty of Computing and Mathematical Sciences and Department of Mathematics and Statistics has been absolutely essential. Cambridge University Press has also provided much encouragement and support, especially Roger Astley and Clare Dennison. Last, but not least, I am grateful for my family's belief in me and support of my work.

Kevin Broughan
December 2016

Acknowledgements

The author gratefully acknowledges the following sources and/or permissions for the non-exclusive use of copyrighted material.

Euclid: Figure 1.1, being an excerpt from Raphael's fresco "School of Athens", Visions of America/Superstock/Getty Images.

L. Euler: Figure 1.2, Apic/Hulton Archive/Getty Images.

C. F. Gauss: Figure 1.3, Bettmann/Getty Images.

J. L. Dirichlet: Figure 1.4, Stringer/Hulton Archive/Getty Images.

B. Riemann: Figure 1.5, Author: Konrad Jacobs. Source: Archives of the Mathematisches Forschungsinstitut Obcrwolfach.

D. Hilbert: Figure 1.6, Ullstein bild/Getty Images.

Zeta flow: Figure 2.2, being figure 2, p. 990, of K. A. Broughan and A. R. Barnett, *The holomorphic flow of the Riemann zeta function*, Mathematics of Computation **73** (2004), 287–1004. Permission of the American Mathematical Society.

Xi flow: Figure 2.3, being figure 1, p. 1274, of K. A. Broughan, *The holomorphic flow of Riemann's function* $\xi(s)$, Nonlinearity **18** (2005), 1289–1294. Copyright © IOP Publishing and London Mathematical Society. Permission of the American Mathematical Society. All rights reserved.

J. E. Littlewood: Figure 4.1, used by permission of the Master and Fellows of Trinity College Cambridge.

J.-L. Nicolas: Figure 5.1, used by permission of J.-L. Nicolas.

P. Erdős: Figure 6.1, Author: Kay Piene. Source: Ragni Piene and Archives of the Mathematisches Forschungsinstitut Oberwolfach.

T. S. Trudgian: Figure 7.2, used by permission of T. S. Trudgian.

J. C. Lagarias: Figure 7.7, used by permission of J. C. Lagarias.

Y. Choie: Figure 8.1, used by permission of Y. Choie.

Section 8.5 was derived from a previously published article: K. A. Broughan and T. Trudgian, *Robin's inequality for 11-free integers*, Integers **15** (2015), A12 (5pp.).

1
Introduction

1.1 Chapter Summary

This chapter is discursive. It begins with an overview of the early history of the Riemann hypothesis (RH) and the evolution of ideas relating to the Ramanujan–Robin inequality. Then in Section 1.3 there is a summary of the contents of the entire volume, first in brief and then in more detail. The section also describes the tables in Appendix A and the software RHpack in Appendix B.

There is a section on notational conventions and special notations, most of which are quite standard. A guide to the reader and two problems complete the chapter.

1.2 Early History

Here the main players in the evolution of the Riemann hypothesis are noted: Euclid, Euler, Gauss, Dirichlet and, last but not least, Riemann himself. The first is Euclid of Alexandria (Figure 1.1) who lived around 300 BCE. His *Elements* includes a proof that there are an infinite number of primes, and that they are the fundamental building blocks of numbers, through the unique factorization of integers. How are the primes distributed? On the face of it, the only pattern appears to be that all are odd, except 2. Do they appear completely at random, or are they uniformly distributed in some sense?

Don Zagier gives a good description of this random/uniform dichotomy: "The first fact is the prime numbers belong to the most arbitrary and ornery objects studied by mathematicians: they grow like weeds among the natural numbers, seeming to obey no other law than that of chance, and nobody can predict where the next one will sprout."

"The second fact is even more astonishing, for it states just the opposite: that the prime numbers exhibit stunning regularity, that there are laws governing their behaviour, and that they obey these laws with almost military precision."

Figure 1.1 Euclid (about 325–265 BCE).

Euler (1707–1783; Figure 1.2) was an amazing mathematician, who wrote a text on calculus, including the first treatment of trigonometric functions, and invented many parts of mathematics that are important today, including the calculus of variations, graph theory and divergent series. In number theory, quadratic reciprocity and Euler products are two of his many contributions, as well as his extensive work on the zeta function. In fact, Euler gave birth to the Riemann zeta function, writing down its definition for the first time. He studied the following sums and product:

$$\zeta(2) = 1 + \frac{1}{2^2} + \frac{1}{3^2} + \frac{1}{4^2} + \cdots = \frac{\pi^2}{6},$$
$$\zeta(4) = 1 + \frac{1}{2^4} + \frac{1}{3^4} + \frac{1}{4^4} + \cdots = \frac{\pi^4}{90},$$
$$\zeta(s) = 1 + \frac{1}{2^s} + \frac{1}{3^s} + \frac{1}{4^s} + \cdots,$$
$$\zeta(s) = \prod_p \left(1 - \frac{1}{p^s}\right)^{-1}.$$

For Euler, the variable s was an integer, and the product over all primes. He tried in vain to find a closed expression for $\zeta(3)$ having the same style as that for $\zeta(2)$ and $\zeta(4)$.

Figure 1.2 L. Euler, 1707–1783.

The next main player was Gauss (1777–1855; Figure 1.3), although he does not appear to have used the zeta function. He was the giant of nineteenth-century mathematics and physics. For example, he used calculus to compute correctly the position of Ceres after it had passed behind the Sun. Although he spent most of his life in his observatory in Göttingen in Germany, he contributed to statistics, non-Euclidean geometry, curvature, geodesy, electromagnetism and complex numbers, as well as to number theory, for which he had an abiding passion.

Gauss was given a book of log tables, at around the age of 14, that included a table of primes. He extended the table, counting the number of primes up to a real positive variable x, now called $\pi(x)$. He considered the average number of primes in each interval of numbers $[1,2,3,\ldots,N]$, and in this way arrived at his prime number conjecture, an inspired guess that was proved about 100 years later:

$$\pi(x) \sim \frac{x}{\log x} \quad \text{which means} \quad \frac{\pi(x)}{x/\log x} \to 1, \quad x \to \infty.$$

It was this conjecture that inspired many mathematicians during the nineteenth century.

Dirichlet (1805–1859; Figure 1.4) was a fine teacher and had a great influence on Bernhard Riemann, who attended his number theory lectures in Berlin during 1847–1849. According to Jacobi, Dirichlet was not only creative, but knew how to make a robust proof: "Only Dirichlet, not I, nor

Figure 1.3 J. C. F. Gauss, 1777–1855.

Figure 1.4 J. L. Dirichlet, 1805–1859.

Cauchy, not Gauss, knows what a perfectly rigorous proof is, but we learn it only from him. When Gauss says he has proved something it is very likely. When Cauchy says it, it's a fifty-fifty bet. When Dirichlet says it, it is certain." Dirichlet showed that there were an infinite number of primes in arithmetic progressions whenever the step and initial value are coprime. He used groups, characters and zeta functions. He was a wonderfully inspiring lecturer, and joined Riemann in Göttingen in 1855.

According to Felix Klein: "Riemann was bound to Dirichlet by the strong inner sympathy of a like mode of thought. Dirichlet loved to make things clear to himself with an intuitive underpinning. Along with this he would give acute, logical analyses of foundational questions and avoid long computations as much as possible." His manner suited Riemann, who adopted it and worked in many ways according to Dirichlet's methods.

So we come to the central character in this account, Bernhard Riemann (1826–1866; Figure 1.5). Riemann was impoverished for most of his life. He was shy and withdrawn, sickly and a hypochondriac. Other than the two years in Berlin, he spent most of his working life in the still very beautiful walled German town of Göttingen, but he changed mathematics forever. As well as numbers, formulae and concepts, the idea of mathematical spaces and of the

Figure 1.5 B. Riemann, 1826–1866.

relationships between them can be traced back to Riemann. He went beyond Dirichlet by supporting his profound ideas with extensive calculations and manipulations.

Riemann contributed to real analysis (the Riemann integral), complex analysis, (Cauchy–Riemann equations), potential theory and geometry (Riemannian manifolds). He richly deserves to be called a genius and a great mathematician.

An eight-page paper Riemann published in the Notices of the Berlin Academy in 1859, titled *On the number of primes less than a given magnitude* [140], contained an outline of a possible proof of Gauss's prime number conjecture. It started from ideas of Cauchy and Dirichlet. For example, he extended the domain of $\zeta(s)$ to the whole of the complex plane other than the point $s = 1$. The paper did not contain proofs, but radically changed analytic number theory. It took 30 years for mathematicians to begin to appreciate what Riemann's ideas really meant, and that his assertions were provable. Over 70 years later the analytic and computational underpinnings for the paper, through Siegel and his exploration of Riemann's hand-written notes, became clear. Riemann states in the paper what is now known as the Riemann hypothesis, or RH. Here is a translation of what he wrote:

One finds in fact about this many zeros of the zeta function within these bounds on the critical line, and it is **very likely that all of the zeros are on the critical line**. *One would of course like to have a rigorous proof of this, but I have put aside the search for such a proof after some fleeting vain attempts, because it is not necessary for the immediate objective of my investigation.* (For information about the term "critical line" see Section 2.2 (7).)

Indeed it was shown later that to complete the proof of the prime number theorem, the main objective of the paper, one did not need to prove the RH, only to show that the zero-free region for the zeta function includes all of the line $\Re s = 1$.

At the 1900 International Congress of Mathematicians, ICM1900, held in Paris, the great mathematician David Hilbert (Figure 1.6) gave an address in which he listed the 23 most outstanding problems of the day he considered worthy of the efforts of mathematicians, and fundamental for the further advancement of the subject. Problem number eight was the "Riemann hypothesis", which conjectured that the complex zeros of $\zeta(s)$ all had real part $\frac{1}{2}$.

The final event in this brief history, and significant for the work reported in this volume, is one of the consequences of assuming the hypothesis is true. In 1914, soon after the commencement of World War I when resources where scarce, the great Indian mathematician Srinivasa Ramanujan published an article in the Proceedings of the London Mathematical Society titled *Highly composite numbers*. At over 50 pages long it must have represented a considerable publishing challenge. As it turned out, there was a large amount

Figure 1.6 David Hilbert, 1862–1943.

of additional material on related topics written by Ramanujan which was not included in the published work. This was discovered in more recent times, and a typeset version with notes was eventually published in 1997. It includes an inequality, derived by Ramanujan, who assumed the Riemann hypothesis in this case, relating to the sum-of-divisors of an integer arithmetic function $\sigma(n)$. This can be stated as follows: for all $n \in \mathbb{N}$ sufficiently large we have

$$\frac{\sigma(n)}{n} < e^{\gamma} \log\log n.$$

Here γ is Euler's constant,

$$\gamma = \lim_{n\to\infty}\left(\left(\sum_{j=1}^{n}\frac{1}{j}\right) - \log(1+n)\right).$$

The nice surprise is that this inequality, with an explicit lower bound for n replacing "sufficiently large", and other related arithmetic inequalities and equalities, is equivalent to the Riemann hypothesis. The original inequality involving $\sigma(n)$ is the most famous, and the equivalence is known as "Robin's theorem". A detailed history of its evolution is given in [128] with quotes from Erdős, Hardy, Rankin, Berndt and Nicolas.

1.3 Volume One Summary

Except for Chapter 10, the text contains a mostly linear progression of ideas. In Chapter 2, key properties of the Riemann zeta function, which will be needed, are outlined and two key parameters developed. This is then followed in Chapter 3 by explicit estimates for functions of primes, given in such a way that their derivations can potentially be improved. Then in Chapter 4 some classical equivalences of the Riemann hypothesis are proved in full. These are used in the work that follows. In Chapter 5 a set of equivalences, for the most part in the form of inequalities, involving the Euler totient function $\varphi(n)$, are derived. Chapter 6 provides preparatory material for Chapter 7 by developing the properties of two types of so-called "abundant" numbers, wherein it is the number of divisors that are abundant. These are the numbers that appear as possible counterexamples to inequalities. In Chapter 7 an inequality based on values of $\sigma(n)$, the sum-of-divisors function, is shown to be equivalent to the hypothesis. In Chapter 8 the focus shifts to numbers not satisfying the inequality, with several results showing the numbers are very constrained, so sit in a narrow class. It is expected that this class will be further constrained. Chapter 7 already contained an alternative inequality, equivalent to the Riemann hypothesis, and Chapter 9 continues in this mode, expressing equivalent formulations for the hypothesis in terms of so-called "extraordinary" numbers and "extremely abundant" numbers.

The final chapter (Chapter 10) breaks the sequence of ideas, by giving ten other equivalent statements for the Riemann hypothesis, mostly proved in full. In one form or another, inequalities play a role in these formulations. The idea of this chapter is to reveal the ubiquitous nature of the hypothesis, and be the source of new ideas.

Further summary details are given in this section in the paragraphs below.

This first chapter, in Section 1.2, gives a sketch of the genesis of the Riemann hypothesis. Interestingly, it was not posed as a problem, conjecture or even hypothesis by its principal author, Bernhard Riemann, but we know that the issues therein were first publicly discussed in 1859. This date also marks the origin of what we now call the Riemann zeta function as a function of a complex variable, $\zeta(s)$.

Chapter 2 summarizes some basic properties of the Riemann zeta function, and gives an intuitive idea of its behaviour. Appendix H "The gamma function" and Appendix I "Riemann zeta function" in Volume Two [32] give more background and proofs. The fundamental parameters H and R are introduced and used throughout the text. The symbol H represents a y-value up to which all zeros of $\zeta(s)$ with positive imaginary part have been demonstrated to have their real part equal to 1/2. Comments are made, when needed, on which value of H has been chosen in a particular circumstance. The value of H that could be used is expected to increase in time, and theorems and algorithms are presented so new values can be adopted easily.

The value of R describes a simple form for a zero-free region of $\zeta(s)$, namely for $\zeta(\beta+i\gamma)=0$ we have

$$\beta < 1 - \frac{1}{R\log\gamma}.$$

It is expected that the value of R will decrease in time. We give the derivation of Landau, and quote more recent improvements.

Chapter 3 is devoted to numerical estimates, especially of the arithmetical functions $\theta(x)$ and $\psi(x)$. Note that there are additional estimates of products of functions of primes in Section 5.2. Estimates in this chapter are primarily derived without using RH. The material depends not only on the work of Rosser and Schoenfeld in their separate and joint papers of 1941, 1962, 1963 and 1975, but also on von Mangold's theorems of 1905.

Chapter 4 gives derivations of some well-known explicit classical equivalences to RH. It may be skipped by anyone familiar with introductory material. The proof of an important theorem of Landau, used in many cases in the text where RH is assumed to be false, is included. The chapter concludes with overview material on zero-free regions and a summary of heights up to which the hypothesis has been shown to hold, with the corresponding numbers of critical zeros.

The work of Rosser and Schoenfeld is remarkable in that it contains sharp explicit results produced before computer-based methods became ubiquitous. In this chapter we take some advantage of the much greater processing speed now available to extend the ranges for which computer verification of inequalities is practicable, and the greatly improved height below which all the zeros of $\zeta(s)$, with positive imaginary part, have real part $\sigma=\frac{1}{2}$. Improving these estimates is the subject of current research, since they have a strong influence on the equivalences to RH.

Moving on to Chapter 5, Nicolas proved in 1983 that RH is true if and only if for every non-prime primorial q we have

$$\frac{q}{\varphi(q)\log\log(q)} > e^{\gamma}.$$

Here a primorial is the product of all primes up to a given prime, so the kth primorial, say q, is $q = N_k := p_1 \cdots p_k$, where p_i is the ith prime starting with $p_1 = 2$.

Nicolas improved this result in 2012, and found four statements equivalent to the Riemann hypothesis. Let $N_k = 2\cdot 3\cdots p_k$ be the kth primorial and let

$$c(n) := \left(\frac{n}{\varphi(n)} - e^{\gamma}\log\log(n)\right)\sqrt{\log n}$$

and define

$$\beta := \sum_{\rho} \frac{1}{\rho(1-\rho)} = 2 + \gamma - \log \pi - 2\log 2 = 0.046191\ldots,$$

where ρ ranges through the non-trivial zeros of $\zeta(s)$ with increasing absolute value of the imaginary part. Then RH is equivalent to each of the following:

(1) $\limsup_{n\to\infty} c(n) = e^{\gamma}(2+\beta) = 3.644415\ldots.$
(2) For all $n \geq N_{120569} = 2\cdot 3 \cdots 1591\,883$ we have $c(n) < e^{\gamma}(2+\beta)$.
(3) For all $n \geq 2$, $c(n) \leq c(N_{66}) = c(2\cdot 3\cdots 317) = 4.0628356921$.
(4) For all $k \geq 1$ we have $c(N_k) \geq c(N_1) = c(2) = 2.208589\ldots.$

From this one shows that the Riemann hypothesis is equivalent to the following inequality holding for all $n \geq N_{120569}$:

$$\frac{n}{\varphi(n)} < e^{\gamma} \log\log n + \frac{e^{\gamma}(2+\beta)}{\sqrt{\log n}}.$$

The chapter includes the derivation of useful estimates for treating the Euler phi function, namely some products and sums of functions of primes. These depend on the relationship between primes and zeta zeros, and form an essential basis, along with other estimates, for developing the equivalences.

In Chapter 6, fundamental results to do with two types of number with many divisors are developed. These are called *superabundant* and *colossally abundant*. The work includes that of Alaoglu, Erdős and Nicolas. These numbers appear as counterexamples to RH, so are used in the chapters that follow. We include a corrected proof of a fundamental theorem of Alaoglu and Erdős.

Chapter 7 is also a fundamental chapter. Grönwall proved in 1913 that

$$G(n) := \frac{\sigma(n)}{n \log\log(n)} \quad \Longrightarrow \quad \limsup_{n\to\infty} \frac{\sigma(n)}{n\log\log n} = e^{\gamma} = 1.78107\ldots.$$

Ramanujan showed, probably in 1915, that if RH is true then for n sufficiently large we must have $G(n) < e^{\gamma}$. Then Robin showed in 1983 that RH is true if and only if $n > 5040$ implies $G(n) < e^{\gamma}$. Also included is the equivalence of Lagarias, which is dependent on the result of Robin, namely if $H_n = 1 + 1/2 + \cdots + 1/n$ is the n harmonic number, then RH is equivalent to the inequality

$$\sigma(n) \leq H_n + \exp(H_n)\log(H_n), \quad n \geq 1.$$

In Chapter 8, properties of numbers that do not satisfy Robin's inequality are explored. The work of Choie, Lichiardopol, Moree and Solé is developed and extended. In particular it is shown that any integer greater than 5040 not satisfying Robin's inequality must be even, fail to be squarefree and not squarefull. The smallest such number must be a so-called Hardy–Ramanujan number and superabundant. Then it is shown, successively, that any counterexample must be divisible by the fifth power of at least one prime, the seventh power and finally the 11th power of at least one prime.

We now move on to Chapter 9. Define a composite integer $n > 1$ to be left abundant or GA1 if it satisfies $G(n/p) \leq G(n)$ for all primes $p \mid n$. Call it right abundant or GA2 if $G(n) \geq G(an)$ for all natural numbers $a > 1$. Then in 2011 Caveney, Nicolas and Sondow showed that the Riemann hypothesis is true if and only if 4 is the only number that is both left and right abundant. They called such numbers *extraordinary*. They also showed that a right abundant number $n > 5040$ exists if and only if the Riemann hypothesis is false, and, if so, such an integer n is even with $n > 10^{8576}$. The chapter concludes with the equivalence of Nazardonyavi and Yakubovich based on "champions" for $G(n)$, which they call *extremely abundant*.

Although Chapter 9 represents the final section of the main body of this text, a range of other equivalences that have caught the attention of the author, and are not in Volume Two [32], have been included in Chapter 10. Proofs are given, as in the earlier chapters, to show depth and illustrate technique: Farey fractions, the Redheffer matrix, the divisibility graph, the Dirichlet eta function, the derivative of the Riemann zeta function, a simple zeta-related inequality, and an inequality for the real part of the logarithmic derivative of the Riemann xi function, $\xi(s)$. The chosen set of equivalences is far from comprehensive. For detailed descriptions of these equivalences, the reader is invited to consult Chapter 10. An improvement of a result of Lagarias, which gives, for every $t \geq 168\pi$, the imaginary part of a zeta zero γ with $|t-\gamma| \leq 1.5$, is included.

In the penultimate section of Chapter 10, an equivalence regarding the maximum multiplicative order, $\operatorname{ord} n$, of any element of the symmetric group S_n is outlined. Not all of the details are included – these are spread over many papers. The development is based on the work of Landau and Shah, and includes the theorem of Massias, Nicolas and Robin, which says that RH is equivalent to the inequality

$$\log \operatorname{ord}(n) < \sqrt{\operatorname{li}^{-1}(n)}, \quad n \text{ sufficiently large},$$

where $\operatorname{li}(x)$ is the logarithmic integral. A new "straight line" proof of the theorem of Shah which is fundamental to the work is included, as is an introduction to the related work of Erdős and Turan on the statistics of S_n.

In the final section of Chapter 10, a simple equivalence, but very influential conjecture, going under the name of the "Hilbert–Pólya conjecture", is introduced.

In Figure 1.7 we give some of the dependences between the chapters in this volume. An arrow from Chapter A to Chapter B means some results in B depend on material in A. Not all dependences have been included, but the network flow diagram provides a general overview.

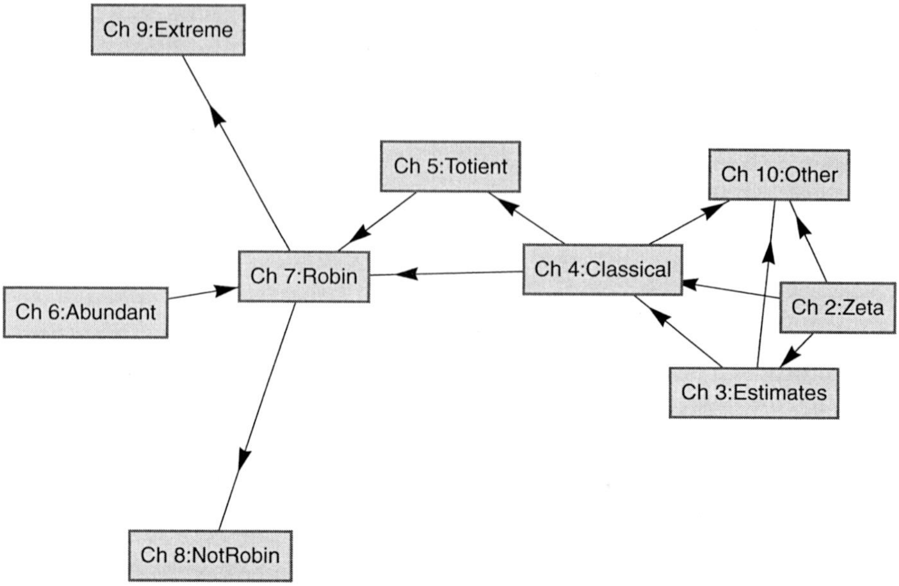

Figure 1.7 Some relationships between chapters.

Appendix A consists of a set of tables giving examples of numbers that appear in the text: numbers less than 5041 that do not satisfy Robin's inequality, superabundant numbers, colossally abundant numbers, the values of the maximum order of an element of the symmetric group, solutions of what is called *Nicolas' reversed inequality*, a list of primes in order which can be used to make colossally abundant numbers, and some extremely abundant numbers, all for small initial values.

Appendix B is a mini-manual for the *Mathematica* software RHpack, a package that is tailored for this book. As well as a description of each RHpack function, instructions are included on how to locate, download and install the package.

1.4 Notation

We use the following notation, sometimes without (further) explanation:

$$n = p_1^{\alpha_1} \cdots p_m^{\alpha_m} \quad \text{with } \alpha_1 > 0 \text{ is the standard prime factorization,}$$

$$d(n) := \sum_{d|n} 1, \quad \text{the number of divisors of } n,$$

$$\sigma(n) := \sum_{d|n} d, \quad \text{the divisor sum,}$$

$$\mu(n) := (-1)^m \quad \text{if } \alpha_j = 1 \text{ for all } j, \text{ else } 0,$$

$$\omega(n) := m, \quad \text{the number of distinct prime divisors of } n,$$

$$
\begin{aligned}
\Omega(n) &:= \sum_{i=1}^{m} \alpha_i, \quad \text{the total number of prime power divisors of } n,\\
\lambda(n) &:= (-1)^{\sum_{i=1}^{m} \alpha_i}, \quad \text{Liouville's function},\\
\Lambda(n) &:= \log p \quad \text{if } n \text{ is a power of } p, \text{ and } 0 \text{ otherwise},\\
\varphi(n) &:= \#\{j : 1 \le j \le n \text{ and } (n,j) = 1\}, \quad \text{Euler's phi function},\\
H_n &:= \sum_{j=1}^{n} \frac{1}{j}, \quad \text{the harmonic number},\\
\nu_p(n) &:= k \quad \text{where } p^k \mid n \text{ and } p^{k+1} \nmid n,\\
\pi(x) &:= \#\{p : p \le x,\ p \text{ is prime}\},\\
\theta(x) &:= \sum_{p \le x} \log p,\\
\psi(x) &:= \sum_{p^n \le x} \log p,\\
M(x) &:= \sum_{n \le x} \mu(n),\\
\mathrm{li}(x) &:= \lim_{\epsilon \to 0} \left(\int_0^{1-\epsilon} \frac{dt}{\log t} + \int_{1+\epsilon}^{x} \frac{dt}{\log t} \right), \quad \text{the logarithmic integral},\\
\mathrm{Li}(x) &:= \int_2^{\infty} \frac{dt}{\log t}, \quad \text{also called the logarithmic integral},
\end{aligned}
$$

$\mathrm{ord}(\sigma)$, the multiplicative order of the element σ of a group.

A note on the use of symbols: Many authors have built explicit rational constants into their explicit results. We have attempted to avoid this practice, since an improvement in a constant will invalidate a result. Constants are called a_1, a_2, …, c_1, c_2, …, etc. and given explicit definitions. There is also a problem in this part of the literature with overloaded symbols. Two examples are the function names $\phi(\cdot)$ and $K(\cdot)$, both of which have a variety of meanings, sometimes in the same context. We have attempted to avoid this practice.

The symbols p and q invariably represent rational prime numbers. The Landau–Vinogradov symbols O, o, $\asymp, \ll$ and $\gg$, with their usual meanings, are also used. Additional notation is introduced as needed.

In addition, e, π and i have their usual meanings, γ is Euler's constant, B is Mertens' constant, β is $2 + \gamma - \log \pi - 2 \log 2$ and H is a known height up to which all complex zeros of $\zeta(s)$ lie on the critical line.

1.5 Background Reading

There are several books about the Riemann hypothesis that set it in its historical and cultural context – see for example Sabbagh [148] and du

Sautoy [151]. Books by Edwards [51], Patterson [131] and Borwein et al. [13] give accessible mathematical introductions, whilst Titchmarsh [167], Ivić [82] and Karatsuba and Voronin [88] are more advanced monographs.

1.6 Unsolved Problems

(1) Investigate why Dirichlet's famous theorem on primes in arithmetic progressions was not included in his original set of lecture notes.

(2) There are Riemann hypotheses in more general or analogous contexts: extended Riemann hypotheses, the generalized Riemann hypothesis and the grand Riemann hypothesis for example. Conjecture and prove an equivalent form in one or more of these settings.

2
The Riemann Zeta Function

2.1 Introduction

The purpose of this chapter is to give some insight into the nature of the Riemann zeta function, and describe some of its properties, which are essential for the development of arithmetic equivalences to RH.

Following this introductory section, the second section of this chapter discusses some of the main properties of the Riemann zeta function, such as how it is defined, the functional equation and the function's zeros, which are all important since they are the subject of the Riemann hypothesis. An idea of how $\zeta(s)$ behaves is given with a presentation of plots of its modulus and argument in small parts of the complex plane, including some of the complex zeros. There is no attempt to be comprehensive. It is interesting to note that few of the properties of $\zeta(s)$ are used in deriving equivalent formulations to the hypothesis.

In Section 2.2, statements of the formulae of von Mangoldt are included, which give explicit representations of errors in a form of the prime number theorem in terms of sums of quantities dependent on the zeros of $\zeta(s)$. This material is presented with references but without proofs, since there are a number of excellent texts that give comprehensive expositions.

In Section 2.3 information is given on heights up to which all zeros of $\zeta(s)$ have been proved, numerically, to be on the critical line. The zero-free region of Landau is derived in Section 2.4. The zeta function properties and zero-free regions are used in this volume in essential ways.

There are quite a few very important results in Section 2.6, which has complete proofs. These include the product formula for $\xi(s)$ given by

$$\xi(s) = \xi(0) \prod_{\rho} \left(1 - \frac{s}{\rho}\right),$$

and the explicit expression for the convergent sum

$$\sum_{\rho} \frac{1}{\rho} = 1 + \frac{\gamma}{2} - \frac{1}{2}\log(4\pi).$$

Both the product and sum are over the non-trivial zeros of $\zeta(s)$. When each ρ is associated with the zero $1-\rho$, the product and sum converge absolutely.

The following RHpack functions (see Appendix B) relate to the material in this chapter: Psi0, PlotVonMangoldtPsi, BacklundB, ZetaZeroHeight, LargestZetaZero and NextZetaZero.

2.2 Basic Properties

In this section a summary of some of the properties of the Riemann zeta function is given. These are well set out and derived in many texts such as the classics [6, 51, 82, 167], and the derivations will not be repeated here. As usual $s = \sigma + it$ is a complex variable and $\zeta(s)$ is the complex value of the Riemann zeta function at $s \in \mathbb{C}$.

(1) $\zeta(s)$ is holomorphic on $\mathbb{C} \setminus \{1\}$. It has a simple pole at $s = 1$ with residue 1.

(2) On the half plane $\sigma > 1$ it can be represented by the absolutely convergent series and Euler product

$$\zeta(s) = \sum_{n=1}^{\infty} \frac{1}{n^s} = \prod_{p \in \mathbb{P}} \left(1 - \frac{1}{p^s}\right)^{-1}.$$

(3) On the right half plane $\sigma > 0$ it can be represented by the conditionally convergent series

$$\zeta(s) = -\sum_{n=1}^{\infty} \frac{(-1)^n}{n^s}.$$

(4) Also on the half plane $\sigma > 0$ it can be represented by the integral

$$\zeta(s) = \frac{s}{s-1} - s\int_1^{\infty} \frac{\{x\}}{x^{s+1}}\,dx$$

where $\{x\}$ is the fractional part of $x \in \mathbb{R}$.

(5) It satisfies a functional equation. This can be written in the form $\xi(s) = \xi(1-s)$ where $\xi(s)$ is the so-called "completed" zeta function, the Riemann xi function, defined by

$$\xi(s) := \frac{s(s-1)}{2}\pi^{-s/2}\Gamma\left(\frac{s}{2}\right)\zeta(s).$$

The function $\xi(s)$ is entire.

(6) The complex zeros of $\zeta(s)$ and $\xi(s)$ are the same. In addition $\zeta(s)$ has zeros at each negative even integer $s \in \{-2, -4, -6, \ldots\}$. It is traditional to call these the "trivial zeros" of $\zeta(s)$. Other than at the trivial zeros, the function is non-zero outside of the vertical open strip $0 < \sigma < 1$. If $\rho = \beta + i\gamma$ is a complex zero of $\zeta(s)$ with $\beta \neq \frac{1}{2}$ then we have a set of four associated zeros:

$$\{\rho = \beta + i\gamma,\ \bar{\rho} = \beta - i\gamma,\ 1-\rho = 1-\beta-i\gamma,\ 1-\bar{\rho} = 1-\beta+i\gamma\}.$$

(7) There are an infinite number of simple zeros of $\zeta(s)$ on the so-called "critical line", $\sigma = \frac{1}{2}$ [76]. The Riemann hypothesis is the conjecture that all of the non-trivial zeros of $\zeta(s)$ lie on this line. This is a long story, beginning with Hardy and Littlewood [74]. Conrey [40] has shown that more than two-fifths of the critical zeros are on this line.

To give the reader some intuitive understanding of how $\zeta(s)$ behaves for small values of $|s|$, three graphical descriptions are given. The first is Figure 2.1, which includes some of the contours for $|\zeta(s)|$. We see the first three zeros on the critical line $\Re s = \frac{1}{2}$ surrounded by closed curves.

The graphic of Figure 2.2 gives the direction modulo π of the complex number $\zeta(s)$ at each s in the domain. It is a phase portrait of the flow $\dot{s} = \zeta(s)$. The first three real zeros at $s = -2$, -4 and -6, where the flow is alternately a source or a sink, and the first three complex zeros, where the flow is a source focus, are shown in the figure.

The passage of $\zeta(s) \to 1$ as $\Re s \to \infty$ is also clear. For further information see the papers [17, 19, 20, 23, 25, 26].

The graphic of Figure 2.3 is a depiction of part of the flow $\dot{s} = \xi(s)$. It shows the symmetry of the flow stemming from the functional equation, the first three complex zeros, each of which is a centre for the flow, and the trending of the separatrices to the positive and negative x-axis. For further information see the papers [21, 22, 26].

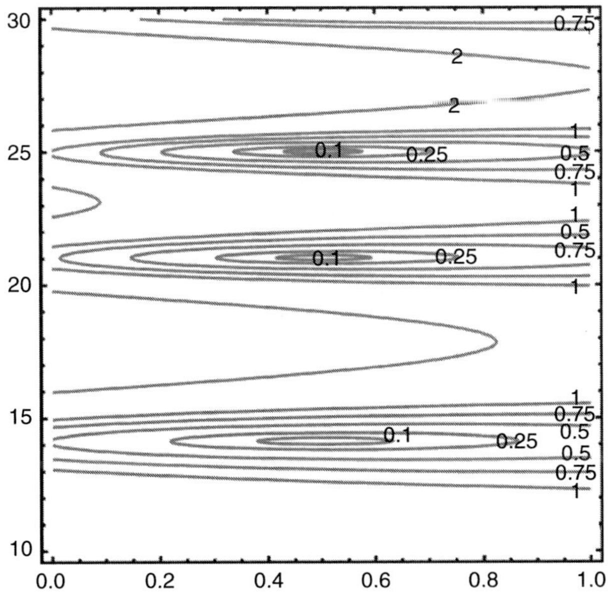

Figure 2.1 The contours of $|\zeta(s)|$.

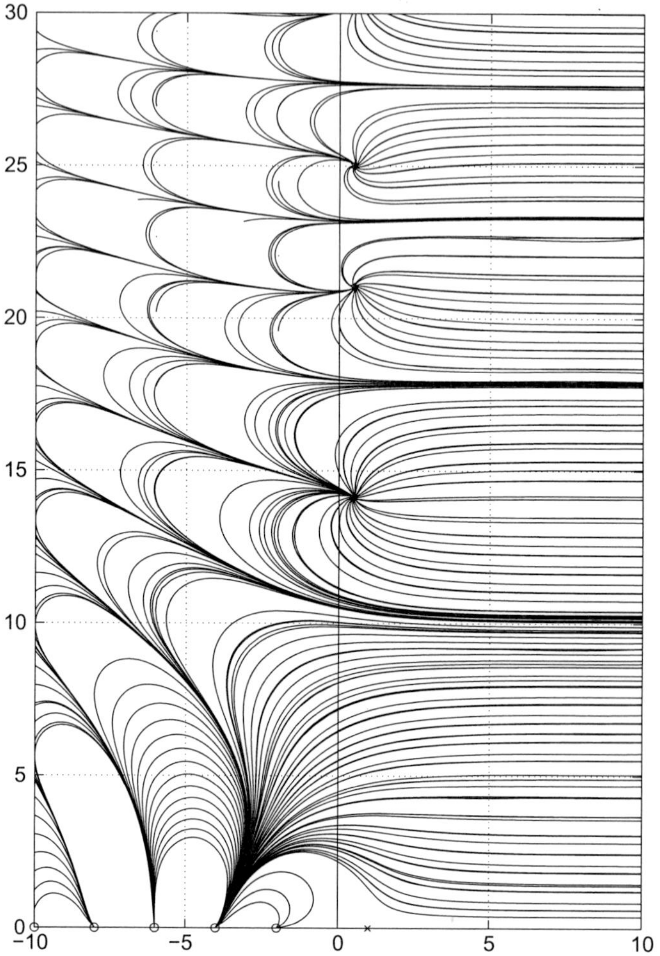

Figure 2.2 The flow $\dot{s} = \zeta(s)$.

(8) Bounds on the modulus of zeta, or its multiplicative inverse, are crucial in estimating integrals using Cauchy's theorem. One we use is the following: assuming the Riemann hypothesis, for every $\epsilon > 0$ and $\delta > 0$ there is a $T_0 > 0$ such that for a positive constant K, dependent on the chosen constants, such that for every $s = \sigma + it$ with $\sigma \geq \frac{1}{2} + \epsilon$ and $t \geq T_0$ we have the bound $|1/\zeta(s)| \leq Kt^{\delta}$. Detailed derivations of bounds of this type are given in Volume Two [32, Appendix I].

(9) We also need bounds on zeta and its derivative that apply on vertical strips $a \leq \sigma \leq b$ where a and b are any given real numbers. The bound is unconditional and the implied constant and power A depend on a and b. There is a constant $A > 0$ such that for every $s = \sigma + it$ with $a \leq \sigma \leq b$ and $|t| \geq \delta > 0$, we have $|\zeta(s)| \ll t^A$ and $|\zeta'(s)| \ll t^A$ [167].

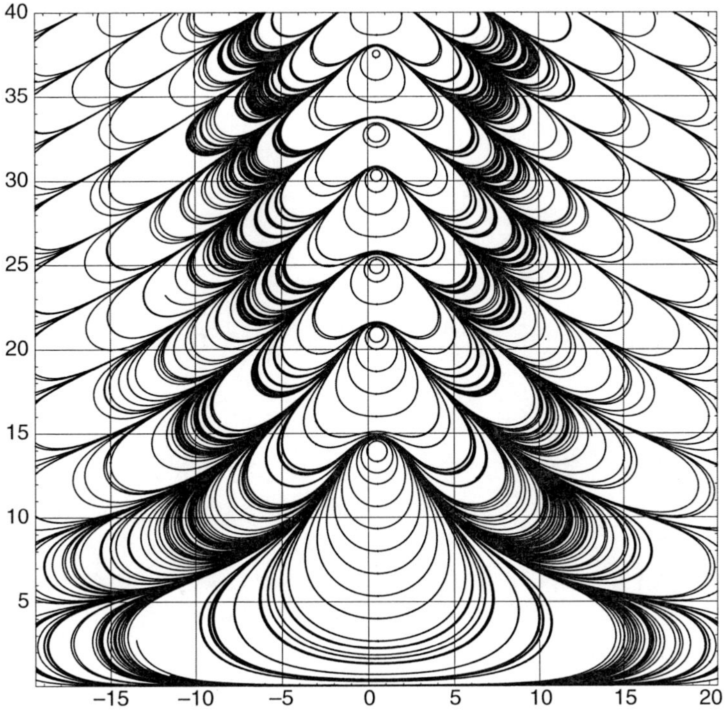

Figure 2.3 The flow $\dot{s} = \xi(s)$.

At times, more precise bounds for the modulus of $\zeta(s)$ on the critical line are needed. These have been the subject of close study over a considerable period. For example, there is the inequality of Cheng and Graham [37]:

$$|\zeta(\tfrac{1}{2} + it)| \le \min(6t^{1/4} + 57,\ 3t^{1/6} \log t), \quad t \ge 3. \tag{2.1}$$

We make particular use of the inequality of Platt and Trudgian [134], which is better than (2.1):

$$|\zeta(\tfrac{1}{2} + it)| \le 0.732 t^{1/6} \log t, \quad t \ge 2. \tag{2.2}$$

Some use is also made of a bound of Richert [139] as made completely explicit by Ford [60], and which gives a bound for the entire right half of the critical strip. It is most useful for σ near 1:

$$|\zeta(\sigma + it)| \le 76.2 |t|^{4.45(1-\sigma)^{3/2}} \log^{2/3} |t|, \quad |t| \ge 3,\ \tfrac{1}{2} \le \sigma < 1 \tag{2.3}$$

(see also Cheng [36]).

We also need to use explicit zero-free regions for $\zeta(s)$, beyond heights at which all zeros are known to be on the critical line. These are studied in some detail in Section 2.3. We use the result of Korobov [96] and Vinogradov

Table 2.1 Some improvements for Backlund's zero counting estimate.

Year	Author	a_1	a_2	a_3	T_0
1919	Backlund [7]	0.137	0.443	4.35	200
1938	Rosser [143]	1.12	0	9.5	1450
1941	Rosser [144]	0.137	0.443	1.588	1467
2014	Trudgian [169]	0.122	0.278	2.510	e

[176]: there is an absolute constant $c > 0$ such that if $s = \sigma + it$ satisfies

$$1 - \sigma \leq \frac{c}{\log^{3/2}|t| \operatorname{loglog}^{1/3}|t|} \tag{2.4}$$

and $|t|$ is sufficiently large, then $\zeta(s) \neq 0$.

(10) Having a value of the height y, denoted by H, such that every complex zero $\rho = \beta + i\gamma$ of $\zeta(s)$ with $0 < \gamma \leq y$ has $\beta = \frac{1}{2}$, is useful in proving explicit statements. A summary history of known results is given in Section 2.3. An approximation to $N(T)$ (see (11)) is derived in Volume Two [32, Appendix I, Theorem I.1].

(11) It is also important in applications to be able to count the number of zeta zeros with $0 < \gamma \leq T$ with specified accuracy. First we need some definitions. Let $N(T)$ be the number of zeros $\rho = \beta + i\gamma$ of $\zeta(s)$ for which $0 < \beta < 1$ and $0 < \gamma \leq T$. Let

$$F(T) := \frac{T}{2\pi} \log \frac{T}{2\pi} - \frac{T}{2\pi} + \frac{7}{8}$$

and

$$B(T) := a_1 \log T + a_2 \operatorname{loglog} T + a_3, \quad T \geq T_0,$$

where a_1, a_2 and a_3 are real numbers – see Table 2.1. We use the result of Backlund [7] as refined by Trudgian:

$$|N(T) - F(T)| \leq B(T), \quad T \geq e.$$

For a longer list of progressive improvements, see [169, table I]. For a description of Backlund's approach, see [51, section 6.7].

(12) Essential tools in this study are the two explicit formulae of von Mangoldt. They express $\psi(x)$, or its integral, as a sum of terms involving x and the complex zeros of $\zeta(s)$, together with some additional terms of minor significance. Recall the definition:

$$\psi(x) := \sum_{p^n \leq x} \log p = \sum_{p \leq x} \left\lfloor \frac{\log(x)}{\log p} \right\rfloor \log p.$$

In spite of their central importance, I do not give their derivation as this is set out fully in [51, sections 3.2 and 3.5]. Let ρ range over the complex zeros of $\zeta(s)$ with increasing absolute value of the imaginary part, and let p represent a rational prime and n a natural number. Then we have:

Theorem 2.1 (von Mangoldt) *For $x > 1$ which is not a power of a prime number we have*

$$\psi(x) = x - \sum_{\rho} \frac{x^\rho}{\rho} + \sum_{n=1}^{\infty} \frac{1}{2nx^{2n}} - \frac{\zeta'(0)}{\zeta(0)} = x - \sum_{\rho} \frac{x^\rho}{\rho} - \frac{1}{2}\log\left(1 - \frac{1}{x^2}\right) - \log 2\pi.$$

If x is a prime power the expression given on the right-hand side is equal to the average $(\psi(x-) + \psi(x+))/2$, so differs from $\psi(x)$ by $O(\log x)$.

Theorem 2.2 (von Mangoldt) *For $x > 1$ we have*

$$\int_0^x \psi(t)\,dt = \frac{x^2}{2} - \sum_{\rho} \frac{x^{\rho+1}}{\rho(\rho+1)} - \sum_{n=1}^{\infty} \frac{x^{-2n+1}}{2n(2n-1)} - \frac{\zeta'(0)}{\zeta(0)}x + \frac{\zeta'(-1)}{\zeta(-1)}.$$

2.3 Zero-Free Regions

A principal tool in this work is a mixture of the classical style of zero-free region for $\zeta(s)$, asymptotically tending to the boundary of the critical strip along the lines of Theorem 2.4, and a height, denoted by H, below which all critical zeros are on the critical line. Intensive use is made of the explicit formulae of von Mangoldt, Theorems 2.1 and 2.2, using these properties to bound infinite sums. We do not always use the best forms that are available, but write results as far as possible in terms of symbolic parameters, so it should be straightforward to introduce improvements.

Although the zero-free region with explicit constants of Kevin Ford [59] is better, asymptotically, than that developed by Landau [105] described below in Section 2.4, it is only when the height is of the order of 10^{70} that it is better. Since checking the Riemann hypothesis to this height is currently unattainable, we will present the method of Landau, which gives a zero-free region that is used in explicit applications, especially in Chapter 3, to derive explicit estimates. See also the notes at the end of Section 2.5 where there is a summary of recent results, including those of Ford, Kadiri, Jang and Kwon [83], and Platt and Trudgian.

Table 2.2 of heights H up to which all imaginary zeros are on the critical line has been compiled using data from Borwein et al. [13, p. 39] and published articles or direct computations for the heights up to 35×10^5. A variety of computational methods were used to establish these zeros and heights. Good examples include [14, 15, 68, 178]. For references consult [13], or for the most recent result, which is that of David Platt, see [132] and [133, section 7]. Below we discuss the seminal 1953 value of Alan Turing.

Table 2.2 Zeros on the critical line and corresponding heights.

Year	Number of zeros	Height H	Author(s)
1856?	3	25.02	B. Riemann
1903	15	65.12	J. P. Gram
1914	79	198.02	R. J. Backlaund
1925	138	299.85	J. I. Hutchinson
1935	195	388.85	E. C. Titchmarsh
1936	1041	1466.51	L. J. Comrie, E. C. Titchmarsh
1953	1104	1539.31	A. M. Turing
1956	15 000	14 040.46	D. H. Lehmer
1956	25 000	21 942.60	D. H. Lehmer
1958	35 337	29 750.17	N. A. Meller
1966	25×10^4	170 570.23	R. S. Lehman
1968	35×10^5	1459 433.86	J. B. Rosser, J. M. Yohe, L. Schoenfeld
1977	4×10^7	—	R. P. Brent
1979	75×10^6	32 585 736.4	R. P. Brent
1982	$2\times 10^8+1$	81 702 130.19	R. P Brent, J. van de Lune, H. J. J. te Riele, D. T. Winter
1983	$3\times 10^8+1$	119 590 809.282	J. van de Lune, H. J. J. te Riele
1986	$15\times 10^8+1$	545 439 823.215	J. van de Lune, H. J. J. te Riele, D. T. Winter
2004	9×10^{11}	—	S. Wedeniwski
2004	10^{13}	—	X. Gourdon
2011	103 800 788 359	30 610 046 000	D. J. Platt

Because the final three items in Table 2.2 are “unpublished” at the time of writing, we have used the 1986 value of van de Lune, te Riele and Winter [111] in our own derivations in Chapter 3. However Platt’s value of H was used, as reported in [133], to show that

$$\pi(10^{24}) = 18\,435\,599\,767\,349\,200\,867\,866,$$

which is his theorem 7.1. This result corresponds to that of Büthe, Franke, Jost and Kleinjung of 2010 in [33], which was derived using RH. Thus there is considerable confidence that Platt’s value of H is correct.

The most influential method to date for verifying RH, up to some height H, has been Turing’s method, which was published in 1953, a year before his most unfortunate death [174, 175]. Almost all methods following Turing share important components with it. There are details in [51, section 8.2]. For an introduction see Booker’s article [12] or that by Hejhal and Odlyzko [78]; see also Trudgian [172]. Here we give a brief outline, with many details skipped.

We saw in Section 2.2 (11) that if we define $E(t) := N(t) - F(t)$ we can write $|E(t)| \le B(t)$. Now $-F(t)$ is a smooth decreasing function and $N(t)$ a step function that increases by 1 at each simple critical line zero, and by at least 2 if t is the ordinate of a pair of zeros off the critical line. We will see below that detecting (and hence counting) zeros on the critical line is straightforward. We have more analytic information on $S(t)$, a good approximation to $E(t)$, defined below, which is contradicted if a pair of off-critical-line zeros occurs. The great strength of Turing's method is that it enables RH to be checked by considering only the values of $\zeta(s)$, and related functions, with s on the critical line.

Some further details: let $\gamma(s) := \pi^{-s/2}\Gamma(s/2)$ and $\Lambda(s) := \gamma(s)\zeta(s)$. Then the functional equation can be written $\Lambda(s) = \Lambda(1-s)$, and this can be used to show that $\Lambda(\frac{1}{2}+it)$ is real for all real t. The zeros of $\Lambda(\frac{1}{2}+it)$ are precisely the zeros of $\zeta(\frac{1}{2}+it)$. Now let $\vartheta(t)$ be the argument of $\gamma(\frac{1}{2}+it)$, with $\vartheta(0) := 0$, or in other words

$$\gamma(\tfrac{1}{2}+it) = |\gamma(\tfrac{1}{2}+it)|e^{i\vartheta(t)}.$$

Then it can be shown that

$$\frac{\vartheta(t)}{\pi} + 1 = F(t) + O\left(\frac{1}{t}\right).$$

The real-valued function of a real variable $S(T)$ is the argument of $\zeta(s)$ on the critical line at the point $\frac{1}{2} + iT$ which is not the ordinate of a zero. The argument is fixed by taking the value found by continuous variation from 0 along the horizontal line from (∞, iT) to $(\frac{1}{2}, iT)$. At a zero the value is defined to be $\lim_{t\to 0+} S(t)$. See Figure 10.4 to get a feeling for how $S(T)$ behaves. It is well known that we can write

$$N(T) = \frac{T}{2\pi}\log\left(\frac{T}{2\pi e}\right) + \frac{7}{8} + S(T) + O\left(\frac{1}{T}\right).$$

See for example Edwards [51, chapters 8 and 9]. It follows that

$$S(t) = N(t) - \left(\frac{\vartheta(t)}{\pi} + 1 + O\left(\frac{1}{t}\right)\right).$$

Now $|\Lambda(s)|$ decays exponentially on the critical line, so, for detecting zeros, it is easier to work with the function

$$Z(t) := \frac{\Lambda(\frac{1}{2}+it)}{|\gamma(\frac{1}{2}+it)|},$$

which also has the same zeros for real t as $\zeta(\frac{1}{2}+it)$. Each simple zero on the critical line for $\zeta(s)$ is accompanied by a sign change in $Z(t)$. This function also has the advantage in computations of requiring fewer terms in a summation than competing methods, via the famous Riemann–Siegel formula:

$$Z(t) = 2\sum_{n=1}^{\lfloor\sqrt{t/2\pi}\rfloor} \frac{\cos(\vartheta(t) - t\log n)}{\sqrt{n}} + O\left(\frac{1}{t^{1/4}}\right).$$

A key ingredient, exploited by Turing, is the theorem of Littlewood, which states that the average value of $S(t)$ is zero:

$$\lim_{T\to\infty} \frac{1}{T} \int_0^T S(t)\,dt = 0.$$

Turing developed an explicit form for this, which was recently improved by Trudgian. For $h > 0$ and $T > 168\pi$ we have

$$\int_T^{T+h} S(t)\,dt \le 2.067 + 0.059 \log(T+h). \tag{2.5}$$

See Lemma 10.29 below.

So, with these preliminaries in hand, we can describe the essence of the method: the function $S(t)$ oscillates about zero in a reasonably controlled manner, with average value zero. Proceed up the critical line counting zeros and "storing" $-S(t)$ using those values and checking (2.5) for h a small multiple of $\log T$. Should a pair of zeros be missed (because they are off the critical line), or a zero on the line be missed, then this would register as an immediate increase in the "central" value of $-S(t)$, and soon become evident as a change in the moving average, by violating (2.5). The region in $\mathbb{C}$ where this occurred could be investigated further to determine whether a zero or multiple zero on the critical line had been missed, or indeed a counterexample to RH had been discovered. This is illustrated in Figures 2.4 and 2.5 for which $-S(t)$ is contrasted with $T(t)$, a function that is the same as $-S(t)$, but has a "fake" missing double zero at $t = 1010$, registered by $S(t)$, but not by $N(t)$.

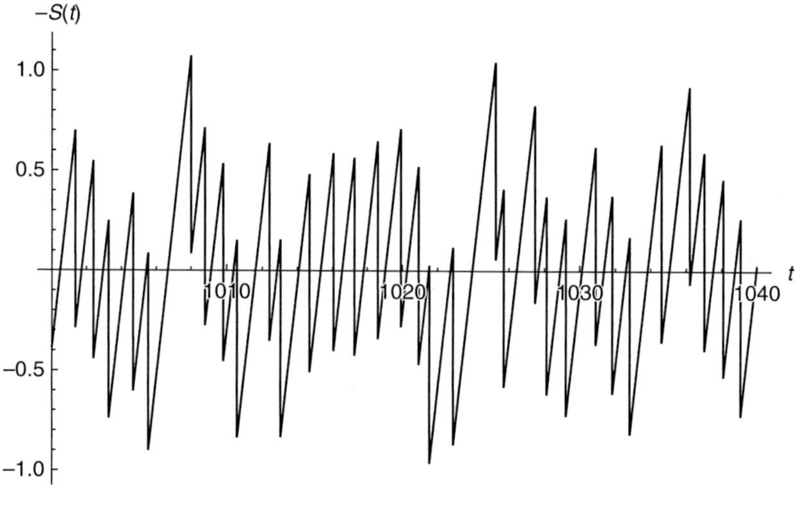

Figure 2.4 The function $-S(t)$ for $1000 \le t \le 1040$.

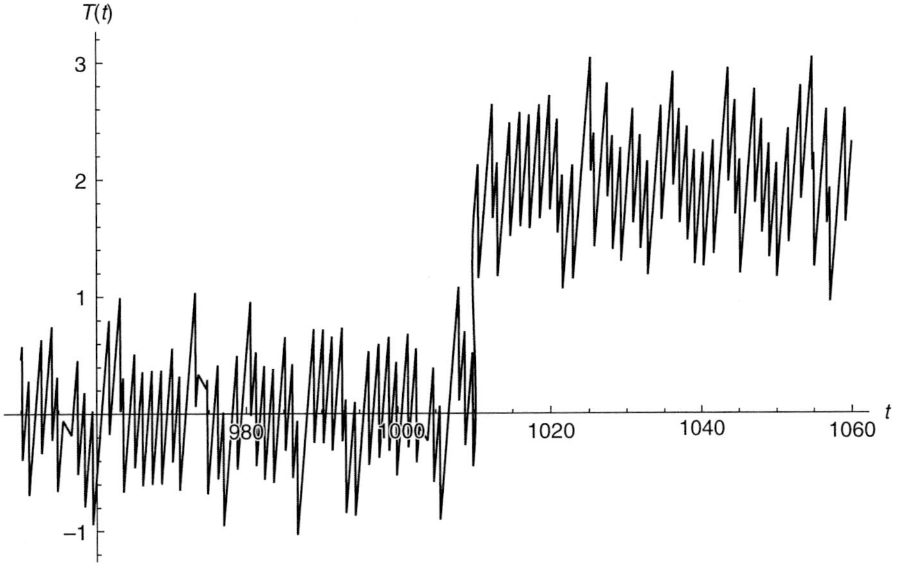

Figure 2.5 The function $T(t)$ on $[950, 1060]$ with a phantom zero at $t = 1010$.

2.4 Landau's Zero-Free Region

Zero-free regions play a vital role in this development. Although Landau's derivation of a zero-free region for $\zeta(s)$ is by now quite classical and has undergone successive improvements (see Section 2.5), it has the advantage of relative simplicity and clear ideas. The progress that has been made has been with considerable effort, but will point to the way forward if RH is shown to be false.

Lemma 2.3 *[105, section 79] If $1 < \sigma \le 2$ and $t \ge 2$ there is a real constant b_1 such that*

$$-\Re\frac{\zeta'(\sigma+it)}{\zeta(\sigma+it)} = \sum_{p,m\ge 1}\frac{\log p\cos(mt\log p)}{p^{m\sigma}}$$
$$< \frac{1}{2}\log t + b_1 - \sum_{\rho}\left(\frac{\sigma-\beta}{(\sigma-\beta)^2+(t-\gamma)^2} + \frac{\beta}{\beta^2+\gamma^2}\right),$$

where the sum is over the imaginary zeros $\rho = \beta + it$ of $\zeta(s)$.

Proof First note that when $\sigma > 1$ we can write

$$\zeta(s) = \prod_p \left(1 - \frac{1}{p^s}\right)^{-1},$$

with the product converging absolutely when σ is bounded away from 1. Taking the negative of the logarithmic derivative gives [6, p. 239]

$$-\Re\frac{\zeta'(s)}{\zeta(s)} = \sum_{p,m} \frac{\log p \cos(mt\log p)}{p^{m\sigma}}.$$

Now consider the Weierstrass product for $\xi(s)$, Theorem 2.11, together with its definition in terms of $\zeta(s)$:

$$\xi(s) = \xi(0)\prod_{\rho}\left(1-\frac{s}{\rho}\right) = (s-1)\pi^{-s/2}\Gamma\left(\frac{s}{2}+1\right)\zeta(s),$$

where the product is over the imaginary zeros of $\zeta(s)$ taken in pairs ρ and $1-\rho$. Take the logarithmic derivative of this identity at points that are not zeta zeros, and use Lemma 2.10, to get

$$\frac{\zeta'(s)}{\zeta(s)} = b - \frac{1}{s-1} - \frac{1}{2}\frac{\Gamma'(s/2+1)}{\Gamma(s/2+1)} + \sum_{\rho}\left(\frac{1}{s-\rho}+\frac{1}{\rho}\right),$$

where the constant $b = \log 2\pi - 1 - \gamma/2$. Then take the negative of the real part of this equation, with $s = \sigma + it$ and $\rho = \beta + i\gamma$, to get

$$\sum_{p,m} \frac{\log p \cos(mt\log p)}{p^{m\sigma}} = -b + \frac{\sigma-1}{(\sigma-1)^2+t^2} + \frac{1}{2}\Re\frac{\Gamma'(\sigma/2+1+ti/2)}{\Gamma(\sigma/2+1+ti/2)}$$
$$-\sum_{\rho}\left(\frac{\sigma-\beta}{(\sigma-\beta)^2+(t-\gamma)^2}+\frac{\beta}{\beta^2+\gamma^2}\right).$$

Now use the bound $|\Gamma'(s)/\Gamma(s)| \le \log t + O(1)$, which can be derived from the equation following (H.4) in Volume Two [32], and which is valid for $0 \le \sigma \le 2$ and $t \ge 2$ to get

$$\sum_{p,m} \frac{\log p \cos(mt\log p)}{p^{m\sigma}} < \frac{1}{2}\log t + b_1 - \sum_{\rho}\left(\frac{\sigma-\beta}{(\sigma-\beta)^2+(t-\gamma)^2}+\frac{\beta}{\beta^2+\gamma^2}\right),$$

which completes the proof of the lemma. □

Theorem 2.4 *[105, section 78] If for all θ we have*

$$a_0 + a_1\cos(\theta) + a_2\cos 2\theta) + \cdots + a_n\cos n\theta \ge 0$$

with each $a_j \ge 0$ and $0 < a_0 < a_1$, and $\rho = \beta + i\gamma$ is a zero of $\zeta(s)$ with $\gamma > 0$, then

$$\beta < 1 - \frac{1}{R\log\gamma},$$

where R is any real number greater than

$$\frac{a_1 + a_2 + \cdots + a_n}{2(\sqrt{a_1}-\sqrt{a_0})^2}.$$

Proof From Lemma 2.3, if we apply the stricter bounds $1 < \sigma \leq 2$ and $t \geq 2$ we have

$$\sum_{p,m} \frac{\log p \cos(mt \log p)}{p^{m\sigma}} < \frac{1}{2}\log t + b_1 - \sum_{\rho}\left(\frac{\sigma-\beta}{(\sigma-\beta)^2+(t-\gamma)^2} + \frac{\beta}{\beta^2+\gamma^2}\right),$$

so in particular we have for some constant b_1 the bound

$$-\Re\left(\frac{\zeta'(\sigma+it)}{\zeta(\sigma+it)}\right) < \frac{1}{2}\log t + b_1.$$

It also follows that for $t = \gamma \geq 2$, if γ is the ordinate of a zero ρ of $\zeta(s)$ and $\sigma \in (1,2]$, then

$$\sum_{p,m} \frac{\log p \cos(m\gamma \log p)}{p^{m\sigma}} < \frac{1}{2}\log \gamma + b_1 - \frac{1}{\sigma-\beta}.$$

Therefore, since for all θ we have

$$-\cos\theta \leq \frac{a_0}{a_1} + \frac{a_2}{a_1}\cos 2\theta + \cdots + \frac{a_n}{a_1}\cos n\theta,$$

we can write

$$\begin{aligned}
\frac{1}{\sigma-\beta} &< \frac{1}{2}\log\gamma + b_1 - \sum_{p,m}\frac{\log p\cos(m\gamma\log p)}{p^{m\sigma}} \\
&\leq \frac{1}{2}\log\gamma + b_1 + \sum_{p,m}\left(\frac{a_0}{a_1} + \frac{a_2}{a_1}\cos(2m\gamma\log p) + \cdots\right. \\
&\qquad \left.\cdots + \frac{a_n}{a_1}\cos(nm\gamma\log p)\right)\frac{\log p}{p^{m\sigma}} \\
&= \frac{1}{2}\log\gamma + b_1 - \frac{a_0}{a_1}\frac{\zeta'(\sigma)}{\zeta(\sigma)} - \frac{a_2}{a_1}\Re\left(\frac{\zeta'(\sigma+2\gamma i)}{\zeta(\sigma+2\gamma i)}\right) - \cdots \\
&\qquad \cdots - \frac{a_n}{a_1}\Re\left(\frac{\zeta'(\sigma+n\gamma i)}{\zeta(\sigma+n\gamma i)}\right).
\end{aligned}$$

Also, since n is fixed and $1 < \sigma \leq 2$, using the Laurent series for $\zeta(s)$ from Volume Two [32, Theorem I.2] to bound $\zeta'(\sigma)/\zeta(\sigma)$ and the bound derived above, there is a constant b_2 such that we have

$$\begin{aligned}
\frac{1}{\sigma-\beta} &< b_2 + \frac{1}{2}\left(1 + \frac{a_2}{a_1} + \cdots + \frac{a_n}{a_1}\right)\log\gamma + \frac{a_0}{a_1}\frac{1}{\sigma-1} \\
&= b_2 + \frac{a_0}{a_1}\frac{1}{\sigma-1} + \frac{a_1+\cdots+a_n}{2a_1}\log\gamma.
\end{aligned}$$

If $\chi := a_0/a_1$ and a real number λ satisfies

$$\lambda > \frac{a_1+\cdots+a_n}{2a_1},$$

it follows that for all sufficiently large γ and $1 < \sigma \le 2$ we have

$$\frac{1}{\sigma - \beta} < \frac{\chi}{\sigma - 1} + \lambda \log \gamma.$$

If we set $\sigma = 1 + g/\log\gamma$, with σ fixed in $(1,2]$, this will define a positive real number g. Applying the inequality we have derived, when g is sufficiently large:

$$\begin{aligned}
\frac{1}{1-\beta+g/\log\gamma} &< \left(\frac{\chi}{g}+\lambda\right)\log\gamma \\
1-\beta+\frac{g}{\log\gamma} &> \left(\frac{1}{\chi/g+\lambda}\right)\frac{1}{\log\lambda} \\
1-\beta &> \left(\frac{1}{\chi/g+\lambda}-g\right)\frac{1}{\log\gamma} = \left(\frac{1-\chi-\lambda g}{\chi/g+\lambda}\right)\frac{1}{\log\gamma}.
\end{aligned}$$

Now choose g so that the coefficient

$$\frac{1-\chi-\lambda g}{\chi/g+\lambda} = \frac{1}{\chi/g+\lambda} - g$$

is as large as possible: we have

$$\begin{aligned}
0 &= \frac{d}{dg}\left(\frac{1}{\chi/g+\lambda} - g\right) \\
&= \frac{\chi}{(\chi+g\lambda)^2} - 1,
\end{aligned}$$

and therefore

$$g = \frac{\sqrt{\chi}-\chi}{\lambda}.$$

Because $0 < a_0 < a_1$, the number g is real and positive. The maximum value of the coefficient is

$$\frac{1}{\chi/g+\lambda} - g = \frac{1-\sqrt{\chi}}{\lambda} - \frac{\sqrt{\chi}-\chi}{\lambda} = \frac{(1-\sqrt{\chi})^2}{\lambda}$$

so that for all sufficiently large γ we have

$$1-\beta > \frac{(1-\sqrt{\chi})^2}{\lambda}\frac{1}{\log\gamma} = \frac{1}{R\log\gamma}$$

with $R > (a_1 + a_2 + \cdots + a_n)/[2(\sqrt{a_1} - \sqrt{a_0})^2]$. This completes the proof. □

Landau used the choices $a_0 = 5$, $a_1 = 8$, $a_2 = 4$ and $a_3 = 3$, with Theorem 2.4 to derive $R = 18.52$.

Table 2.3 Examples for finding values of R.

No.	Year	Author	Form	R
(1)	1898	Hadamard	$3+4\cos(\theta)+\cos(2\theta)$	34.82
(2)	1909	Landau	$5+8\cos(\theta)+4\cos(2\theta)$ $+3\cos(3\theta)$	18.52
(3)	1941	Rosser	$18+30\cos(\theta)+17\cos(2\theta)$ $+6\cos(3\theta)+\cos(4\theta)$	17.72
(4)	1962	Rosser/Schoenfeld	$2(1+\cos(\theta))^2(3+10\cos(\theta))^2$	17.51
(5)	2000	Ford	$(\cos(\theta)+0.225)^2(\cos(\theta)+0.9)^2$	8.463
(6)	2014	Kadiri	$8(\cos(\theta)+0.265)^2(\cos(\theta)+0.91)^2$	5.69693
(7)	2014	Jang/Kwon	$8(\cos(\theta)+a)^2(\cos(\theta)+b)^2$	5.68371
(8)	2014	Platt/Trudgian	$\sum_{k=0}^{16} a_k\cos(k\theta)$	5.573412

2.5 Zero-Free Regions Summary

In this section we present a summary of some of the developments regarding zero-free regions of $\zeta(s)$. Example (1) was the original used by Hadamard and de la Vallée Poussin in their proof of the prime number theorem, (2) appears in Landau's book [105], (3) was employed by Rosser in his 1941 paper, (4) after probably some serious searching by Rosser and Schoenfeld in their 1962 paper, with only a modest improvement in R. Example (5) was found using numerical optimization by Ford [59].

The value from (4) is

$$R = \frac{515}{(\sqrt{546}-\sqrt{322})^2} = 17.51631055012728537\ldots.$$

Using Ford's explicit bound from 2002 [60], namely for $\rho=\beta+i\gamma$ a zeta zero with $\gamma>0$,

$$\beta < 1 - \frac{1}{57.54\log^{2/3}\gamma\log\log^{1/3}\gamma},$$

one could make R arbitrarily small, provided H is sufficiently large. At the time of writing, the known value of H is well out of range.

In their 1975 paper [146], Rosser and Schoenfeld used a different method, which is more analytic than Landau's, to find a better, but quite convenient, error-free region for $\zeta(s)$. It takes a form that is just slightly better than

$$\beta < 1 - \frac{1}{R_1\log(\gamma/17)},$$

where $\rho=\beta+i\gamma$ is a zero of zeta with

$$R_1 := (1-1/\sqrt{5})515/(\sqrt{546}-\sqrt{322})^2 = 9.6827783291110\ldots$$

and $\gamma \geq 21$. This form for the zero-free region, together with a better approximation for the integrals

$$\int_h^\infty \phi_m(t)\log\left(\frac{t}{2\pi}\right)dt,$$

than that which has been used here in Chapter 3, gives, on the face of it, better results than those we computed and use when calculating estimates. However, since we do not need sharper estimates than those obtained by using larger values of H, and in the interests of simplicity of exposition, we have not used R_1.

Finally, note some more recent results: Kadiri [84] showed that we could choose $R = 5.69693$. He used Weil's explicit formula with the method and one of the test functions of Heath-Brown. For background definitions, proofs and applications of the explicit formula, see Volume Two [32, Chapter 9]. The value was slightly improved by Jang and Kwon, who used an alternative test function in the explicit formula and Platt's height $H = 3.0610046 \times 10^{10}$ of 2011 to derive $R = 5.68371$; see [83]. This has been further improved to $R = 5.573412$ by Mossinghoff and Trudgian [121] by finding a well-adapted trigonometric polynomial (of degree 16) using simulated annealing, and carefully analysing the error term of Kadiri. They used Platt's value of H.

Ford used the value for H established by van de Lune, te Riele and Winter [111], and Kadiri used the 2004 height of Wedeniwski. We have chosen to use a value $R = 8.0$ in the computations reported in Chapter 3.

The details of these more recent results are outside the scope of this book. The interested reader could consult the fundamental papers [77, 179], or the references given above.

2.6 The Product Over Zeta Zeros

In this section we show that the completed zeta function $\xi(s)$ has a product representation

$$\xi(s) = \xi(0)\prod_\rho\left(1 - \frac{s}{\rho}\right).$$

This is given in Theorem 2.11. First we have two lemmas, which take the form of evaluations, and then a call-out to Hadamard's product expansion, which is proved in Volume Two [32, Appendix B].

Lemma 2.5 *We have the special values:*

(1) $\dfrac{\Gamma'(1)}{\Gamma(1)} = -\gamma,$

(2) $\dfrac{\Gamma'(3/2)}{\Gamma(3/2)} = 2 - \gamma - \log 4,$

(3) $\dfrac{\zeta'(0)}{\zeta(0)} = \log(2\pi)$.

Proof **(1)** By Lemma H.1 of Volume Two [32], for all z which is not a negative integer, we have the following representation for the logarithmic derivative of $\Gamma(z)$:

$$\frac{\Gamma'(z)}{\Gamma(z)} = -\gamma - \frac{1}{z} + \sum_{n=1}^{\infty}\left(\frac{1}{n} - \frac{1}{n+z}\right),$$

and the evaluation follows on setting $z = 1$.

(2) By the formula in Step (1) we also have

$$\begin{aligned}\frac{\Gamma'(3/2)}{\Gamma(3/2)} &= -\gamma - \frac{2}{3} + \sum_{n=1}^{\infty}\left(\frac{1}{n} - \frac{1}{n+3/2}\right)\\ &= -\gamma - \frac{2}{3} + \frac{8-3\log 4}{3}\\ &= 2-\gamma-\log 4.\end{aligned}$$

Alternatively, the value in part (2) of the lemma follows from Gauss's digamma theorem [92]: for $p,q \in \mathbb{N}$

$$\begin{aligned}\frac{\Gamma'(p/q)}{\Gamma(p/q)} =& -\gamma - \log(2q) - \frac{1}{2}\pi\cot(\pi p/q)\\ &+ 2\sum_{0<n<q/2}\cos((2\pi pn)/q)\log(\sin(\pi n/q)).\end{aligned}$$

(3) A proof using Euler–Maclaurin summation is as follows. By Abel's theorem [6, theorem 4.2] we have for $\sigma > 1$

$$\sum_{n=N}^{\infty}\frac{1}{n^s} = \frac{1}{(s-1)N^{s-1}} + \frac{1}{N^s} - s\int_N^{\infty}\frac{\{x\}}{x^{s+1}}\,dx,$$

and

$$\int_N^{\infty}\frac{dx}{x^s} = \frac{1}{(s-1)N^{s-1}}.$$

Hence

$$\sum_{n=N}^{\infty}\frac{1}{n^s} - \int_N^{\infty}\frac{dx}{x^s} = \frac{1}{N^s} - s\int_N^{\infty}\frac{\{x\}}{x^{s+1}}\,dx.$$

For $\Re s = \sigma > 1$ and positive integer N, evaluating the integral gives

$$\zeta(s) - \frac{1}{s-1} = \sum_{n=1}^{\infty}\frac{1}{n^s} - \int_1^{\infty}\frac{dx}{x^s}$$

$$= \sum_{n=1}^{N-1} \frac{1}{n^s} + \sum_{n=N}^{\infty} \frac{1}{n^s} - \int_1^N \frac{dx}{x^s} - \int_N^{\infty} \frac{dx}{x^s}$$
$$= \sum_{n=1}^{N-1} \frac{1}{n^s} - \int_1^N \frac{dx}{x^s} + \sum_{n=N}^{\infty} \frac{1}{n^s} - \int_N^{\infty} \frac{dx}{x^s}$$
$$= \sum_{n=1}^{N-1} \frac{1}{n^s} - \int_1^N \frac{dx}{x^s} + \frac{1}{N^s} - s\int_N^{\infty} \frac{\{x\}}{x^{s+1}}\,dx.$$

The right-hand side converges to a holomorphic function on the half plane $\sigma > 0$ and gives an analytic continuation of the left-hand side in this half plane. At $s = 1$ the right-hand side converges, as $N \to \infty$, to Euler's constant γ, because we have the representation

$$\gamma = \lim_{N\to\infty}\left(1 + \frac{1}{2} + \cdots + \frac{1}{N} - \log N\right).$$

Therefore the Taylor expansion of $g(s) := (s-1)\zeta(s)$ about $s = 1$ has the form

$$g(s) = (s-1)\zeta(s) = 1 + \gamma(s-1) + O(|s-1|^2) \quad \Longrightarrow \quad \frac{g'(1)}{g(1)} = \frac{\gamma}{1} = \gamma.$$

Next write the functional equation of $\zeta(s)$ in the form [6, theorem 12.7]

$$\zeta(s) = \Gamma(1-s)(2\pi)^{s-1}2\sin(s\pi/2)\zeta(1-s).$$

This implies $(s-1)\zeta(s) = -\Gamma(2-s)(2\pi)^{s-1}2\sin(s\pi/2)\zeta(1-s)$. Take the logarithmic derivative of this equation at $s = 1$ and use part (1) of the lemma to derive

$$\frac{g'(1)}{g(1)} = \gamma = -\frac{\Gamma'(1)}{\Gamma(1)} + \log(2\pi) + \frac{\pi\cos(\pi/2)}{2\sin(\pi/2)} - \frac{\zeta'(0)}{\zeta(0)}$$
$$\Longrightarrow \quad \frac{\zeta'(0)}{\zeta(0)} = \log(2\pi),$$

which completes the derivation. □

Lemma 2.6 *If γ is Euler's constant then*

$$\lim_{s\to 1}\left(\frac{\zeta'(s)}{\zeta(s)} + \frac{1}{s-1}\right) = \gamma.$$

Proof This follows from Step (3) of the proof of Lemma 2.5. □

We also need, in several places below, the well-known product expansion for $\xi(s)$ in terms of the non-trivial zeta zeros. This can be proved directly from properties of the constituent functions (see for example [51, chapter 2]), but we have chosen to give a proof based on Hadamard's expansion, which is proved in Volume Two [32, Theorem B.10]; see also [43, section 12].

Recall the definition: the **order** of an entire function $f(z)$ is the infimum of real values b such that as $z \to \infty$, $f(z) \ll \exp(|z|^b)$. Next we give Hadamard's theorem statement.

Theorem 2.7 (Hadamard factorization) *Let $f(z)$ be an entire function of finite order ρ, having $f(0) \neq 0$ and zeros $z_1, z_2, \ldots$, with multiple zeros appearing multiple times. Then $f(z)$ has the absolutely and uniformly converging on compact subsets product representation*

$$f(z) = e^{P(z)} \prod_{n=1}^{\infty} \left(\left(1 - \frac{z}{z_n} \right) e^{Q(z/z_n)} \right), \tag{2.6}$$

where $P(z)$ is a polynomial of degree less than or equal to ρ, and $Q(z)$ is a polynomial of the form

$$Q(z) = z + \frac{z^2}{2} + \cdots + \frac{z^m}{m},$$

where $m \geq 0$ is the smallest integer such that

$$\sum_{n=1}^{\infty} \frac{1}{|z_n|^{m+1}} < \infty.$$

Recall the definition:

$$\xi(s) := \tfrac{1}{2} s(s-1)\pi^{-s/2}\Gamma(s/2)\zeta(s).$$

Lemma 2.8 *For all $s \subset \mathbb{C}$ we have constants A and B such that*

$$\xi(s) = e^{A+Bs} \prod_{\rho} \left(\left(1 - \frac{s}{\rho} \right) e^{s/\rho} \right),$$

where, if we combine the factors associated with $(1 - s/\rho)$ and $(1 - s/(1-\rho))$, the product converges absolutely and uniformly on compact subsets of $\mathbb{C}$.

Proof **(1)** It follows from properties of the zeros and poles of $\zeta(s)$ and $\Gamma(s)$ that $\xi(s)$ is an entire function. We first show that $\xi(s)$ has order at most 1. By the functional equation we need only consider $\sigma \geq \frac{1}{2}$. In this half plane, for $|s|$ sufficiently large and some constant $c_1 > 0$ we have

$$|\tfrac{1}{2} s(s-1)\pi^{-s/2}| \leq \exp(c_1|s|).$$

By Stirling's approximation (see (H.5) in Theorem H.5 in Volume Two [32]), there is a constant $c_2 > 0$ such that, for $s \to \infty$ in $\sigma \geq \frac{1}{2}$,

$$\Gamma(s/2) \leq \exp(c_2|s|\log|s|).$$

Then using Abel's theorem [6, theorem 4.2] for $\sigma > 1$, and extending the domain of validity of the expression by analytic continuation to $\sigma > 0$,

$$\zeta(s) = \frac{s}{s-1} - s\int_1^{\infty} \frac{\{x\}}{x^{s+1}}\, dx,$$

which gives $|\zeta(s)| \leq c_3|s|$ as $s \to \infty$ in $\sigma \geq \frac{1}{2}$. Combining these estimates with the definition of $\xi(s)$, and the functional equation, gives for each $\epsilon > 0$, as $|s| \to \infty$,

$$|\xi(s)| \leq \exp(|s|^{1+\epsilon}).$$

Hence the order is at most 1. We cannot do better than this in terms of the value of the exponent on the right-hand side, since when $\sigma \to \infty$, $\log\Gamma(\sigma)$ is asymptotic to $\sigma\log\sigma$. Hence the order of $\xi(s)$ is 1.

(2) Next we show that the parameter m in Hadamard's theorem (Theorem 2.7) has the value 1. This is because the $\sum_\rho 1/|\rho|^2$ converges and $\sum_\rho 1/|\rho|$ diverges. To verify the former write, since the number of Riemann zeta zeros up to T, by Lemma 2.9, satisfies $N(T) \ll T\log T$,

$$\sum_\rho \frac{1}{|\rho|^2} = 2\sum_{\gamma>0} \frac{1}{\beta^2+\gamma^2} \leq 2\sum_{\gamma>0} \frac{1}{\gamma^2} \ll \int_1^\infty \frac{d(t\log t)}{t^2} < \infty.$$

To verify the latter statement, by Theorem 10.32 proved below, for each $n \in \mathbb{N}$ there is a zeta zero γ_{j_n} such that $|4n - \gamma_{j_n}| \leq 1.5$. Therefore if $n \neq n'$, we have $\gamma_{j_n} \neq \gamma_{j_{n'}}$ and we can write

$$\sum_\rho \frac{1}{|\rho|} \geq 2\sum_{\gamma>0} \frac{1}{\sqrt{1/4+\gamma^2}} \geq \sum_{\gamma>0} \frac{1}{|\gamma|} \geq \sum_{n=1}^\infty \frac{1}{|\gamma_{j_n}|} \geq \sum_{n=1}^\infty \frac{1}{|4n-2|} = \infty.$$

(3) It follows from Steps (1) and (2) and Hadamard's theorem that there are constants A, B such that for $s \in \mathbb{C}$

$$\xi(s) = e^{A+Bs}\prod_\rho \left(1 - \frac{s}{\rho}\right)e^{s/\rho}. \tag{2.7}$$

This completes the proof. □

In Section 2.2 (11), we commented on available good estimates for zeta zeros in the critical strip. Sometimes in what follows all we need to have is an asymptotic upper bound for the number of zeros up to T as a function of T. This is given in Lemma 2.9, which is a standard application of Jensen's formula. Indeed, we used the bound $N(t) \ll t\log t$ in Step (3) of the previous lemma, and will use it again in Lemma 2.10. Note the argument here is not circular!

Lemma 2.9 *The function $\xi(s)$ has at most $O(R\log R)$ zeros in any disc $B(0,R]$.*

Proof Let $0 < R$, let $n(R)$ be the number of zeros of $\xi(z)$ in $B(0,R]$, and let $M(R) = \sup\{|\xi(z)| : |z| \leq R\}$. By the proof of Lemma 2.8, Step (1), we have $M(R) \leq \exp(c_1 R\log R)$.

Label the zeros (a_n) of $\xi(z)$ in $B(0,R]$ such that

$$|a_1| \le |a_2| \le \cdots.$$

Next we use Jensen's formula, from Volume Two [32, Theorem B.5], to write

$$M(2R) \ge \exp\left(\frac{1}{2\pi}\int_0^{2\pi} \log|\xi(2Re^{i\theta})|\,d\theta\right) = \prod_{n=1}^{n(2R)} \frac{2R}{|a_n|} \ge \prod_{n=1}^{n(R)} \frac{2R}{|a_n|} \ge 2^{n(R)}.$$

Therefore $n(R) \le \log M(2R)/\log 2$. Since we have $M(R) \le e^{c_1 R\log R}$ so $n(R) \le c_3 R\log R$, and therefore $n(R) = O(R\log R)$. This completes the proof. □

We also need, frequently in the developments that follow in later chapters, several expressions and estimates for sums over critical zeta zeros. Statement (a) in Lemma 2.10 was, apparently, known to Riemann. We saw in the proof of Lemma 2.8 that the convergence was conditional. In the second sum, as usual $1/\rho$ is associated with $1/(1-\rho)$, in which form the series is absolutely convergent.

Lemma 2.10 *We have*

(a) $\displaystyle\sum_\rho \frac{1}{\rho} = 1 + \frac{\gamma}{2} - \frac{1}{2}\log 4\pi,$

(b) $\displaystyle\sum_\rho \frac{1}{\rho(1-\rho)} = 2 + \gamma - \log 4\pi,$

(c) $\displaystyle\sum_\rho \frac{1}{|\rho|^2} \le 0.046191441,$

where the sums are over all of the non-trivial zeros of $\zeta(s)$.

Proof By the definition of $\zeta(s)$ and Lemma 2.8, for some real U, V we can write

$$\xi(s) = \frac{1}{2}s(s-1)\pi^{-s/2}\Gamma\left(\frac{s}{2}\right)\zeta(s) = e^{U-Vs}\prod_\rho\left(1-\frac{s}{\rho}\right)e^{s/\rho} = \xi(1-s), \tag{2.8}$$

where the product is over the non-trivial zeros of $\zeta(s)$.

Logarithmic differentiation gives

$$\begin{aligned}\frac{\xi'(s)}{\xi(s)} &= \frac{\zeta'(s)}{\zeta(s)} + \frac{1}{s-1} - \frac{1}{2}\log\pi + \frac{1}{2}\frac{\Gamma'(s/2+1)}{\Gamma(s/2+1)}\\ &= -V + \sum_\rho\left(\frac{1}{s-\rho}+\frac{1}{\rho}\right).\end{aligned} \tag{2.9}$$

By Lemmas 2.5 (2) and 2.6, respectively, we know that

$$-\frac{1}{2}\frac{\Gamma'(3/2)}{\Gamma(3/2)} = \frac{\gamma}{2} - 1 + \log 2 \quad\text{and}\quad \lim_{s\to 1}\left(\frac{\zeta'(s)}{\zeta(s)} + \frac{1}{s-1}\right) = \gamma,$$

so using (2.9) we can write

$$\begin{aligned}-V &= \lim_{s\to 1}\left(\frac{\zeta'(s)}{\zeta(s)} + \frac{1}{s-1}\right) - \frac{1}{2}\log\pi + \frac{1}{2}\frac{\Gamma'(3/2)}{\Gamma(3/2)} \\ &= 1 + \frac{\gamma}{2} - \frac{1}{2}\log(4\pi).\end{aligned}$$

But by (2.8) we have

$$V = -\frac{\xi'(0)}{\xi(0)} = \frac{\xi'(1)}{\xi(1)},$$

so using (2.9) we get

$$-V + \sum_{\rho}\left(\frac{1}{-\rho} + \frac{1}{\rho}\right) = V - \sum_{\rho}\left(\frac{1}{1-\rho} + \frac{1}{\rho}\right).$$

Because $\zeta(\rho) = 0$ if and only if $\zeta(1-\rho) = 0$, there are matching terms in these sums that cancel, leading to $2V = 2\sum_{\rho} 1/\rho$ and therefore

$$V = \sum_{\rho}\frac{1}{\rho} = 1 + \frac{\gamma}{2} - \frac{1}{2}\log(4\pi). \tag{2.10}$$

This completes the proof of part (*a*). Part (*b*) is immediate since

$$\sum_{\rho}\frac{1}{\rho(1-\rho)} = \sum_{\rho}\left(\frac{1}{\rho} + \frac{1}{1-\rho}\right) = 2\sum_{\rho}\frac{1}{\rho}.$$

Now choose a height H up to which all of the zeta zeros are on the critical line and define

$$S_H = \sum_{|\gamma|\le H}\frac{1}{1/4+\gamma^2} \quad \text{and} \quad S_\infty = \sum_{|\gamma|>H}\frac{1}{\beta^2+\gamma^2}.$$

We choose in this derivation $H = 600\,269.6771$, giving 10^6 zeros with positive imaginary part. Then

$$\sum_{\rho}\frac{1}{|\rho|^2} \le S_H + S_\infty.$$

Note also that

$$|\gamma| > H \quad \Longrightarrow \quad \frac{\beta}{\gamma^2} < \left(1 + \frac{1}{H^2}\right)\frac{\beta}{\beta^2+\gamma^2}$$

and

$$\frac{\beta-1}{\gamma^2} < \left(1 + \frac{1}{H^2}\right)\frac{\beta-1}{(\beta-1)^2+\gamma^2},$$

and therefore

$$S_\infty < \sum_{|\gamma|>H} \frac{1}{\gamma^2} = \sum_{|\gamma|>H} \left(\frac{\beta}{\gamma^2} + \frac{1-\beta}{\gamma^2} \right)$$
$$< \left(1 + \frac{1}{H^2}\right) \sum_{|\gamma|>H} \left(\frac{\beta}{\beta^2+\gamma^2} + \frac{1-\beta}{(1-\beta)^2+\gamma^2} \right).$$

But

$$2V = \sum_\rho \frac{1}{\rho} + \frac{1}{\bar\rho} = \sum_\rho \frac{2\beta}{|\rho|^2} \quad \implies \quad S_\infty < \left(1+\frac{1}{H^2}\right)(2V - S_H),$$

and therefore

$$\sum_\rho \frac{1}{|\rho|^2} < S_H + \left(1+\frac{1}{H^2}\right)(2V - S_H) < 0.046191441,$$

which completes the proof of part (c). □

Theorem 2.11 *For all $s \in \mathbb{C}$ we have*

$$\xi(s) = \xi(0) \prod_\rho \left(1 - \frac{s}{\rho}\right),$$

where if we combine the factors $(1 - s/\rho)$ and $(1 - s/(1-\rho))$ the product converges absolutely and uniformly on compact subsets of $\mathbb{C}$. In addition $\xi(0) = \frac{1}{2}$.

Proof **(1)** By Lemma 2.8 there are constants A, B such that for $s \in \mathbb{C}$

$$\xi(s) = e^{A+Bs} \prod_\rho \left(1 - \frac{s}{\rho}\right) e^{s/\rho}. \tag{2.11}$$

Use the definition of ξ and take the logarithmic derivative of both sides of

$$e^{A+Bs} \prod_\rho \left(1 - \frac{s}{\rho}\right) e^{s/\rho} = \frac{1}{2} s(s-1)\pi^{-s/2}\Gamma(s/2)\zeta(s)$$

to get

$$\frac{\zeta'(s)}{\zeta(s)} = B + \frac{1}{2}\log\pi - \frac{1}{s-1} - \frac{1}{2}\frac{\Gamma'(s/2+1)}{\Gamma(s/2+1)} + \sum_{\rho\in S}\left(\frac{1}{s-\rho} + \frac{1}{\rho}\right).$$

By (2.10) of Lemma 2.10 we get the value of B:

$$B = -V = -\sum_\rho \frac{1}{\rho} = \frac{1}{2}\log(4\pi) - 1 - \frac{\gamma}{2}.$$

(2) From the definition of $\xi(s)$ and the functional equation $\xi(s) = \xi(1-s)$ we can write

$$\xi(0) = \xi(1) = \frac{1}{2}\pi^{-1/2}\Gamma(1/2)\lim_{s\to 1}(s-1)\zeta(s) = \frac{1}{2}.$$

Thus, by (2.11), we have $e^A = \xi(0) = \frac{1}{2}$.

(3) Next we consider the convergence of the product. With $\rho = \beta + i\gamma$ let

$$\prod_{\rho}\left(1 - \frac{s}{\rho}\right) = \prod_{\gamma>0}\left(1 - \frac{s(1-s)}{\rho(1-\rho)}\right).$$

This converges absolutely provided $\sum_{\gamma>0} 1/|\rho(1-\rho)| < \infty$. To see this, complete the square

$$\sum_{\gamma>0}\frac{1}{|\rho(1-\rho)|} = \sum_{\gamma>0}\frac{1}{\left|(\rho-\frac{1}{2})^2 + \frac{1}{4}\right|} < \sum_{\gamma>0}\frac{1}{|\rho-\frac{1}{2}|^2} \ll \int_1^\infty \frac{d(t\log t)}{t^2} < \infty.$$

(4) Because $V = -B$ we are now able to write

$$\xi(s) = \xi(0)\exp\left(s\left(B + \sum_{\rho}\frac{1}{\rho}\right)\right)\prod_{\rho}\left(1 - \frac{s}{\rho}\right) = \xi(0)\prod_{\rho}\left(1 - \frac{s}{\rho}\right).$$

This completes the proof. □

Finally, as a corollary to Theorem 2.11, we can derive a useful expression for the logarithmic derivative of $\zeta(s)$.

Lemma 2.12 *[51, section 3.2] Let S be the zeros of $\zeta(s)$ with $0 < \Re s < 1$. Then for all $s \neq 1$ and not in S in $\mathbb{C}$ we have*

$$\frac{\zeta'(s)}{\zeta(s)} = \log(2\pi) - 1 - \frac{\gamma}{2} - \frac{1}{s-1} - \frac{1}{2}\frac{\Gamma'(s/2+1)}{\Gamma(s/2+1)} + \sum_{\rho\in S}\left(\frac{1}{s-\rho} + \frac{1}{\rho}\right).$$

Proof From the definition of $\xi(s)$ and Theorem 2.11 we have

$$\Gamma(s/2+1)\pi^{-s/2}(s-1)\zeta(s) = \xi(s) = \xi(0)\prod_{\rho}\left(1 - \frac{s}{\rho}\right),$$

where the product is over the non-trivial zeros of $\zeta(s)$. If s is not a zeta zero we can take the logarithmic derivative to get

$$-\frac{\zeta'(s)}{\zeta(s)} = -\frac{1}{2}\log\pi + \frac{1}{s-1} + \frac{1}{2}\frac{\Gamma'(s/2+1)}{\Gamma'(s/2+1)} - \sum_{\rho}\frac{1}{s-\rho}.$$

Evaluating this at $s = 0$ and using Lemma 2.5 (1) gives

$$\frac{\zeta'(0)}{\zeta(0)} = \frac{1}{2}\log\pi + 1 + \frac{\gamma}{2} - \sum_{\rho}\frac{1}{\rho}.$$

Adding these two equations gives

$$-\frac{\zeta'(s)}{\zeta(s)} + \frac{\zeta'(0)}{\zeta(0)} = 1 + \frac{\gamma}{2} + \frac{1}{s-1} + \frac{1}{2}\frac{\Gamma'(s/2+1)}{\Gamma'(s/2+1)} - \sum_{\rho}\left(\frac{1}{s-\rho} + \frac{1}{\rho}\right).$$

Finally substitute for $\zeta'(0)/\zeta(0)$ from Lemma 2.5 (3) to complete the proof of the lemma. □

2.7 Unsolved Problems

(1) Prove that $\zeta(3)$ is a rational multiple of π^3.

(2) Further improve the result of Backlund which gives an estimate for the number of zeros of $\zeta(s)$ in the positive critical strip up to height T.

(3) Find a sum of cosines giving a smaller value of R, searching through functions of the form $f(\theta) = \prod_{1 \le i \le n}(a_i + b_i \cos\theta)^2$ or otherwise.

(4) Determine, using Dirichlet's pigeonhole principle or otherwise, whether the value of R can be made arbitrarily close to $\frac{1}{2}$, i.e. for all $\epsilon > 0$ there is a positive sum of cosines such that $\frac{1}{2} < R < \frac{1}{2} + \epsilon$.

(5) Improve the proof of Rosser and Schoenfeld of [146, theorem 1] to find a smaller value of the denominator $R_1 \log(t/17)$ with $|t| \ge t_0 := 21$, by either reducing the value of R_1 or replacing 17 with a larger value, at possibly the cost of using a much larger t_0.

(6) Study the works given in the references at the end of Section 2.3, and use the methods therein, with improved test functions, to find a smaller value of R.

3
Estimates

3.1 Introduction

This chapter includes the derivation of estimates that are used throughout the volume. The aim here is not to be comprehensive and derive all estimates of a given type, or to present the best that are, in some sense, currently available. Almost all of the results that are needed for deriving equivalences included in the volume are deduced from first principles, in a manner which could be followed step by step, should the reader so wish.

The main focus is on numerical estimates, especially of the arithmetical functions $\psi(x)$ and $\theta(x)$. Note that there are additional estimates of products of functions of primes in Section 5.2. In the main, the estimates in this chapter are derived without using RH, and so are significantly weaker than Theorems 4.5 or 4.6, which give good estimates for $\psi(x)$ and $\theta(x)$, which could be regarded as best possible. The material is based on the work of Rosser and Schoenfeld, scattered through their separate or joint papers of 1941, 1962 and 1975, but also depends on von Mangoldt's theorems of 1905.

The work of Rosser and Schoenfeld is remarkable in that they produced sharp explicit results before computer-based methods became easy to use and ubiquitous. In this chapter we take some advantage of the much greater processing speed available in the early twenty-first century to extend the ranges for which computer verification of inequalities is practicable. We also are able to take advantage of the much greater height, denoted by H, below which all the zeros of $\zeta(s)$, with positive imaginary part, have real part $\sigma = \frac{1}{2}$. Results are expressed in terms of key parameters, regarded as constants, but which are likely to change in time with further improvements.

Recall that $\theta(x) := \sum_{n \le x} \log p$ and $\psi(x) := \sum_{p^m \le x} \log p$. In Section 3.2 tables of values of b and $\epsilon_b > 0$ are derived such that

$$x \ge e^b \quad \Longrightarrow \quad |\psi(x) - x| < \epsilon_b x,$$

where e is the exponential number. This represents a recalculation of the table developed by Rosser and Schoenfeld in 1962 [145, table I], with improved parameters and some more minor improvements. The range is increased from $b = 5000$ to $b = 10000$, but the vital tool is an algorithm to compute ϵ_b given b, rather than the table.

In Section 3.3 some details are given of the simple strategy used to derive exact results from inequalities using floating-point arithmetic. In Section 3.4 estimates are given for $\theta(x)$ of the form

$$|\theta(x) - x| < \epsilon \frac{x}{\log x},$$

with x in a range depending on ϵ, and also the useful bounds $0.8x \leq \theta(x) \leq 1.02x$ valid for $x \geq 121$. The inequality $\theta(x) < 1.000081x$, used by some authors, was able to be proved, but only by assuming RH. This proof is included in the final section, as are some useful bounds for $\psi(x) - \theta(x)$.

Many authors simply assume these estimates, especially those from [145]. However, the work presented here, especially that in Chapters 5 and 7, is totally dependent on their validity, so sufficient detail of derivations is given in this chapter for readers to reproduce results, or to set about improving them.

The RHpack (see Appendix B) functions ComputeKm, PsiSmallXEpsilon, PsiLargeXEpsilon, PsiLargeXEpsilon2 and VerifyThetaX relate to the material in this chapter.

3.2 Constructing Tables of Bounds for $\psi(x)$

The first lemma, Lemma 3.1, gives two bounds for integrals that are used in the derivation of the ϵ_b, the first bound rather complex and the second simpler. We used the more complex form in computations, but the simpler form could find theoretical uses. Note that in their 1975 paper, Rosser and Schoenfeld use modified Bessel functions to find better upper bounds for the integrals, and their method was implemented by Dusart as reported in [49]. Those results are not needed for the RH equivalences developed in later chapters.

The parameter R is described in Section 2.3: if $\rho = \beta + i\gamma$ is an imaginary zero of $\zeta(s)$, with $\gamma > 0$, then

$$\beta < 1 - \frac{1}{R \log \gamma}.$$

Lemma 3.1 *[144, lemma 19] If $m \geq 1$ is a natural number with real x satisfying*

$$0 < 2 \log x < Rm \log^2 h,$$

then for $h > 2\pi e$, if we define

$$\phi_m(t) := t^{-m-1} e^{-(\log x)/(R \log t)},$$

we have

$$\int_h^\infty \phi_m(t)\log\left(\frac{t}{2\pi}\right)dt < \frac{1+m\log\left(\dfrac{h}{2\pi}\right)}{m^2h^m x^{1/(R\log h)}\left(1-\dfrac{(1+m\log(h/2\pi))\log x}{\log(h/2\pi)Rm^2\log^2 h}\right)}$$
$$=: \mathcal{B}_1(m,x,h).$$

If however we have $0<4\log x<Rm\log^2 h$, *again with* $h\geq 2e\pi$, *then the bound takes a simpler form:*

$$\int_h^\infty \phi_m(t)\log\left(\frac{t}{2\pi}\right)dt < \frac{2\log\left(\dfrac{he}{2\pi}\right)}{mh^m x^{1/(R\log h)}}.$$

Proof First integrate $I(h) := \int_h^\infty \phi_m(t)\log(t/2\pi)\,dt$ by parts, noting that

$$\int \frac{\log\left(\dfrac{t}{2\pi}\right)}{t^{m+1}}\,dt = -\frac{1+m\log\left(\dfrac{t}{2\pi}\right)}{m^2t^m},$$

to get

$$I(h) = \frac{1+m\log\left(\dfrac{h}{2\pi}\right)}{m^2h^m x^{1/(R\log h)}} + \int_h^\infty \frac{\left(1+m\log\left(\dfrac{t}{2\pi}\right)\right)\log x}{Rm^2\log^2 t\log\left(\dfrac{t}{2\pi}\right)}\log\left(\frac{t}{2\pi}\right)\phi_m(t)\,dt.$$

Therefore

$$I(h) < \frac{1+m\log\left(\dfrac{h}{2\pi}\right)}{m^2h^m x^{1/(R\log h)}} + \left(\frac{\left(1+m\log\left(\dfrac{h}{2\pi}\right)\right)\log x}{Rm^2\log^2 h\log\left(\dfrac{h}{2\pi}\right)}\right)I(h).$$

Solving for $I(h)$ gives the stated upper bound. Note that the condition on $\log x$ ensures the denominator of the upper bound is positive. The more stringent condition ensures the bracketed term in the denominator is greater than $\frac{1}{2}$ and the numerator less than $m\log(he/2\pi)$. This completes the proof of the lemma. □

In Chapter 4 a simplified version of the following development is given, sufficient for what is needed to prove one of the equivalences to RH presented there. Here, in order that Tables 3.2 and 3.3 might be derived, the additional integer parameter m, used in Lemma 3.1, is needed. In addition, we replace

$\psi(x) - x$ with an approximating function $\eta(x)$ which simplifies the estimation problem. Let

$$\eta(x) := \psi(x) - x + \log(2\pi) + \frac{1}{2}\log\left(1 - \frac{1}{x^2}\right).$$

The basic definition needed here is an extension of $E(x,h)$, which will be defined in the proof of Lemma 4.3, to all integers $m \geq 1$:

$$E_m(x,h) := \int_0^h \cdots \int_0^h \eta(x + y_1 + \cdots + y_m)\, dy_1 \cdots dy_m.$$

Lemma 3.2 *[146, lemma 8, p. 256] If $x > 1$, m is a positive integer and $0 < \delta < (1 - 1/x)/m$, we have*

$$-\epsilon_1 x \leq \eta(x) \leq \epsilon_2 x,$$

where

$$\epsilon_1 := \frac{E_m(x, -x\delta)}{(-1)^{m+1} x^{m+1} \delta^m} + \frac{m\delta}{2} \quad \textit{and} \quad \epsilon_2 := \frac{E_m(x, x\delta)}{x^{m+1}\delta^m} + \frac{m\delta}{2}.$$

Proof **(1)** First we need two definitions:

$$f_{m,n,\alpha}(x,h,z) := \frac{E_m(x,h)}{h^n} + \frac{nh^\alpha}{2} - zh^{\alpha-1},$$

$$f_m(x,h,z) := f_{m,m,1}(x,h,z) = \frac{E_m(x,h)}{h^m} + \frac{mh}{2} - z.$$

Also assume $x > 1$ and $x + mh > 1$. The variable h could be negative; indeed we will later choose $h = \pm\delta x$. In the properties below, (2) and (3) are immediate and (4) can be deduced by induction on $n > 1$ using (3) and (4) applied to the integrand on the right.

(2) $\displaystyle\int_0^h f_{m,1,\alpha}(x,h,z)\, dz = E_m(x,h).$

(3) $\displaystyle\int_0^h f_{m,n,\alpha}(x,h,y_1 + \cdots + y_n)\, dy_n = f_{m,n-1,\alpha+1}(x,h,y_1 + \cdots + y_{n-1}).$

(4) $\displaystyle E_m(x,h) = \int_0^h \cdots \int_0^h f_{m,n,\alpha}(x,h,y_1 + \cdots + y_n)\, dy_n \cdots dy_1.$

(5) If $0 < h$ there is a z with $0 \leq z \leq mh$ such that $\eta(x+z) \leq f_m(x,h,z)$; if not we would have by (4)

$$\begin{aligned} E_m(x,h) &= \int_0^h \cdots \int_0^h \eta(x + y_1 + \cdots + y_m)\, dy_1 \cdots dy_m \\ &> \int_0^h \cdots \int_0^h f_m(x,h,y_1 + \cdots + y_m)\, dy_1 \cdots dy_m = E_m(x,h), \end{aligned}$$

which is false.

(6) If $0 > h$ there is a z with $0 \geq z \geq mh$ such that $\eta(x+z) \geq f_m(x,h,z)$; if not we would have

$$\begin{aligned}(-1)^m E_m(x,h) &= \int_h^0 \cdots \int_h^0 \eta(x+y_1+\cdots+y_m)\,dy_1\cdots dy_m \\ &< \int_h^0 \cdots \int_h^0 f_m(x,h,y_1+\cdots+y_m)\,dy_1\cdots dy_m \\ &= (-1)^m E_m(x,h).\end{aligned}$$

(7) If δ and x satisfy $0 < \delta < (1-1/x)/m$, then $-\epsilon_1 x \leq \eta(x) \leq \epsilon_2 x$. To see this let $h = \delta x$ with $\delta > 0$. By (5) and the definition of ϵ_2, there is a $z \in [0,h]$ such that

$$\eta(x+z) \leq \frac{E_m(x,\delta x)}{\delta^m x^m} + \frac{m\delta x}{2} - z \leq x(1+\epsilon_2).$$

Therefore, since $z \geq 0$ and $-\frac{1}{2}\log(1-1/(x+z)^2) \leq -\frac{1}{2}\log(1-1/x^2)$ we get

$$\psi(x) \leq \psi(x+z) \leq x(1+\epsilon_2) - \log(2\pi) - \frac{1}{2}\log\left(1-\frac{1}{x^2}\right).$$

In other words, $\eta(x) \leq \epsilon_2 x$. The proof of the lower bound follows in a similar manner. □

Now define for $m \geq 1$

$$V(m,x) := \sum_\rho \frac{x^{\beta-1}}{|\gamma|^{m+1}},$$

where the sum is over all complex zeros $\rho = \beta + i\gamma$ of $\zeta(s)$.

Lemma 3.3 *[144, theorems 15 and 16] Let $m \geq 1$, $x > 1$, $0 < \delta < (1-1/x)/m$, and suppose also that*

$$\delta \geq 2V(m,x)^{1/(m+1)}.$$

Then if the real number θ satisfies

$$\theta \geq \frac{\delta}{2}\left(\frac{b_m(\delta)}{2^m} + m\right),$$

where $b_m(\delta) := \sum_{j=0}^m \binom{m}{j}(1+j\delta)^{m+1}$, we have $\max(\epsilon_1,\epsilon_2) < \theta$.

Proof **(1)** To begin we use Theorem 2.2 which is for $x > 1$

$$\int_0^x \psi(t)\,dt = \frac{x^2}{2} - \sum_\rho \frac{x^{\rho+1}}{\rho(\rho+1)} - \sum_{n=1}^\infty \frac{x^{-2n+1}}{2n(2n-1)} - \frac{\zeta'(0)}{\zeta(0)}x + \frac{\zeta'(-1)}{\zeta(-1)}.$$

First integrate the definition of $\eta(x)$ using $\zeta'(0)/\zeta(0) = \log(2\pi)$ from Lemma 2.5:

$$\begin{aligned}
\int_0^h \eta(x+z)\,dz &= \int_0^h \psi(x+z)\,dz - \int_0^h (x+z)\,dz \\
&\quad + \frac{1}{2}\int_0^h \log\left(1 - \frac{1}{(x+z)^2}\right) dz + h\log(2\pi) \\
&= \int_x^{x+h} \psi(u)\,du - hx - \frac{1}{2}h^2 + \frac{1}{2}\int_x^{x+h} \log\left(1-\frac{1}{u^2}\right) du \\
&\quad + h\log(2\pi) \\
&= \sum_\rho \frac{x^{\rho+1} - (x+h)^{\rho+1}}{\rho(\rho+1)}.
\end{aligned}$$

Thus

$$\int_0^h \eta(x+z)\,dz = \sum_\rho \frac{x^{\rho+1} - (x+h)^{\rho+1}}{\rho(\rho+1)}.$$

We **claim** further integrations lead to

$$E_m(x, \pm x\delta) = \sum_\rho \frac{x^{\rho+m}}{\rho(\rho+1)\cdots(\rho+m)} \left(\sum_{j=0}^{m} (-1)^{j+m+1} \binom{m}{j} (1 \pm j\delta)^{\rho+m} \right).$$

To see this, use the absolute convergence of $\sum_\rho 1/|\rho|^2$, and integrate twice to get

$$\begin{aligned}
&\int_0^h \int_0^h \eta(x+y_1+y_2)\,dy_1\,dy_2 \\
&= \int_0^h \sum_\rho \frac{(x+y_2)^{\rho+1} - (x+y_2+h)^{\rho+1}}{\rho(\rho+1)}\,dy_2 \\
&= \sum_\rho \frac{1}{\rho(\rho+1)} \int_0^h \left[(x+y_2)^{\rho+1} - (x+y_2+h)^{\rho+1}\right] dy_2 \\
&= \sum_\rho \frac{1}{\rho(\rho+1)(\rho+2)} \left[(x+y_2)^{\rho+2} - (x+y_2+h)^{\rho+2}\right]_0^h \\
&= \sum_\rho -\frac{x^{\rho+2} - 2(x+h)^{\rho+2} + (x+2h)^{\rho+2}}{\rho(\rho+1)(\rho+2)}.
\end{aligned}$$

Continuing in this manner and using induction, after m integrations followed by a substitution $h = \pm\delta x$, we obtain the given formula for $E_m(x, \pm\delta x)$.

(2) Now we derive the bound

$$|E_m(x, \pm x\delta)| < x^{m+1}((1+\delta)^{m+1} + 1)V(m, x).$$

Take absolute values of the result of Step (1) using the upper bound

$$\left|\frac{x^{\rho+m}}{\rho(\rho+1)\cdots(\rho+m)}\right| < \frac{x^{\beta+m}}{|\gamma|^{m+1}}$$

and the bound $|(1 \pm j\delta)^{\rho+m}| < (1+j\delta)^{m+1}$, to get

$$\begin{aligned}|E_m(x, \pm x\delta)| &= \left|\sum_\rho \frac{x^{\rho+m}}{\rho(\rho+1)\cdots(\rho+m)}\left(\sum_{j=0}^{m}(-1)^{j+m+1}\binom{m}{j}(1 \pm j\delta)^{\rho+m}\right)\right| \\ &< x^{m+1}\sum_{j=0}^{m}\binom{m}{j}(1+j\delta)^{m+1}\sum_\rho \frac{x^{\beta-1}}{|\gamma|^{m+1}}.\end{aligned}$$

Hence

$$\begin{aligned}|E_m(x, \pm x\delta)| &\le x^{m+1} b_m(\delta)\sum_\rho \frac{x^{\beta-1}}{|\gamma|^{m+1}} \\ &= x^{m+1} b_m(\delta) V(m,x).\end{aligned}$$

(3) Since we are given $\delta \ge 2V(m,x)^{1/(m+1)}$ we have

$$\frac{V(m,x)}{\delta^{m+1}} \le \frac{1}{2^{m+1}},$$

and therefore

$$\begin{aligned}\epsilon_2 &= \frac{E_m(x, x\delta)}{x^{m+1}\delta^m} + \frac{m\delta}{2} \\ &< \frac{x^{m+1} b_m(\delta)}{x^{m+1}\delta^m}\sum_\rho \frac{x^{\beta-1}}{|\gamma|^{m+1}} + \frac{m\delta}{2} \\ &= \frac{\delta}{2}\left(2b_m(\delta)\frac{V(m,x)}{\delta^{m+1}} + m\right) \\ &\le \frac{\delta}{2}\left(\frac{b_m(\delta)}{2^m} + m\right).\end{aligned}$$

The proof for ϵ_1 is similar. □

Recall the definition $\phi_m(\gamma) = \exp(-\log x/(R\log\gamma))/|\gamma|^{m+1}$. Also note that we are designating H as a fixed height below which all complex zeros of $\zeta(s)$ are on the critical line, and

$$V(m,x) := \sum_\rho \frac{x^{\beta-1}}{|\gamma|^{m+1}},$$

where the sum is over all non-trivial zeros $\rho = \beta + i\gamma$ of $\zeta(s)$.

The next lemma shows how the value of H and an explicit zero-free region using the parameter R can be used to obtain an upper bound for $V(m,x)$.

Lemma 3.4 *For $x > 1$ and $m \geq 1$ we have*

$$V(m,x) \leq \frac{1}{\sqrt{x}} \sum_{\rho} \frac{1}{|\gamma|^{m+1}} + \sum_{H<\gamma} \phi_m(\gamma).$$

Proof Splitting the sum in the definition of $V(m,x)$ we get

$$V(m,x) = \sum_{\beta \leq \frac{1}{2}} \frac{x^{\beta-1}}{|\gamma|^{m+1}} + \sum_{\beta > \frac{1}{2}} \frac{x^{\beta-1}}{|\gamma|^{m+1}}.$$

For the first sum

$$\sum_{\beta \leq \frac{1}{2}} \frac{x^{\beta-1}}{|\gamma|^{m+1}} \leq \frac{1}{\sqrt{x}} \sum_{\beta \leq \frac{1}{2}} \frac{1}{|\gamma|^{m+1}} \leq \frac{1}{\sqrt{x}} \sum_{\rho} \frac{1}{|\gamma|^{m+1}},$$

and for the second

$$\sum_{\beta > \frac{1}{2}} \frac{x^{\beta-1}}{|\gamma|^{m+1}} = 2 \sum_{\beta > \frac{1}{2}, \gamma > 0} \frac{x^{\beta-1}}{|\gamma|^{m+1}}.$$

In addition, by Theorem 2.4, for $\gamma > H$ we have

$$\frac{x^{\beta-1}}{\gamma^{m+1}} < \frac{\exp\left(-\frac{\log x}{R \log \gamma}\right)}{\gamma^{m+1}} = \phi_m(\gamma),$$

and so

$$\sum_{\beta > \frac{1}{2}} \frac{x^{\beta-1}}{|\gamma|^{m+1}} < 2 \sum_{\beta > \frac{1}{2}, \gamma > 0} \phi_m(\gamma) \leq \sum_{\gamma > H} \phi_m(\gamma).$$

This completes the proof. □

Next we obtain an upper bound for the sums of the $\phi_m(\gamma)$. Recall the improvement to Backlund's result, given in Section 2.2, for the difference between the number $N(T)$ of zeta zeros with positive imaginary part up to height T and its approximation $F(T) = (T/2\pi)\log(T/2\pi e) + 7/8$, namely

$$B(T) := a_1 \log T + a_2 \log\log T + a_3,$$

where, using the best available values at the time of writing, $a_1 = 0.122$, $a_2 = 0.278$ and $a_3 = 2.510$ for $T \geq e$.

Lemma 3.5 *[145, theorem 27] If $2 \leq h$ then*

$$\sum_{h<\gamma} \phi_m(\gamma) < P(h) e^{-(\log x)/(R \log h)} + Q(h) \int_h^{\infty} \phi_m(t) \log\left(\frac{t}{2\pi}\right) dt,$$

where

$$P(h) := 2\frac{a_1 \log h + a_2 \log\log h + a_3}{h^{m+1}}$$

and

$$Q(h) := \frac{1}{2\pi} + \frac{a_1 \log h + a_2}{h \log h \log(h/2\pi)}.$$

Proof By Lemma 4.2 (which relies simply on Lemma 4.1, a generalized form of Abel summation) and Backlund's bound $B(T)$, we can write

$$\sum_{h<\gamma} \phi_m(\gamma) < \int_h^{\infty} \left(\frac{1}{2\pi} \log\left(\frac{t}{2\pi}\right) + \left(a_1 + \frac{a_2}{\log t}\right) \frac{1}{t} \right) \phi_m(t)\, dt + 2B(h)\phi_m(h).$$

But for $h \le t$ we have

$$\frac{h}{t} \le \frac{\log\left(\frac{t}{2\pi}\right)}{\log\left(\frac{h}{2\pi}\right)} \quad \text{and} \quad \frac{1}{t \log\left(\frac{t}{2\pi}\right)} \le \frac{1}{h \log\left(\frac{h}{2\pi}\right)}.$$

Therefore

$$\begin{aligned} \sum_{h<\gamma} \phi_m(\gamma) &< \left(\frac{1}{2\pi} + \frac{a_1 \log h + a_2}{h \log h \log(h/2\pi)} \right) \int_h^{\infty} \phi_m(t) \log\left(\frac{t}{2\pi}\right) dt + 2B(h)\phi_m(h) \\ &= P(h) e^{-(\log x)/(R \log h)} + Q(h) \int_h^{\infty} \phi_m(t) \log\left(\frac{t}{2\pi}\right) dt. \end{aligned}$$

This completes the proof. □

For $m \ge 1$ let $k_m(h) := \sum_{\rho,\gamma>h} 1/\gamma^{m+1}$. Now we can combine Lemmas 3.4 and 3.5 to obtain an upper bound for $V(m,x)$. The reader might compare this with [145, lemma 6]. Recall that we have defined $\mathcal{B}_1(m,x,H)$ in the proof of Lemma 3.1:

$$\mathcal{B}_1(m,x,H) := \frac{1 + m \log\left(\frac{h}{2\pi}\right)}{m^2 h^m x^{1/(R \log h)} \left(1 - \frac{(1 + m \log(h/2\pi)) \log x}{\log(h/2\pi) R m^2 \log^2 h}\right)}.$$

Lemma 3.6 *Let* $0 < 2 \log x < R m \log^2 H$. *Then*

$$V(m,x) < \left(\frac{k_m(0)}{\sqrt{x}} + \frac{P(H)}{x^{1/(R \log H)}} + Q(H) \mathcal{B}_1(m,x,H) \right) =: \mathcal{B}_2(m,x,H).$$

Next recall the definition:

$$\eta(x) := \psi(x) - x + \log(2\pi) + \frac{1}{2} \log\left(1 - \frac{1}{x^2}\right), \quad x > 1.$$

Finally, putting all of these derivations together with Lemma 3.3, we are able to prove the following theorem, which will enable us to derive Tables 3.2 and 3.3.

Theorem 3.7 *[144, theorem 21] If positive real numbers δ, ϵ, a and x and integer $m \geq 1$ are such that*

$$\begin{aligned} \epsilon(\delta) &:= \frac{\delta}{2}\left(\frac{[(1+\delta)^{m+1}+1]^m}{2^m}+m\right), \\ a &> 1+m\delta a, \\ 2\log a &< Rm\log^2 H, \\ \delta &\geq 2\mathcal{B}_2(m,a,H)^{1/(m+1)}, \end{aligned}$$

then for all $x \geq a$ we have $-\epsilon x < \eta(x) < \epsilon x$.

Proof Note that for $x \geq a$ we have $V(m,x) \leq V(m,a)$ so

$$V(m,x) \leq V(m,a) < \mathcal{B}_2(m,a,H) \leq \left(\frac{\delta}{2}\right)^{m+1}$$

and we can use Lemma 3.2 with $\epsilon_i = \epsilon(\delta)$ and the rather gross estimate $b_m(\delta) < [(1+\delta)^{m+1}+1]^m$. □

In deriving Tables 3.2 and 3.3, we used Rosser's 1941 value of R and a reasonably large value of H for 2014, namely $H = 5.4 \times 10^8$; see the 1986 entry of Table 2.2 from van de Lune, te Riele and Winter. The values of $k_m(0)$, which occur in Table 3.1, were obtained in a manner similar to those in [144, p. 225], which were based on [143, lemma 8]. The bound for $k_1(0)$ comes from Lemma 2.10. Those with $m > 1$ are computed by evaluating the first 100 terms $1/\gamma^{m+1}$ numerically and bounding the remainder using the bound for $k_{m-1}(0)$, which had been computed previously.

Finally, we present only a subset of the values of ϵ_b that are required. A copy of the *Mathematica* program used to compute the successive $k_m(0)$ values, ComputeKm, is part of RHpack – see Appendix B.

Of course, Theorem 3.7 gives an ϵ that works for the function η and table I of [145] is in terms of $\psi(x)$, so a small adjustment needs to be made. Indeed, we have shown that

$$-\epsilon x \leq \eta(x) \leq \epsilon x \quad \Longrightarrow \quad 1-\epsilon-\frac{\log 2\pi}{x}+\frac{1}{2x^2} \leq \frac{\psi(x)}{x} \leq 1+\epsilon-\frac{\log 2\pi}{x}+\frac{1}{x^2}.$$

Hence if $\log x \geq \log a = b$ we need to replace ϵ with

$$\epsilon' := \max\left(\epsilon+\frac{\log 2\pi}{e^b}-\frac{1}{2e^{2b}},\ \epsilon-\frac{\log 2\pi}{e^b}+\frac{1}{e^{2b}}\right).$$

Thus, with this revised ϵ denoted ϵ_b we obtain Tables 3.2 and 3.3 where the row b gives the value of b, the value of m giving the smallest ϵ, and the

Table 3.1 Upper bounds used for $k_m(0)$, $1 \le m \le 28$.

m	$k_m(0)$ bound	m	$k_m(0)$ bound
1	0.046300	2	0.0016700
3	0.000074400	4	4.46301×10^{-6}
5	2.88354×10^{-7}	6	1.93507×10^{-8}
7	1.32606×10^{-9}	8	9.19838×10^{-11}
9	6.42733×10^{-12}	10	4.5113×10^{-13}
11	3.17539×10^{-14}	12	2.23908×10^{-15}
13	1.58067×10^{-16}	14	1.11669×10^{-17}
15	7.89294×10^{-19}	16	5.58061×10^{-20}
17	3.94653×10^{-21}	18	2.79132×10^{-22}
19	1.97443×10^{-23}	20	1.3967×10^{-24}
21	9.88051×10^{-26}	22	6.98986×10^{-27}
23	4.94499×10^{-28}	24	3.49838×10^{-29}
25	2.47499×10^{-30}	26	1.75098×10^{-31}
27	1.23877×10^{-32}	28	8.76397×10^{-34}

Table 3.2 Values of b, m and ϵ_b.

b	m	ϵ_b
10	1	0.03604387616
11	1	0.02792600226
12	1	0.02166958397
13	1	0.01683159959
14	1	0.013082730087
15	1	0.010173897079
16	1	0.007914814691
17	1	0.006159239110
18	1	0.004794353324
19	1	0.003732913461
20	1	0.002907340379
25	1	0.0008427406872
30	1	0.0002748595135
35	1	0.0001515932079
40	1	0.0001348313906
45	2	0.0001315791411
50	2	0.00008556326873
55	2	0.00003720592016
60	2	0.000016278904663
65	2	$7.594348788 \times 10^{-6}$
75	2	$4.457990402 \times 10^{-6}$
80	2	$4.377481479 \times 10^{-6}$
85	2	$4.330519430 \times 10^{-6}$
90	2	$4.286798206 \times 10^{-6}$
95	3	$2.5990044920 \times 10^{-6}$
100	3	$1.4581963574 \times 10^{-6}$

Table 3.3 More values of b, m and ϵ_b.

b	m	ϵ_b
200	5	$2.3461827847 \times 10^{-7}$
300	8	$1.1237935640 \times 10^{-7}$
400	11	$8.578404834 \times 10^{-8}$
500	14	$7.743154524 \times 10^{-8}$
600	15	$7.3869596744 \times 10^{-8}$
700	14	$7.1000004141 \times 10^{-8}$
800	14	$6.8129091697 \times 10^{-8}$
900	13	$6.5258532748 \times 10^{-8}$
1000	13	$6.2436274496 \times 10^{-8}$
2000	7	$3.4825165615 \times 10^{-8}$
3000	3	$1.1625210647 \times 10^{-8}$
4000	3	$2.563076921 \times 10^{-9}$
5000	4	$1.0662477419 \times 10^{-9}$
6000	4	$3.162913973 \times 10^{-10}$
7000	5	$2.252110357 \times 10^{-10}$
8000	5	$8.153215009 \times 10^{-11}$
9000	6	$7.617681319 \times 10^{-11}$
10000	7	$7.415233861 \times 10^{-11}$

corresponding value of ϵ_b so that

$$|\psi(x) - x| \leq \epsilon_b x, \quad x \geq e^b.$$

We also present plots of the value of ϵ_b as a function of b in three ranges, corresponding to the tables going up to $b = 10^4$. Figures 3.1, 3.2 and 3.3 show how the optimal value of m changes in the mid-range.

The program used to compute the ϵ_b is PsiSmallXEpsilon, which is part of RHpack – see Appendix B.

A recent article by Faber and Kadiri [57, corollary 1.2] uses the Platt value $H = 3.061 \times 10^{10}$ to derive, for example,

$$|\psi(x) - x| \leq 5.3688 \times 10^{-4}\, x, \quad x \geq e^{20}.$$

The estimates for $\psi(x)$ derived here are sufficient for our purposes.

3.3 Exact Verification Using Computation

In verifying estimates over a finite range, both in this chapter and in what follows, we often use a computational method, not always giving details. These computations are very useful since they sometimes simplify proofs, and sometimes make covering the infinite part of variable ranges easier.

We give here a description of the approach that has been taken. It might be regarded as a very simple form of interval arithmetic.

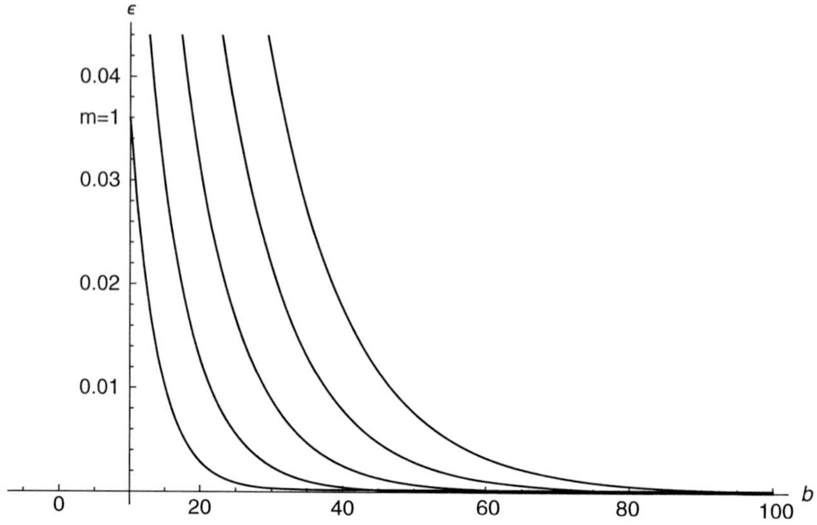

Figure 3.1 The values of ϵ_b for $10 \le b \le 100$ and $1 \le m \le 5$.

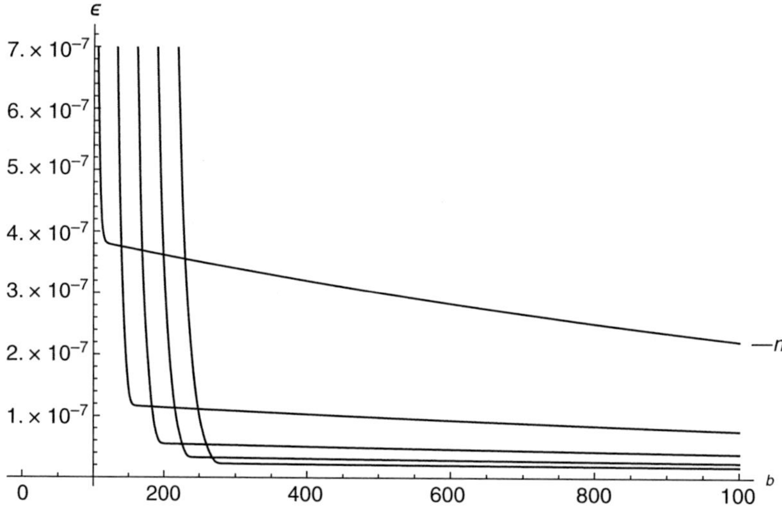

Figure 3.2 The values of ϵ_b for $100 \le b \le 1000$ and $3 \le m \le 7$.

(1) Suppose we wish to test the validity of the bound

$$\sum_{i=1}^{n} a_i < f(n), \quad 1 \le n \le N_0, \tag{3.1}$$

where (a_i) is a given real sequence and $f(n)$ is a real-valued function. The values of each element of the sequence and of $f(n)$ are only known

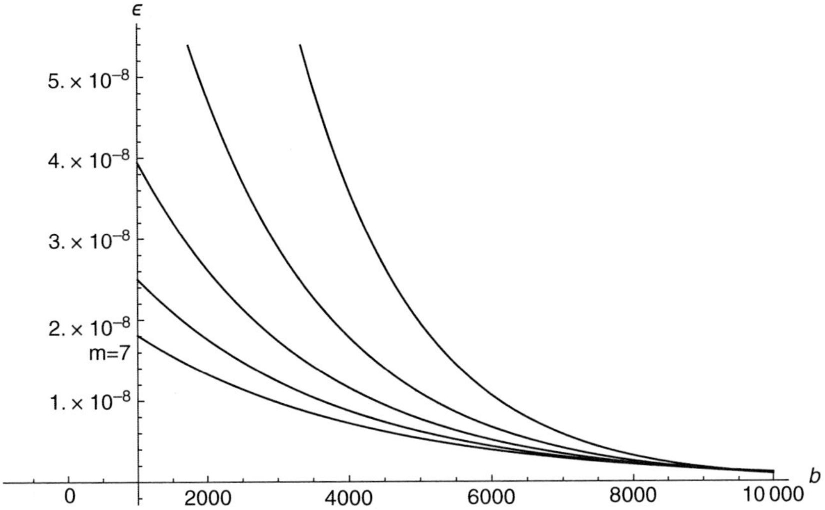

Figure 3.3 The values of ϵ_b for $10^3 \le b \le 10^4$ and $3 \le m \le 7$.

approximately. We set a target for success

$$\sum_{i=1}^{n} a_i + \epsilon_0 < f(n), \qquad 1 \le n \le N_0.$$

To achieve that target we need to approximate $f(n)$ and the terms of (a_i) with a "local accuracy" $\epsilon > 0$ so we can calculate rationals (b_i) and $g(n)$ such that $|a_i - b_i| \le \epsilon$ and $|f(n) - g(n)| \le \epsilon$ for all $1 \le i, n \le N_0$. Then we need to have, for each $1 \le n \le N_0$,

$$\epsilon_0 \le f(n) - \sum_{i=1}^{n} a_i \le g(n) + \epsilon - \sum_{i=1}^{n} b_i + n\epsilon,$$

so we should test $\epsilon_1 := \epsilon_0 - (n+1)\epsilon \le g(n) - \sum_{i=1}^{n} b_i$, and that should be sufficient to verify (3.1) exactly.

For example, if we wanted to show that $\theta(x) < x$ up to $x = 10^{11}$ with accuracy 10^{-10} we could set the local accuracy to $\epsilon = 10^{-23}$ and define $a_n = \log n$ for prime n and zero otherwise.

(2) Continuing with this example, the equation $\theta(n) < n$ will have been verified at integer values n of x in the given range. Since $\theta(x)$ is constant between integers and $f(x) = x$ is increasing, evaluation at integers is sufficient to show it is true for all real numbers x up to N_0.

(3) In case the inequality is similar, for example, to $0.8x < \theta(x)$ for $0 < x \le 10^6$ say, then the process adopted is similar, except we verify the equation, using the approach of Step (1), $0.8(n+1) < \theta(n)$ at integers n. Then, since both sides are increasing, if $n < x < n+1$ and n is an integer, we have

$$0.8x < 0.8(n+1) < \theta(n) \le \theta(x),$$

and the equation has been verified for all real x in the given range.

3.4 Estimates for $\theta(x)$

The next theorem provides estimates for $\theta(x)$ and $\psi(x)$ in an unbounded range. It depends on a function $\epsilon(x)$, defined in the statement, and depends on x being sufficiently large.

The proof, given by Rosser and Schoenfeld in 1962 in [145] as theorem 11 (with a different conclusion from that stated here), but based in part on the proofs from 1938 in [143], is by dividing the range into sections and using a variety of lemmas. In problem (1) at the end of this chapter we suggest that a different and simpler proof should be possible, with potentially a different and smaller form for the upper bound.

Here we have used an improved value for the lower bound for the range of values of R, from $R \ge 18$ to $R \ge 8$, at the cost of a small increase in the leading coefficient. This enables us to use $R = 8$ in calculations; see Section 2.5.

Recall the definitions of $k_m(h)$, $N(T)$ and $F(T)$:

$$k_m(h) := \sum_{\rho,\gamma>h} \frac{1}{\gamma^{m+1}},$$

$$N(T) := \#\{\rho : 0 < \Im\rho \le T\},$$

$$F(T) := \frac{T}{2\pi}\log\left(\frac{T}{2\pi e}\right) + \frac{7}{8},$$

and $|N(T) - F(T)| < 1.12\log T + 9.5$ for $T \ge 1450$, which is Rosser's improvement of Backlund's bound from 1938 [143, lemma 5]. For other bounds and ranges see part (11) of Section 2.2.

Theorem 3.8 *Let*

$$\epsilon(x) := \sqrt{2\log x}\exp\left(-\sqrt{\frac{\log x}{R}}\right),$$

where $8 \le R \le 18$. *Then for all* $x \ge e^{140}$ *we have*

$$|\psi(x) - x| < \epsilon(x)x \quad and \quad |\theta(x) - x| < \epsilon(x)x.$$

Proof **(1)** First we have some definitions. Let

$$S_8 := \sum_{1469 \le \gamma \le e^8} \frac{1}{\gamma^2} < 3.32 \times 10^{-4} < (0.274)8e^{-8},$$

$$S_n := \sum_{e^{n-1} < \gamma \le e^n} \frac{1}{\gamma^2}, \quad n \ge 9,$$

$$\epsilon_1(x) := \frac{k_1(0)}{\sqrt{x}} + \sum_{n=9}^{\infty} S_n x^{-1/Rn} + \frac{1.9}{x} + \frac{3}{x^2},$$

$$\epsilon_2(x) := \sum_{n=9}^{\infty} S_n \exp\left(-\frac{\log x}{Rn}\right),$$

$$\epsilon_3(x) := \frac{k_1(0)}{\sqrt{x}} + \frac{1.9}{x} + \frac{3}{x^2} \quad \text{so} \quad \epsilon_1 = \epsilon_2 + \epsilon_3.$$

(2) Next we **claim** that for S_r, when $r \geq 9$ we have $S_r < (0.274) r \exp(-r)$. To see this, define

$$\Delta := F(e^r) - F(e^{r-1}),$$

so $\Delta = e^{r-1}(e-1)(r-a)/(2\pi)$ where $a := \log 2\pi + 1 - 1/(e-1) > 2.25$. We also define for r fixed $g(x) := F(xe^{r-1})$, so $g(x)$ is increasing and concave upwards on $[1, e]$, and so

$$g\left(1 + \frac{e-1}{g(e) - g(1)} n\right) \leq g(1) + n, \quad 1 \leq n \leq \Delta,$$

which translates to

$$F\left(e^{r-1} + \frac{2\pi n}{r-a}\right) \leq F(e^{r-1}) + n.$$

Therefore, since $N(e^r) \leq F(e^r) + 9.5 + 1.12r$, if we let M be n plus that upper bound for $N(e^r)$ we must have

$$e^{r-1} + \frac{2\pi n}{r-a} < \gamma_M$$

for $0 \leq n \leq \Delta$. But, considering the lower bound for $N(e^{r-1})$,

$$m < F(e^{r-1}) - 9.5 - 1.12(r-1) \quad \Longrightarrow \quad \gamma_m < e^{r-1}.$$

Hence

$$\begin{aligned} S_r &= \sum_{e^{r-1} < \gamma \leq e^r} \frac{1}{\gamma^2} \\ &< \frac{19 + 2.24r}{e^{2r-2}} + \sum_{n=1}^{\Delta} \frac{1}{\left(e^{r-1} + \frac{2\pi n}{r-a}\right)^2} \\ &< \frac{2.24(r+9)}{e^{2r-2}} + \frac{1}{e^{2r-2}} \sum_{n=1}^{\Delta} \frac{1}{\left(1 + \frac{(e-1)n}{\Delta}\right)^2} \end{aligned}$$

$$< \frac{2.24(r+9)}{e^{2r-2}} + \frac{1}{e^{2r-2}} \int_0^{\Delta} \frac{dt}{\left(1 + \frac{e-1}{\Delta} t\right)^2}$$

$$= \frac{2.24(r+9)}{e^{2r-2}} + \frac{\Delta}{e^{2r-1}}$$

$$= \frac{2.24(r+9)}{e^{2r-2}} + \frac{(r-a)(e-1)}{2\pi e^r}$$

$$< 16.6(r+9)e^{-2r} + 0.274(r-2.25)e^{-r} < 0.274re^{-r},$$

where the last inequality is true for $r \geq 9$. This completes the proof of the claim.

(3) In this step we focus on the term that depends on the zeta zeros ρ where $\rho = \beta + i\gamma$, namely $\epsilon_2(x)$. Using the result of Step (2) we have

$$\epsilon_2(x) = \sum_{n=9}^{\infty} S_n \exp\left(-\frac{\log x}{Rn}\right) < (0.274) \sum_{n=9}^{\infty} ne^{-n} x^{-1/Rn}.$$

Next we **claim** that for all $y > 0$ and $n \geq 1$ we have

$$\frac{n}{e^{n+y/n}} < \left(\sqrt{y} + \frac{1}{2} + \frac{1}{8\sqrt{y}}\right) e^{-2\sqrt{y}}. \tag{3.2}$$

To see this, let

$$f(y,n) := \frac{n}{\sqrt{y}} \exp\left(-n - \frac{y}{n} + 2\sqrt{y}\right)$$

and regard y and n as real variables. Note that $f(y, \sqrt{y}) = 1$ and let $n_1 = \sqrt{y}$. We get

$$\frac{\partial f(y,n)}{\partial n} = \frac{e^{-(n-\sqrt{y})^2/n}}{n\sqrt{y}}(n - n^2 + y),$$

which for $n \geq 1$ is zero at the unique value $n_2 := (1 + \sqrt{4y+1})/2$. Also, at $n = n_2$ we get

$$\frac{\partial^2 f(y,n)}{\partial n^2} = -\frac{4[2y(\sqrt{4y+1}+2) + \sqrt{4y+1}+1]\exp\left(-\frac{(-2\sqrt{y}+\sqrt{4y+1}+1)^2}{2(\sqrt{4y+1}+1)}\right)}{\sqrt{y}(\sqrt{4y+1}+1)^3},$$

which is negative. Thus for fixed y, n_2 gives the maximum value for $f(y,n)$. Then, setting $h := n_2 - n_1$ and using convexity and a linear approximation we get

$$f(y, n_1 + h) \leq f(y, n_1) + h\frac{\partial f(y, \sqrt{y})}{\partial n}$$

$$= 1 + \frac{1}{\sqrt{y}}\left(\frac{1+\sqrt{4y+1}}{2} - \sqrt{y}\right) = \frac{1}{2\sqrt{y}} + \left(1 + \frac{1}{4y}\right)^{1/2}$$
$$< 1 + \frac{1}{2\sqrt{y}} + \frac{1}{8y}.$$

Therefore

$$\frac{n}{e^{n+y/n}} = f(y,n)\sqrt{y}e^{-2\sqrt{y}} < \left(\sqrt{y} + \frac{1}{2} + \frac{1}{8\sqrt{y}}\right)e^{-2\sqrt{y}},$$

which completes the proof of the **claim**.

Setting $y := (\log x)/R$, it follows that for $n \geq 1$, $8 \leq R \leq 18$ and $x \geq e^{140}$, we have

$$ne^{-n}x^{-1/Rn} \leq \left(\sqrt{\frac{\log x}{R}} + \frac{1}{2} + \frac{1}{8\sqrt{(\log x)/R}}\right)e^{-2\sqrt{(\log x)/R}}$$
$$< 1.2\sqrt{\frac{\log x}{R}}e^{-2\sqrt{(\log x)/R}}. \tag{3.3}$$

Returning to the sum for $\epsilon_2(x)$ given in Step (1), we can now write

$$\epsilon_2(x) = (0.274)\sum_{9\leq n\leq 2\sqrt{(\log x)/R}} ne^{-n}x^{-1/Rn} + (0.274)\sum_{n>2\sqrt{(\log x)/R}} ne^{-n}x^{-1/Rn}$$
$$< (0.274)2\sqrt{\frac{\log x}{R}}(1.2)\sqrt{\frac{\log x}{R}}\exp\left(-2\sqrt{\frac{\log x}{R}}\right)$$
$$+ (0.274)\sqrt{\frac{\log x}{R}}\int_{2\sqrt{(\log x)/R}}^{\infty} te^{-t}\,dt$$
$$< (2\times 0.274)\left(\frac{\log x}{R}\right)\left(1.2 + \frac{1}{\sqrt{(\log x)/R}} + \frac{1}{2(\log x)/R}\right)\exp\left(-2\sqrt{\frac{\log x}{R}}\right)$$
$$< (0.8895)\frac{\log x}{R}\exp\left(-2\sqrt{\frac{\log x}{R}}\right),$$

where we have used the integral $\int_b^\infty t\exp(-t)\,dt = (b+1)e^{-b}$, and the final inequality follows for $x \geq \exp(140)$. Finally in this step, using the bound we have derived for $\epsilon_2(x)$ and that for $\epsilon_3(x)$ from Step (1), for $x \geq e^{140}$ we get, since in addition $R \geq 8$,

$$\epsilon_1(x) = \epsilon_2(x) + \epsilon_3(x) < \frac{1}{8}\log x \exp\left(-2\sqrt{\frac{\log x}{R}}\right). \tag{3.4}$$

(4) By Theorem 2.2 we have for $x > 1$

$$\psi_1(x) := \int_0^x \psi(t)\,dt = \frac{x^2}{2} - \sum_\rho \frac{x^{\rho+1}}{\rho(\rho+1)} - \sum_{n=1}^\infty \frac{x^{-2n+1}}{2n(2n-1)} - \frac{\zeta'(0)}{\zeta(0)}x + \frac{\zeta'(-1)}{\zeta(-1)}.$$

Therefore

$$\left|\psi_1(x) - \frac{x^2}{2}\right| \le 2x^2\epsilon_1(x) \quad\Longrightarrow\quad \frac{x^2}{2} - 2x^2\epsilon_1(x) \le \psi_1(x) \le \frac{x^2}{2} + 2x^2\epsilon_1(x).$$

Therefore, if $h > 0$, because $\epsilon_1(x) > \epsilon_1(x+h)$ we have

$$\begin{aligned}\psi_1(x+h) - \psi_1(x) &< \tfrac{1}{2}(x+h)^2 - \tfrac{1}{2}x^2 + 2(x+h)^2\epsilon_1(x+h) + 2x^2\epsilon_1(x)\\ &\le xh + \tfrac{1}{2}h^2 + 2\epsilon_1(x)(2x^2 + 2xh + h^2).\end{aligned}$$

Discarding the positive last term in the first lower bound below

$$\begin{aligned}\psi_1(x) - \psi_1(x-h) &\ge \tfrac{1}{2}x^2 - 2x^2\epsilon_1(x) - \tfrac{1}{2}(x-h)^2 + 2(x-h)^2\epsilon_1(x-h)\\ &> xh - \tfrac{1}{2}h^2 - 2x^2\epsilon_1(x)\\ &> xh - \tfrac{1}{2}h^2 - 2\epsilon_1(x)(2x^2 + 2xh + h^2).\end{aligned}$$

Thus, because $\psi(x)$ is non-decreasing we have

$$\begin{aligned}x - \frac{1}{2}h - \frac{2\epsilon_1(x)}{h}(2x^2 + 2xh + h^2) &\le \frac{\psi_1(x) - \psi_1(x-h)}{h} \le \psi(x)\\ &\le \frac{\psi_1(x+h) - \psi_1(x)}{h} \le x + \frac{1}{2}h + \frac{2\epsilon_1(x)}{h}(2x^2 + 2xh + h^2).\end{aligned}$$

Finally, set $h := 4x\sqrt{\epsilon_1(x)}$ and use (3.4) to derive

$$\begin{aligned}|\psi(x) - x| &\le x\sqrt{\epsilon_1}\left(3 + 4\sqrt{\epsilon_1(x)} + 8\epsilon_1(x)\right)\\ &< 4x\sqrt{(\log x)/8}\exp\left(-\sqrt{\frac{\log x}{R}}\right)\\ &= x\sqrt{2\log x}\exp\left(-\sqrt{\frac{\log x}{R}}\right),\end{aligned}$$

for $x \ge e^{140}$, completing the proof of the estimate for $\psi(x)$.

(5) Bringing the estimates together,

$$\theta(x) - x \le \psi(x) - x \le 3.983x\sqrt{(\log x)/8}\exp(-\sqrt{(\log x)/R}) < \epsilon(x)x.$$

Then, using Lemma 3.10 which gives $\psi(x) - \theta(x) < 1.02\sqrt{x} + 3\sqrt[3]{x}$ $(x > 0)$, we get

$$\begin{aligned}\theta(x) - x &\ge \psi(x) - x - 1.02\sqrt{x} - 3\sqrt[3]{x}\\ &\ge -3.983x\sqrt{(\log x)/8}\exp(-\sqrt{(\log x)/R}) - 1.02\sqrt{x} - 3\sqrt[3]{x}\\ &\ge -4x\sqrt{(\log x)/8}\exp(-\sqrt{(\log x)/R}) = -x\epsilon(x),\end{aligned}$$

with the last inequality holding for $x \ge e^{81}$. Therefore

$$|\theta(x) - x| < \epsilon(x)x, \quad x \ge e^{140}.$$

This completes the proof. □

Table 3.4 Values of $b = \log x$, and rounded $\log \epsilon(x)$ for $R = 8$ and $R = 18$.

b	$\log \epsilon(x)$, $R = 8$	$\log \epsilon(x)$, $R = 18$	b	$\log \epsilon(x)$, $R = 8$	$\log \epsilon(x)$, $R = 18$
1000	−7	−3	10000	−30	−18
2000	−11	−6	20000	−44	−28
3000	−15	−8	30000	−55	−35
4000	−17	−10	40000	−65	−41
5000	−20	−12	50000	−73	−47
6000	−22	−13	60000	−80	−52
7000	−24	−15	70000	−87	−56
8000	−26	−16	80000	−94	−61
9000	−28	−17	90000	−100	−65
10000	−30	−18	100000	−105	−68

Table 3.5 Larger values of b and $\log \epsilon(x)$.

$b = \log x$	$\log \epsilon(x)$, $R = 8$	$\log \epsilon(x)$, $R = 18$
100000	−105	−68
200000	−151	−99
300000	−186	−122
400000	−216	−142
500000	−243	−160
600000	−266	175
700000	−288	−190
800000	−309	−204
900000	−328	−216
1000000	−346	−228

In Tables 3.4 and 3.5 we give values of $\log \epsilon(x)$ for a range of values of $b = \log x$. The second and fifth columns of Table 3.4 are computed rounding the result of Theorem 3.8, as are the $R = 8$ plots in Figures 3.4 and 3.5. The third and sixth columns are computed using the Rosser–Schoenfeld expression for $\epsilon(x)$, namely $\sqrt{\log x}\exp(-\sqrt{(\log x)/R})$, as are the $R = 18$ plots in Figures 3.4 and 3.5.

As an application of these values, first consider Table 3.5 at line $b = 10^6$. Since $\exp(-346) < 5.43 \times 10^{-151}$, we get

$$|\psi(x) - x| \leq x \times 5.43 \times 10^{-151}, \quad x \geq e^{10^6}.$$

At the other extreme, from Table 3.4, line $b = 1000$,

$$|\psi(x) - x| \leq x \times (0.00092), \quad x \geq e^{10^3}.$$

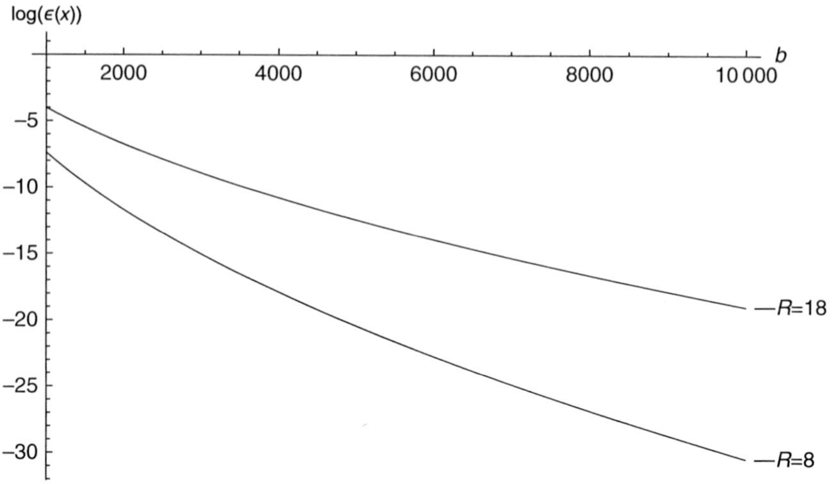

Figure 3.4 A plot of b, where $x = e^b$, vs $\log \epsilon(x)$ for $10^3 \le b \le 10^4$.

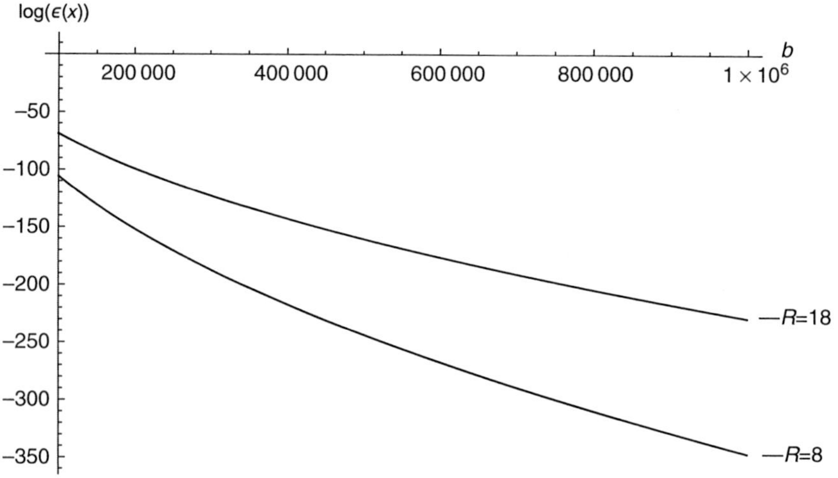

Figure 3.5 A plot of b, where $x = e^b$, vs $\log \epsilon(x)$ for $10^5 \le b \le 10^6$.

This compares badly with the value obtained in Section 3.2; see Table 3.3, which gives

$$|\psi(x) - x| \le x \times (4.45 \times 10^{-9}), \quad x \ge e^{10^3}.$$

There is one more preliminary lemma in this chapter, viz. Landau's lemma (Lemma 3.9). We have made some small changes in his proof and result, but the interesting parts of a fascinating proof are his. Other than this, the remaining material consists of essential estimates, with applications spread throughout the rest of the book. For example, Lemma 3.14 below is used in Lemmas 3.10, 5.6, 7.7, 8.13 and 9.16, and in Theorems 5.25, 7.11 and 7.13.

Lemma 3.9 *[105, pp. 89–90] For all $x \geq 1$ we have*

$$\psi(x) < \alpha x + (\log x + 2)(3 \log x)$$

where

$$\alpha := \left(\frac{6}{5}\right)\left(\frac{\log 2}{2} + \frac{\log 3}{3} + \frac{\log 5}{5} - \frac{\log 30}{30}\right) < 1.105551.$$

Proof **(1)** For $x \geq 1$ define

$$T(x) := \sum_{1 \leq n \leq x} \log n = \log(\lfloor x \rfloor !).$$

Then

$$\begin{aligned} T(x) &= \sum_{n=1}^{x-1} \log n + \log\lfloor x \rfloor \leq \sum_{n=1}^{x-1} \int_n^{n+1} \log t\, dt + \log\lfloor x \rfloor \\ &\leq \int_1^x \log t\, dt + \log x = x \log x - x + 1 + \log x, \\ T(x) &\geq \sum_{n=2}^{x} \log n \geq \sum_{n=2}^{x} \int_{n-1}^{n} \log t\, dt \\ &= \int_1^{\lfloor x \rfloor} \log t\, dt = \int_1^x \log t\, dt - \int_{\lfloor x \rfloor}^x \log t\, dt \\ &\geq \int_1^x \log t\, dt - \log x > x \log x - x + 1 - \log x. \end{aligned}$$

Hence there is a θ_0, which depends on x, with $|\theta_0| \leq 1$ such that we have the estimate

$$T(x) = x \log x - x + 1 + \theta_0 \log x.$$

Define $U(x) := T(x) - T(x/2) - T(x/3) - T(x/5) + T(x/30)$ and evaluate $U(x)$ using the estimate for $T(x)$ to obtain the expression, for some θ_1 with $|\theta_1| \leq 1$ and $x \geq 30$,

$$\begin{aligned} U(x) &= x\left(\frac{\log 2}{2} + \frac{\log 3}{3} + \frac{\log 5}{5} - \frac{\log 30}{30}\right) + \theta_1(5 \log x) \\ &= \beta x + \theta_1(5 \log x), \end{aligned} \tag{3.5}$$

where $\beta = 0.921292\ldots$. Equation (3.5) was proved for $x \geq 30$. A simple computation shows it is also true for $1 \leq x \leq 30$. Note that the combination of T evaluations has been chosen to make this bound possible.

(2) Next we **claim** we can write $T(x)=\sum_{n=1}^{x}\psi(x/n)$. To see this derive

$$T(x)=\log(\lfloor x\rfloor!)=\sum_{p\le x}\log p\left(\left\lfloor\frac{x}{p}\right\rfloor+\left\lfloor\frac{x}{p^2}\right\rfloor+\cdots\right)$$
$$=\sum_{p\le x}\log p\sum_{m=1}^{\infty}\left\lfloor\frac{x}{p^m}\right\rfloor=\sum_{m=1}^{\infty}\sum_{p\le\sqrt[m]{x}}\log p\left\lfloor\frac{x}{p^m}\right\rfloor,$$

and so

$$T(x)=\sum_{m=1}^{\infty}\sum_{n=1}^{x}\sum_{p\le\sqrt[m]{x}}\log p=\sum_{m=1}^{\infty}\sum_{n=1}^{x}\theta\left(\sqrt[m]{\frac{x}{n}}\right)$$
$$=\sum_{n=1}^{x}\sum_{m=1}^{\infty}\theta\left(\sqrt[m]{\frac{x}{n}}\right)=\sum_{n=1}^{x}\psi\left(\frac{x}{n}\right).$$

Then substitute this expression for $T(x)$ into the definition of $U(x)$ to get

$$U(x)=\sum_{n=1}^{\infty}\left(\psi\left(\frac{x}{n}\right)-\psi\left(\frac{x}{2n}\right)-\psi\left(\frac{x}{3n}\right)-\psi\left(\frac{x}{5n}\right)+\psi\left(\frac{x}{30}\right)\right)=\sum_{n=1}^{\infty}c_n\psi\left(\frac{x}{n}\right),$$

where the coefficients c_n are determined by the remainder when n is divided by 30 according to the following rules:

$c_n=-1$ for remainders $6,10,12,15,18,20,24,0$,
$c_n=0$ for remainders $2,3,4,5,8,9,14,16,21,22,25,26,27,28$,
$c_n=1$ for remainders $1,7,11,13,17,19,23,29$.

(3) Now expand $U(x)$ in increasing order of denominators of x to give

$$U(x)=\psi(x)-\psi\left(\frac{x}{6}\right)+\psi\left(\frac{x}{7}\right)-\psi\left(\frac{x}{10}\right)+\psi\left(\frac{x}{11}\right)-\psi\left(\frac{x}{12}\right)+\cdots,$$

and we note that the signs of the successive terms alternate and their sizes decrease. A minor miracle! Hence

$$\psi(x)-\psi\left(\frac{x}{6}\right)\le U(x),$$

and therefore, since by what we have shown above in (3.5)

$$\psi(x)-\psi\left(\frac{x}{6}\right)<\beta x+5\log x.$$

Hence, for $n\ge 0$ with $x\ge 6^n$ and $x\ge 1$ we have

$$\psi\left(\frac{x}{6^n}\right)-\psi\left(\frac{x}{6^{n+1}}\right)<\beta\frac{x}{6^n}+5\log\frac{x}{6^n}.$$

Now sum these expressions over n:

$$\psi(x) = \sum_{n=0}^{\log x/\log 6} \left(\psi\left(\frac{x}{6^n}\right) - \psi\left(\frac{x}{6^{n+1}}\right)\right)$$
$$\leq \beta x \sum_{n=0}^{\infty} \frac{1}{6^n} + \left(\frac{\log x}{\log 6} + 1\right)(5 \log x)$$
$$< \frac{6}{5}\beta x + 3 \log x (\log(x) + 2).$$

This completes the proof. □

We need the following Lemma 3.10 in Chapter 4 where it is used in the proof of Theorem 4.5. It plays an essential role in the development of the ideas in the book, being used more than 10 times.

Lemma 3.10 *[145, theorem 18]*

(a) For all x with $0 < x \leq 6.66 \times 10^{11}$ we have $\theta(x) < x$.
(b) For $6.66 \times 10^{11} < x$ we have $\theta(x) \leq 1.0009x$.
(c) For all $x > 0$ we have $\psi(x) - \theta(x) < \theta(\sqrt{x}) + 3\sqrt[3]{x}$.
(d) For $x \geq 121$ we have $0.8x \leq \theta(x)$.

Proof (a) This is proved by computation; see Section 3.3.

(b) This follows directly from the $b = 25$ line of Table 3.2.

(c) A direct computation shows that for integers x in the range $1 \leq x \leq 10^5$, and thus for all real x in that range, we have

$$\psi(x) - \theta(x) < \theta(\sqrt{x}) + 3\sqrt[3]{x}. \tag{3.6}$$

We now extend the range for (3.6) to $10^5 \leq x < 10^{15}$. We **claim** that for $x \geq 1$ we have $\theta(x) < 1.106207x$. This is immediate for $1 \leq x < 10^6$ by the right-hand member of (b), and for $x \geq 10^6$ we get, using Lemma 3.9,

$$\theta(x) \leq \psi(x) \leq x\left(1.105551 + \frac{(3 \log x)(\log x + 2)}{x}\right) < 1.106207x.$$

Using this result, $\theta(x) < 1.106207x,\ x > 0$, and the identity

$$\psi(x) = \sum_{n=1}^{\infty} \theta(\sqrt[n]{x}), \tag{3.7}$$

we can write

$$\psi(x) - \theta(x) - \theta(\sqrt{x}) \leq 1.2x^{1/3} \sum_{n=3}^{\log x/\log 2} x^{1/n-1/3} < 3x^{1/3}$$

when $x \in [10^5, 10^{18}]$.

We can now complete the proof of (3.6) in the range $x \geq 10^{18}$. Using (3.7) and (*a*) and (*b*) again, we can write for $x \geq 10^{18}$

$$\begin{aligned}\psi(x) - \theta(x) - \theta(\sqrt{x}) &= \theta(\sqrt[3]{x}) + \sum_{n\geq 4} \theta(\sqrt[n]{x}) \\ &\leq \theta(\sqrt[3]{x}) + \theta(\sqrt[4]{x})(\log x/\log 2 - 3) \\ &< 1.0009\sqrt[3]{x}\left(1 + \frac{\log x/\log 2 - 3}{\sqrt[12]{x}}\right) < 3\sqrt[3]{x}.\end{aligned}$$

This completes the proof of (*c*).

(*d*) A numerical check shows that for integers x in the range $0 < x \leq 10^{10}$ we have

$$x - 2.05282\sqrt{x} < \theta(x) < x. \tag{3.8}$$

Thus, for $0 < x \leq 10^{10}$, by (3.8) we have $x - 2.05282\sqrt{x} < \theta(x)$, and so for $121 \leq x \leq 10^{10}$

$$0.813337x \leq x\left(1 - \frac{2.05282}{\sqrt{x}}\right) < \theta(x).$$

By (*c*) we have $\psi(x) - 1.0009\sqrt{x} - 3\sqrt[3]{x} < \theta(x)$ for $x > 0$, and by Table 3.2, $x(1 - 0.0361) < \psi(x)$ for $x > 10^{10}$. Therefore, for $x \geq 10^{10}$

$$0.8x < x\left(1 - 0.0361 - \frac{1.0009}{\sqrt{x}} - \frac{3}{\sqrt[6]{x}}\right) < \theta(x)$$

which completes the proof. □

A note regarding part (*a*) is needed here. It is known, see Platt and Trudgian [135, theorem 1], that $\theta(x) < x$ for $2 < x < 1.39 \times 10^{17}$, improving on a result by Schoenfeld [153, p. 360] (see also [142, (4), p. 194]). In addition Platt and Trudgian showed that for all $x > 0$ we have $\theta(x) < (1 + 7.5 \times 10^{-7})x$. The paper [135] also contains the result (corollary 2), which is then an easy consequence,

$$\psi(x) - \theta(x) < (1 + 7.5 \times 10^{-7})\sqrt{x} + 3\sqrt[3]{x}, \quad x > 0.$$

The note [153, p. 360] states that for all $0 < x$ we have $\theta(x) < 1.000081x$. This was not proved by Schoenfeld in that paper, but is the subject of a final note, with details to follow. However, the details have not been found. We have a simple derivation of the estimate, using RH, which is all that is needed.

Lemma 3.11 *If we assume RH, then for* $0 < x$ *we have* $\theta(x) < 1.000081x$.

Proof Since by Lemma 3.10, $\theta(x) < x$ for $0 < x \leq 10^{11}$, the upper bound for $\theta(x)$ holds in that range. By Theorem 4.6, which depends on RH, we have

$$\theta(x) - x \le \frac{\sqrt{x}\log^2 x}{8\pi} \quad \Longrightarrow \quad \theta(x) \le x\left(1 + \frac{\log^2 x}{8\pi\sqrt{x}}\right) < 1.000081x$$

for $x \ge 10^{11}$. □

Using the same approach as for Lemma 3.10 (*d*), we obtain the following list of lower bounds for $\theta(x)$:

Lemma 3.12 *[145, theorem 10] We have $\theta(x) \ge \beta x$ for $x \ge \alpha$ when (α,β) are in the list:*

$$\{(89\,387, 0.995), (32\,057, 0.990), (11\,927, 0.985), (7481, 0.980), (5381, 0.975), (3457, 0.970), (2657, 0.965), (1481, 0.960), (1433, 0.955), (1427, 0.950), (853, 0.945), (809, 0.94), (599, 0.93), (557, 0.92), (349, 0.91), (227, 0.89), (149, 0.86)\}.$$

3.5 More Estimates

We also need the estimates for $\psi(x) - \theta(x)$ which are given in Lemma 3.13. Compare these with [145, theorems 13, 14, 19 and 24].

Lemma 3.13

(*a*) *For $1423 \le x \le 10^8$ we have $x - 2\sqrt{x} < \theta(x)$.*
(*b*) *For $121 < x \le 10^{16}$ we have $\sqrt{x} < \psi(x) - \theta(x)$.*
(*c*) *For $x \ge 10^{13}$ we have $\psi(x) - \theta(x) < 1.021\sqrt{x}$.*
(*d*) *For $121 \le x$ we have $0.98\sqrt{x} < \psi(x) - \theta(x)$.*
(*e*) *For $0 < x$ we have $\psi(x) - \theta(x) < 1.42620\sqrt{x}$.*

Proof (*a*) This may be verified by computation – see Section 3.3.

(*b*) For $x \le 10^5$ we can verify the equation numerically. For $10^5 \le x \le 10^8$ use $x - 2.1\sqrt{x} < \theta(x)$, which is a consequence of the proof of Lemma 3.10 – see (3.8). Finally when $10^8 \le x \le 10^{16}$ we have $\sqrt{x} \le 10^8$ so we can write

$$\begin{aligned}\psi(x) - \theta(x) &\ge \sum_{n\ge 2} \theta(\sqrt[n]{x}) \ge (\sqrt{x} - 2.1\sqrt[4]{x}) + (\sqrt[3]{x} - 2.1\sqrt[6]{x}) + (\sqrt[4]{x} - 2.1\sqrt[8]{x}) \\ &\ge \sqrt{x} + \sqrt[3]{x}\left(1 - \frac{1.1}{\sqrt[12]{x}} - \frac{2.1}{\sqrt[6]{x}} - \frac{2.1}{x^{5/24}}\right) \\ &> \sqrt{x}\end{aligned}$$

since $x \ge 10^8$.

(*c*) For $10^{13} \le x \le 10^{22}$ use $\theta(\sqrt{x}) < \sqrt{x}$ and (3.6), namely $\psi(x) - \theta(x) < \theta(\sqrt{x}) + 3\sqrt[3]{x}$. For $x > 10^{22}$ use $\theta(x) < 1.02x$, from Lemma 3.10(*b*), so that in that case we have

$$\psi(x) - \theta(x) < \sqrt{x}\left(1.02 + \frac{3}{\sqrt[6]{x}}\right) < 1.021\sqrt{x}.$$

(*d*) Up to 10^{16} the result follows by (*b*). For $x > 10^{16}$ use the pair $(0.98, 7481)$ from Lemma 3.12 and write, since $\sqrt{x} \geq 7481$ in this range,

$$\psi(x) - \theta(x) \geq \theta(\sqrt{x}) \geq 0.98\sqrt{x}.$$

(*e*) Up to 10^6 the result follows by a computation – see Section 3.3. For $x \geq 10^6$ use $\psi(x) - \theta(x) < 1.02\sqrt{x} + 3\sqrt[3]{x}$. All five parts of the lemma have now been verified. □

We now give the final lemma, Lemma 3.14. Part (*a*) is the most useful. The others are included with particular constants chosen to meet the needs of particular applications.

Lemma 3.14 *[145, theorems 4 and 29–31]*

(*a*) *For* $19421 \leq x$ *we have* $|\theta(x) - x| < 0.125x/\log x$.
(*b*) *For* $1451 \leq x \leq e^{375}$ *we have* $|\theta(x) - x| < 0.31x/\log x$.
(*c*) *For* $809 \leq x \leq e^{575}$ *we have* $|\theta(x) - x| < 0.40x/\log x$.
(*d*) *For* $569 \leq x$ *we have* $|\theta(x) - x| < 0.47x/\log x$.
(*e*) *For* $563 \leq x$ *we have* $|\theta(x) - x| < 0.5x/\log x$ *with, in addition, for* $0 < x$ *we get* $\theta(x) < x(1 + 0.5/\log x)$.

Proof **(1)** If $1 < x \leq 10^{11}$: each of (*a*)–(*e*) is verified by computation – see Section 3.3.

(2) Next let $10^{11} \leq x \leq e^{5000}$. We use Lemma 3.10(*a*), (*b*) and (*c*) to get for $x > 0$

$$\psi(x) - \theta(x) < 1.0009\sqrt{x} + 3\sqrt[3]{x}, \quad x > 0,$$

and then use for b in Table 3.2 or 3.3

$$|\psi(x) - x| < \epsilon_b x, \quad x \geq e^b,$$

so therefore

$$(1 - \epsilon_b)x - 1.0009\sqrt{x} - 3\sqrt[3]{x} < \psi(x) - 1.0009\sqrt{x} - 3\sqrt[3]{x} < \theta(x).$$

If b_j and b_{j+1} are in the tables with $b_j < b_{j+1}$ and ϵ_j corresponds to $a_j = e^{b_j}$, then for $a_j \leq x \leq a_{j+1}$ we have

$$x\left(1 - \frac{1}{8\log x}\right) \leq x\left(1 - \frac{1}{8\log a_{j+1}}\right) \leq x\left(1 - \epsilon_j - \frac{1.02}{\sqrt{a_j}} - \frac{3}{a_j^{2/3}}\right) < \theta(x)$$

if

$$\epsilon_j + \frac{1.02}{\sqrt{a_j}} + \frac{3}{a_j^{2/3}} \leq \frac{1}{8\log a_{j+1}},$$

for $x \geq 10^{11}$. Now $e^{25} < 10^{11}$ and we were able to verify the inequality for ϵ_j using four intervals for b, namely

$$[25,30],\ [30,40],\ [40,100] \text{ and } [100,5000].$$

(3) For $x \geq e^{5000}$ use Theorem 3.8. In this range, with the notation in that theorem and with $R \geq 8$ we have

$$\epsilon(x) = \sqrt{2\log x}\exp\left(-\sqrt{\frac{\log x}{R}}\right) < \frac{1}{8\log x}.$$

This completes the proof of (a) and therefore of (b)–(e) for $x \geq 10^{11}$. □

3.6 Unsolved Problems

(1) Write a program to reproduce the tables given in Section 3.2, then heuristically determine how ϵ varies with b, with or without fixed m. Derive analytical estimates to replace the tables.
(2) Examine the analytical estimates to derive the bound given in Theorem 3.8 and derive a better bound (which is for example asymptotically smaller).
(3) Extend the finite ranges for any of the results given here which were obtained through computation, using improved computation or otherwise. See Section 3.3.
(4) Replace the approach using tables with something analytic and easier to apply.
(5) Write a survey article or book giving an account of work on explicit estimates and their applications in (a part of) number theory.

4
Classical Equivalences

4.1 Introduction

In this chapter a number of classical equivalences to the Riemann hypothesis are presented, several with complete proofs. In Section 4.2 the explicit result of Schoenfeld is proved in detail, namely that RH is equivalent to the inequality

$$|\psi(x) - x| \leq \frac{\sqrt{x}\log^2 x}{8\pi},$$

being true for all $x \geq 74$. We call this the **Schoenfeld criterion**. It is directly related to the size of the error in different forms of the prime number theorem. Note that RH is equivalent to the weaker form

$$|\psi(x) - x| \ll \sqrt{x}\log^2 x \quad \text{or even} \quad |\psi(x) - x| \ll_\epsilon x^{1/2+\epsilon}.$$

Then in Section 4.3 material under the heading "Oscillation theorems" is given. This material is presented without proof, except for two results which are used explicitly in later chapters. A fundamental theorem of Landau, Theorem 4.12, is used in many places in the volume to derive results when the Riemann hypothesis is assumed to be false. In addition, a famous result of Littlewood, Theorem 4.13, shows in particular that both positive and negative gaps between $\psi(x)$ and x occur infinitely often as $x \to \infty$, with gap sizes being at least a constant times $\sqrt{x}\log\log\log x$, even assuming RH.

In Section 4.4 a number of results are given wherein the hypothesis is equivalent to the order of the error in arithmetic sums. These have been known for a long time and the list is not at all comprehensive. Some failed attempts to show RH is true have been based on them. It is expected that a range of equivalences along this same line, some of them new, might be derived.

A proof of the **Littlewood criterion** is given in Section 4.4, viz. that RH is equivalent to the estimate that, for all $\epsilon > 0$, as $x \to \infty$ we have

$$\sum_{n \le x} \mu(n) \ll x^{1/2+\epsilon}.$$

The following RHpack (see Appendix B) functions relate to the material in this chapter: ComputeTheta, ComputeThetaValues, M and Psi.

4.2 The Prime Number Theorem and Its RH Equivalences

The original prime number theorem, as proved by de la Vallée-Poussin in 1889, is the statement that there exists an absolute constant $a > 0$ such that we have

$$\pi(x) = \mathrm{Li}(x) + O\left(x e^{-a\sqrt{\log x}}\right), \quad x \to \infty,$$

where for $x \ge 2$ we have $\mathrm{Li}(x) := \int_2^x dt/\log t$, called the logarithmic integral.

This was improved by Walfisz [177, chapter 5] without using any special hypotheses. He showed that there is a constant $a > 0$ such that as $x \to \infty$ we have

$$\pi(x) = \mathrm{Li}(x) + O\left(x e^{-a \log^{3/5} x/(\log\log x)^{1/5}}\right), \quad x \to \infty.$$

In 1901–1902 von Koch [93, 94] showed RH implied that a much better error term in the prime number theorem could be assumed. Indeed von Koch proved that the Riemann hypothesis is equivalent to the estimate

$$\pi(x) = \mathrm{Li}(x) + O(\sqrt{x}\log x), \quad x \to \infty,$$

where the implied constant is absolute. The implied constant in the error term was made explicit by Schoenfeld [153, 154] in 1976, who showed the RH implies that for all $x \ge 2657$ we have

$$|\pi(x) - \mathrm{li}(x)| < \frac{\sqrt{x}\log x}{8\pi},$$

where the alternative form of the logarithmic integral $\mathrm{li}(x)$ is defined by the sum of improper integrals for $x > 1$,

$$\mathrm{li}(x) := \lim_{\epsilon \to 0} \int_0^{1-\epsilon} \frac{dt}{\log t} + \lim_{\epsilon \to 0} \int_{1+\epsilon}^{x} \frac{dt}{\log t},$$

and is such that $\mathrm{Li}(x) = \mathrm{li}(x) - \mathrm{li}(2)$ for $x \ge 2$ with $1 < \mathrm{li}(2) < 1.046$.

Schoenfeld [153, 154] also found an explicit bound for the error in the estimate $\psi(x) \sim x$ in the case of the function $\psi(x)$, where $\psi(x) := \sum_{p^n \le x} \log p$ counts primes with a weighted sum. This function, and its close companion $\theta(x) := \sum_{p \le x} \log p$, are fundamental components of this study. They are named after the famous Russian mathematician Chebyshev (1821–1894), who made many contributions to mathematics, including orthogonal polynomials and the theory of probability. He is well known for his proof of a weaker form of the prime number theorem.

Here is the statement of Schoenfeld's theorem: the Riemann hypothesis implies that for all $x \geq 73.2$ we have

$$|\psi(x) - x| < \frac{\sqrt{x}\log^2 x}{8\pi}.$$

It also implies that for all $x \geq 599$ we have

$$|\theta(x) - x| < \frac{\sqrt{x}\log^2 x}{8\pi}.$$

These two classical results have a famous corollary which gives the best possible form of the prime number theorem error. For $x \geq 2657$ we have

$$|\pi(x) - \mathrm{li}(x)| < \frac{\sqrt{x}\log x}{8\pi}.$$

The proof requires quite a few lemmas which are set out below. We use a minimum of generality to make the proofs as simple as possible. The corollary is proved following the theorem.

First, using the same method as in Abel's identity, [6, theorem 4.2] we derive the following well-known stronger result:

Lemma 4.1 *Let $a < \gamma_1 < g_2 < \cdots < \gamma_n \leq b$ be real numbers with $n \geq 2$, and $a_j \in \mathbb{C}$, $1 \leq j \leq n$. For $a \leq x \leq b$ let*

$$A(x) = \sum_{\gamma_i \leq x} a_i,$$

and let $f(x)$ be a real- or complex-valued function on $[a,b]$ with an integrable derivative. Then

$$\sum_{j=1}^{n} a_j f(\gamma_j) = A(b)f(b) - A(a)f(a) - \int_a^b A(t)f'(t)\,dt.$$

Proof First note that

$$\int_{\gamma_j}^{\gamma_{j+1}} A(t)f'(t)\,dt = A(\gamma_j)\int_{\gamma_j}^{\gamma_{j+1}} f'(t)\,dt = A(\gamma_j)f(\gamma_{j+1}) - A(\gamma_j)f(\gamma_j),$$

so therefore, noting that $A(\gamma_n) = A(b)$ and shifting the index in the first sum:

$$\int_a^b A(t)f'(t)\,dt$$

$$
\begin{aligned}
&= \int_a^{\gamma_1} A(t)f'(t)\,dt + \int_{\gamma_1}^{\gamma_n} A(t)f'(t)\,dt + \int_{\gamma_n}^{b} A(t)f'(t)\,dt \\
&= A(a)(f(\gamma_1)-f(a)) + A(\gamma_n)(f(b)-f(\gamma_n)) \\
&\quad + \sum_{j=1}^{n-1}(A(\gamma_j)f(\gamma_{j+1}) - A(\gamma_j)f(\gamma_j)) \\
&= A(b)f(b) - A(a)f(a) + A(a)f(\gamma_1) - f(\gamma_n)A(\gamma_n) \\
&\quad - \sum_{j=1}^{n-1} A(\gamma_j)f(\gamma_j) + \sum_{j=2}^{n} A(\gamma_{j-1})f(\gamma_j) \\
&= A(b)f(b) - A(a)f(a) + A(a)f(\gamma_1) - f(\gamma_n)A(\gamma_n) \\
&\quad + \sum_{j=2}^{n-1}(A(\gamma_{j-1}) - A(\gamma_j))f(\gamma_j) + A(\gamma_{n-1})f(\gamma_n) - A(\gamma_1)f(\gamma_1) \\
&= A(b)f(b) - A(a)f(a) + A(a)f(\gamma_1) - f(\gamma_n)A(\gamma_n) - \sum_{j=1}^{n} a_j f(\gamma_j) \\
&\quad + A(\gamma_{n-1})f(\gamma_n) - A(\gamma_1)f(\gamma_1) \\
&= A(b)f(b) - A(a)f(a) - \sum_{j=1}^{n} a_j f(\gamma_j).
\end{aligned}
$$

This completes the proof of the lemma □

Remark By suitable indexing and setting $a_j = 1$ with (γ_j) a distinct increasing sequence of real numbers meeting the real interval $(a,b]$ in at least two points, so $A(x)$ counts the number of the γ_j not greater than x, we obtain the useful corollary:

$$\sum_{a<\gamma_j\le b} f(\gamma_j) = A(b)f(b) - A(a)f(a) - \int_a^b A(t)f'(t)\,dt.$$

By specializing to the case of the positive imaginary parts of zeta zeros, using Lemma 4.1 we obtain a simplified upper bound for a functional sum. First recall some definitions from Section 2.2 (11):

$$
\begin{aligned}
N(T) &:= \#\{\rho : \zeta(\rho) = 0,\ 0 < \Im\rho \le T\}, \\
F(T) &:= \frac{T}{2\pi}\log\left(\frac{T}{2\pi e}\right) + \frac{7}{8}, \\
B(T) &:= a_1 \log T + a_2 \log\log T + a_3.
\end{aligned}
$$

Lemma 4.2 *[146, lemma 7, p. 255] Let $1 < a < b$ and the real function $f(y)$ be such that $f'(y)$ is continuous with $f'(y) \le 0$ and $f(y) > 0$ for $a \le y \le b$. Then*

$$\sum_{a<\gamma_j\le b} f(\gamma_j) \le \frac{1}{2\pi}\int_a^b f(y)\log\frac{y}{2\pi}\,dy + \left(a_1 + \frac{a_2}{\log a}\right)\int_a^b \frac{f(y)}{y}\,dy + 2B(a)f(a).$$

Proof By the remark following the proof of Lemma 4.1 we have

$$\sum_{a<\gamma_j\le b} f(\gamma_j) = N(b)f(b) - N(a)f(a) - \int_a^b N(y)f'(y)\,dy.$$

Now Backlund's explicit estimate for $N(T)$ (see Section 2.2) can be written, for all $T \ge 2$,

$$|N(T) - F(T)| < B(T) \implies N(T) < F(T) + B(T) \quad \text{and} \quad -N(T) < B(T) - F(T).$$

Therefore, since $f'(y)$ is negative, we can upper bound the sum,

$$\sum_{a<\gamma_j\le b} f(\gamma_j) < -\int_a^b (F(y)+B(y))f'(y)\,dy + (B(b)+F(b))f(b) + (B(a)-F(a))f(a).$$

Integrating by parts and using

$$F'(y) = \frac{1}{2\pi}\log\frac{y}{2\pi}$$

and

$$B'(y) = \left(a_1 + \frac{a_2}{\log y}\right)\frac{1}{y} \le \left(a_1 + \frac{a_2}{\log a}\right)\frac{1}{y}$$

for $a \le y \le b$, we get

$$\sum_{a<\gamma_j\le b} f(\gamma_j) < \frac{1}{2\pi}\int_a^b f(y)\log\frac{y}{2\pi}\,dy + \left(a_1 + \frac{a_2}{\log a}\right)\int_a^b \frac{f(y)}{y}\,dy + 2B(a)f(a).$$

which completes the proof. □

Now we define two sums depending on two real positive parameters for δ and T:

$$S_1(\delta) := (2+\delta)\sum_{0<\gamma\le T}\frac{1}{|\rho|},$$

$$S_2(\delta) := 2\frac{(2+2\delta+\delta^2)}{\delta}\sum_{T<\gamma}\frac{1}{|\rho(\rho+1)|},$$

where $\rho = \beta + i\gamma$ represents a zero of $\zeta(s)$. These sums are used as an intermediate step to obtain the bounds we need for $|\psi(x)-x|$ and $|\theta(x)-x|$. The derivation in Lemma 4.3 is a simplified version of that used in Lemma 3.2.

Lemma 4.3 *[146, lemma 8, p. 256] Assume the Riemann hypothesis is true. If $x > 1$ and $0 < \delta < 1 - 1/x$ we have*

$$\left|\psi(x) - x + \log(2\pi) + \frac{1}{2}\log\left(1 - \frac{1}{x^2}\right)\right| \leq \sqrt{x}(S_1(\delta) + S_2(\delta)) + \frac{\delta x}{2}.$$

Proof **(1)** First we need to define some functions:

$$\begin{aligned}
\eta(x) &:= \psi(x) - x + \log(2\pi) + \frac{1}{2}\log\left(1 - \frac{1}{x^2}\right),\\
E(x,h) &:= \int_0^h \eta(x+y)\,dy,\\
f_{n,\alpha}(x,h,z) &:= \frac{E(x,h)}{h^n} + \frac{nh^\alpha}{2} - zh^{\alpha-1},\\
f(x,h,z) &:= f_{1,1}(x,h,z) = \frac{E(x,h)}{h} + \frac{h}{2} - z,\\
\epsilon_1 &:= \frac{E(x,-x\delta)}{x^2\delta} + \frac{\delta}{2},\\
\epsilon_2 &:= \frac{E(x,x\delta)}{x^2\delta} + \frac{\delta}{2}.
\end{aligned}$$

Also assume $x > 1$ and $x + h > 1$. The variable h could be negative, indeed we will choose later to have $h = \pm\delta x$. In the properties below, (2) and (3) are immediate and (4) can be deduced by induction on $n \geq 1$.

(2) $\displaystyle\int_0^h f_{1,\alpha}(x,h,z)\,dz = E(x,h)$.

(3) $\displaystyle\int_0^h f_{n,\alpha}(x,h,y_1 + \cdots + y_n)\,dy_n = f_{n-1,\alpha+1}(x,h,y_1 + \cdots + y_{n-1})$.

(4) $\displaystyle E(x,h) = \int_0^h \cdots \int_0^h f_{n,\alpha}(x,h,y_1 + \cdots + y_n)\,dy_n \cdots dy_1$.

(5) If $0 < h$ there is a z with $0 \leq z \leq h$ such that $\eta(x+z) \leq f(x,h,z)$: if not, we would have by (2)

$$E(x,h) = \int_0^h \eta(x+y)\,dy > \int_0^h f(x,h,y)\,dy = E(x,h),$$

which is false.

(6) If $0 > h$ there is a z with $0 \geq z \geq h$ such that $\eta(x+z) \geq f(x,h,z)$: if not, we would have

$$-E(x,h) = \int_h^0 \eta(x+y)\,dy < \int_h^0 f(x,h,y)\,dy = -E(x,h).$$

(7) If $< \delta < 1 - 1/x$ then $-\epsilon_1 x \leq \eta(x) \leq \epsilon_2 x$: let $h = \delta x$ with $\delta > 0$. By (5) there is a $z \in [0,h]$ such that

$$\eta(x+z) \leq \frac{E(x,\delta x)}{\delta x} + \frac{\delta x}{2} - z \leq x(1 + \epsilon_2).$$

Therefore since $z \geq 0$ and $\frac{1}{2}\log(1-1/(x+z)^2) \leq -\frac{1}{2}\log(1-1/x^2)$ we get

$$\psi(x) \leq \psi(x+z) \leq x(1+\epsilon_2) - \log(2\pi) - \frac{1}{2}\log\left(1-\frac{1}{x^2}\right).$$

In other words, $\eta(x) \leq \epsilon_2 x$. The proof of the lower bound is similar.

(8) Next we **claim** with $\rho = \frac{1}{2} + i\gamma$ being a zeta zero that

$$\int_0^h \eta(x+z)\,dz = \sum_\rho \frac{x^{\rho+1}-(x+h)^{\rho+1}}{\rho(\rho+1)}.$$

To see this, use Theorem 2.2, which for $x > 1$ is

$$\int_0^x \psi(t)\,dt = \frac{x^2}{2} - \sum_\rho \frac{x^{\rho+1}}{\rho(\rho+1)} - \sum_{n=1}^{\infty} \frac{x^{-2n+1}}{2n(2n-1)} - \frac{\zeta'(0)}{\zeta(0)}x + \frac{\zeta'(-1)}{\zeta(-1)}.$$

First integrate the definition of $\eta(x)$:

$$\begin{aligned}
E(x,h) &= \int_0^h \eta(x+z)\,dz \\
&= \int_0^h \psi(x+z)\,dz - \int_0^h (x+z)\,dz \\
&\quad + \frac{1}{2}\int_0^h \log\left(1-\frac{1}{(x+z)^2}\right)dz + h\log(2\pi) \\
&= \int_x^{x+h} \psi(u)\,du - hx - \frac{1}{2}h^2 + \frac{1}{2}\int_x^{x+h}\log\left(1-\frac{1}{u^2}\right)du \\
&= \sum_\rho \frac{x^{\rho+1}-(x+h)^{\rho+1}}{\rho(\rho+1)}.
\end{aligned}$$

(9) Now we split the sum from (8) so that

$$|E(x,h)| \leq 2\sum_{0<\gamma\leq T}\left|\frac{x^{\rho+1}-(x+h)^{\rho+1}}{\rho(\rho+1)}\right| + 2\sum_{T<\gamma}\left|\frac{x^{\rho+1}-(x+h)^{\rho+1}}{\rho(\rho+1)}\right|.$$

Let $h = \pm\delta x$ and recall we are assuming RH is true. In the first sum use the bound

$$\left|\frac{x^{\rho+1}-(x+h)^{\rho+1}}{(\rho+1)}\right| \leq x^{3/2}\int_0^\delta (1+y) = x^{3/2}\frac{\delta(2+\delta)}{2}.$$

In the second sum use the upper bound $|1-(1\pm\delta)^{1+\rho}| \leq (1+(1+\delta)^2)$ so that

$$\left|\frac{x^{\rho+1}-(x+h)^{\rho+1}}{\rho(\rho+1)}\right| \leq x^{3/2}(2+2\delta+\delta^2)\frac{1}{|\rho(\rho+1)|}.$$

Therefore

$$|E(x, \pm x\delta)| \le x^{3/2}(2+\delta)\delta \sum_{0<\gamma\le T} \frac{1}{|\rho|} + 2x^{3/2}(2+2\delta+\delta^2) \sum_{T<\gamma} \frac{1}{|\rho(\rho+1)|}.$$

Finally using this bound, the definitions of $E(x,h)$, ϵ_i, $S_i(\delta)$ and Step (7), we get

$$|\eta(x)| \le \sqrt{x}(S_1(\delta) + S_2(\delta)) + \frac{\delta x}{2},$$

and the proof of the lemma is complete. □

Now let $T = 2(2 + 2\delta + \delta^2)/(\delta(2 + \delta))$ so $T \to \infty$ as $\delta \to 0+$. Set $D :=$ 158.84998, just less than the y-coordinate of the 58th zeta zero. We also in Lemma 4.4 and in Theorem 4.5 use the choices $a_1 = 0.137$ and $a_2 =$ 0.443, which are sufficient for our purposes, but not the best available. See Section 2.2 and the unsolved problems at the end of this chapter.

Lemma 4.4 *[146, lemma 9, p. 262] If $T_1 \ge D$ and $\delta > 0$ then*

$$S_1(\delta) + S_2(\delta) < \frac{2+\delta}{4\pi}\left(\left(\log\frac{T_1}{2\pi} + 1\right)^2 + 1\right).$$

Proof A computation gives

$$\sum_{\gamma\le D} \frac{1}{\sqrt{\frac{1}{4}+\gamma^2}} < 0.8113925.$$

Next take $f(y) = 1/y$ in Lemma 4.2 to get

$$\sum_{D<\gamma\le T} \frac{1}{\gamma} \le \frac{1}{4\pi}\left(\log^2\frac{T}{2\pi} - \log^2\frac{D}{2\pi}\right) + \left(0.137 + \frac{0.443}{\log D}\right)\left(\frac{1}{D} - \frac{1}{T}\right) + \frac{2B(D)}{D}.$$

Therefore

$$\begin{aligned}\frac{S_1(\delta)}{2+\delta} &< 0.8113925 + \frac{1}{4\pi}\log^2\frac{T}{2\pi} - \frac{0.224212}{T} + \frac{0.224212}{D} + \frac{2B(D)}{D} \\ &\quad - \frac{1}{4\pi}\log^2\frac{D}{2\pi} \\ &\le \frac{1}{4\pi}\log^2\frac{T}{2\pi} - \frac{0.224212}{T} - 0.791068.\end{aligned}$$

Again using Lemma 4.2, this time with $f(y) = 1/y^2$, we get

$$\begin{aligned}\sum_{T<\gamma} \frac{1}{|\rho(\rho+1)|} &< \sum_{T<\gamma} \frac{1}{\gamma^2} \\ &\le \frac{1}{2\pi}\int_T^\infty \frac{1}{y^2}\log\frac{y}{2\pi}\,dy + \left(0.137 + \frac{0.443}{\log T}\right)\int_T^\infty \frac{dy}{y^3} + \frac{2B(T)}{T^2}\end{aligned}$$

$$= \frac{1}{2\pi T}\left(\log\frac{T}{2\pi}+1\right)+\frac{1}{2T^2}\left(0.137+\frac{0.443}{\log T}\right)+\frac{2B(T)}{T^2}.$$

Then, making the choice $T = T_1$ where $(2+\delta)T_1 = 2(2+2\delta+\delta^2)/\delta$ we get

$$S_1(\delta)+S_2(\delta) < \frac{2+\delta}{4\pi}\left(\left(\log\frac{T_1}{2\pi}+1\right)^2+1\right).$$

□

We are now able to present Schoenfeld's proof of one side of the equivalence, wherein we assume the Riemann hypothesis is true.

Theorem 4.5 *[153, theorem 10, p. 337] The Riemann hypothesis implies that for all $x \ge 23\times 10^8$ we have*

$$|\psi(x)-x| < \frac{\sqrt{x}\log x(\log x-2)}{8\pi}$$

and

$$|\theta(x)-x| < \frac{\sqrt{x}\log x(\log x-2)}{8\pi}.$$

Proof Let $x \ge \xi \ge 828\,000$. In the proof, constants δ and α_i with $1 \le i \le 6$ will be defined progressively, with each of them depending on ξ in the given range, which finally will be chosen explicitly to enable the proof to succeed. Let

$$\alpha_1 := \frac{\log^2\xi}{\pi\sqrt{\xi}} \quad \text{and} \quad \alpha_2 := \frac{\log^2\xi}{\pi\sqrt{\xi}} < 0.0126. \tag{4.1}$$

Then

$$\delta := \frac{\log x}{\pi\sqrt{x}} \le \frac{\alpha_1}{\log x} \le \alpha_2.$$

Now let $T_1 := 2(2+2\delta+\delta^2)/(\delta(2+\delta))$ and $D := 158.84998$ so there are precisely 57 zeros $\rho = \frac{1}{2}+i\gamma$ of $\zeta(s)$ with $0<\gamma\le D$ and $T_1 > D$. Then if we set

$$\delta_1 := \frac{\alpha_2(1+\alpha_2)}{2+\alpha_2} < 0.00634 \quad \text{and} \quad \alpha_3 := 2\log\log\xi - 2(1+\delta_1) > 2.841,$$

we get

$$\log\frac{T_1}{2\pi}+1 = \log\left(\frac{\sqrt{x}}{\log x}\frac{2+2\delta+\delta^2}{2+\delta}\right)+1 \le \log\left(\frac{\sqrt{x}}{\log x}(1+\delta_1)\right)+1$$
$$\le \frac{1}{2}(\log x - 2\log\log x + 2(1+\delta_1)) \le \frac{1}{2}(\log x - \alpha_3).$$

Now define

$$\alpha_4 := 2\alpha_3 - \frac{\alpha_3^2 + 4.152828}{\log \xi}.$$

By Lemmas 4.3 and 4.4, since

$$\frac{2+\delta}{4\pi} < \frac{1}{2\pi}\left(1 + \frac{\alpha_1}{2\log x}\right),$$

we can write

$$\begin{aligned}\frac{|\psi(x) - x|}{x} &< \frac{1}{2\pi\sqrt{x}}\left(1 + \frac{\alpha_1}{2\log x}\right)\left(\left(\log\frac{T_1}{2\pi} + 1\right)^2 + 1.038207\right)\\ &\quad + \frac{\log x}{2\pi\sqrt{x}} + \frac{1}{x}\left(\log(2\pi) + \frac{1}{2}\log\left(1 - \frac{1}{x^2}\right)\right)\\ &< \frac{1}{8\pi\sqrt{x}}\left(1 + \frac{\alpha_1}{2\log x}\right)(\log^2 x - 2\alpha_3\log x + \alpha_3^2 + 4.152828)\\ &\quad + \frac{\log x}{2\pi\sqrt{x}} + \frac{\log(2\pi)}{x}.\end{aligned}$$

Thus

$$\begin{aligned}\frac{|\psi(x) - x|}{x} &< \frac{1}{8\pi\sqrt{x}}\left(1 + \frac{\alpha_1}{2\log x}\right)(\log^2 x - \alpha_4\log x)\\ &\quad + \frac{\log x}{8\pi\sqrt{x}}\left(4 + \frac{8\pi\log(2\pi)}{\sqrt{x}\log x}\right).\end{aligned}$$

Next, note that $\alpha_3 < 2\log\log\xi < \log\xi$, so α_4 increases with α_3, and therefore, for this range,

$$\alpha_4 > 2 \times 2.841 - \frac{2.841^2 + 4.152828}{\log\xi} > 0.$$

Therefore, using the upper bound which was derived for $|\psi(x) - x|/x$ and defining

$$\alpha_5 := \alpha_4 - \frac{\alpha_1}{2} - 4 - \frac{8\pi\log(2\pi)}{\sqrt{\xi}\log\xi},$$

we get

$$\frac{|\psi(x) - x|}{x} < \frac{\log x(\log x - \alpha_5)}{8\pi\sqrt{x}}.$$

Finally define

$$\alpha_6 := \alpha_5 - \frac{8.16\pi}{\log\xi} - \frac{24\pi}{\xi^{1/6}\log\xi}.$$

To get the analogous inequality for $\theta(x)$, use parts (a), (b) and (d) of Lemma 3.10, which gives for $0 < x$, the bound $\psi(x) - \theta(x) < 1.02\sqrt{x} + 3\sqrt[3]{x}$. Then

$$\frac{|\theta(x) - x|}{x} < \frac{|\psi(x) - x|}{x} + \frac{\psi(x) - \theta(x)}{x} \leq \frac{\log x}{8\pi\sqrt{x}}(\log x - \alpha_6).$$

Let $\xi = 10^8$ to compute $\alpha_5 > \alpha_6 > 2$, which competes the proof of the theorem. □

Remark The inequality $|\theta(x) - x| \leq \sqrt{x}\log^2 x/(8\pi)$ succeeds for 231 integers n in the range $[2, 598]$. The inequality $|\psi(x) - x| \leq \sqrt{x}\log^2 x/(8\pi)$ succeeds at integers in the list

$$\{13, 19, 20, 23, 24, 25, 26, 27, 29, 30, 31, 32, 33, 34, 35, 38, 39,$$
$$41, 42, 43, 44, 45, 46, 47, 48, 49, 50, 51, 52, 53, 54, 55, 56, 57,$$
$$58, 59, 60, 61, 62, 63, 64, 65, 66, 67, 68, 69, 70, 71, 72, 73\}$$

and fails at all other integers in $[2, 73]$.

Theorem 4.6 *[153, theorem 1] The Riemann hypothesis implies that for all $x \geq 74$ we have*

$$|\psi(x) - x| < \frac{\sqrt{x}\log^2 x}{8\pi}$$

and for all $x \geq 599$ we have

$$|\theta(x) - x| < \frac{\sqrt{x}\log^2 x}{8\pi}.$$

Proof In the range $x \geq 23 \times 10^8$ the result follows from Theorem 4.5. For $90\,000 \leq x < 23 \times 10^8$ we checked the inequalities through making a computation at integers

$$n + 1 - \frac{1}{8\pi}\sqrt{n+1}\log(n+1)(\log(n+1) - 2) < \theta(n)$$
$$< n + \frac{1}{8\pi}\sqrt{n}\log n(\log n - 2).$$

Below 90 000 we checked the inequality

$$n + 1 - \frac{1}{8\pi}\sqrt{n+1}\log^2(n+1) < \theta(n) < n + \frac{1}{8\pi}\sqrt{n}\log^2 n.$$

The derivation in the case of $\psi(x)$ is similar. □

More information about this process of verifying estimates over finite ranges numerically is given in Chapter 3.

Corollary 4.7 *[153, corollary 1, p. 339] Assume the Riemann hypothesis is true. Then for all $x \geq 2657$ we have*

$$|\pi(x) - \mathrm{li}(x)| \leq \frac{\sqrt{x}\log x}{8\pi}.$$

Proof First let x and ξ be real numbers such that $x \geq 23 \times 10^8 \geq \xi > 1$. Then by Abel's identity [6, theorem 4.2] we have

$$\pi(x) - \pi(\xi) = \frac{\theta(x)}{\log x} - \frac{\theta(\xi)}{\log \xi} + \int_\xi^x \frac{\theta(t) - t}{t\log^2 t}\,dt + \int_\xi^x \frac{dt}{\log^2 t}.$$

Integrating $1/\log t$ by parts we have also

$$\int_\xi^x \frac{dt}{\log^2 t} = \frac{\xi}{\log \xi} - \frac{x}{\log x} + \int_\xi^x \frac{dt}{\log t} = \frac{\xi}{\log \xi} - \frac{x}{\log x} + \mathrm{li}(x) - \mathrm{li}(\xi).$$

Therefore we get

$$\pi(x) - \mathrm{li}(x) = \frac{\theta(x) - x}{\log x} + \int_\xi^x \frac{\theta(t) - t}{t\log^2 t}\,dt - \eta(\xi),$$

where $\eta(x) := \mathrm{li}(x) - \pi(x) - (x - \theta(x))/\log x$. Therefore, using Theorem 4.5,

$$\begin{aligned}
|\pi(x) - \mathrm{li}(x)| &\leq \left|\frac{\theta(x) - x}{\log x}\right| + \left|\int_\xi^x \frac{\theta(t) - t}{t\log^2 t}\,dt\right| + |\eta(\xi)| \\
&\leq \frac{\sqrt{x}(\log x - 2)}{8\pi} + \frac{1}{8\pi}\int_\xi^x \frac{dt}{\sqrt{t}} + |\eta(\xi)| \\
&\leq \frac{\sqrt{x}\log x}{8\pi} - \frac{\sqrt{\xi}}{4\pi} + |\eta(\xi)|.
\end{aligned}$$

At $\xi = 10^8$, $0 < \eta(\xi) < 88.5 < 10^4/(4\pi)$ so $|\pi(x) - \mathrm{li}(x)| < \sqrt{x}\log x/(8\pi)$ when $x \geq 23 \times 10^8$.

In the range $n \in [2657, 23 \times 10^8]$ we checked the inequality $\pi(n) < \mathrm{li}(n) + \sqrt{n}\log n/(8\pi)$ numerically at integers and then the inequality $\pi(n) > \mathrm{li}(n+1) - \sqrt{n+1}\log(n+1)/(8\pi)$, also at integers, demonstrating that the theorem is true for all $x \geq 2657$. □

Remark A computation shows that the inequality $|\pi(n) - \mathrm{li}(n)| < \sqrt{n}\log n/(8\pi)$ is also true for a substantial number of integers n with $2 \leq n \leq 2656$. Indeed 1434 successes were found, splitting into about 40 subranges.

We now give an example of a well-known technique which is used a few other times in this volume in case the Riemann hypothesis is false. Define $\theta := \sup\{\sigma : \zeta(\sigma + i\gamma) = 0\}$ so RH is equivalent to $\theta = \frac{1}{2}$, and always $\frac{1}{2} \leq \theta \leq 1$.

Theorem 4.8 *Let $\epsilon > 0$ be given. Then $\psi(x) - x = \Omega_\pm(x^{\theta-\epsilon})$, where the implied constant is arbitrary and when Ω_t is defined in Section 4.3.*

Proof First note that, by Lemma 5.28, for $\Re s > 1$ we have

$$-\frac{\zeta'(s)}{\zeta(s)} = s\int_2^\infty \frac{\psi(t)}{t^{s+1}}\,dt.$$

Now let $0 < \alpha < \theta \le 1$ and let $C \neq 0$ be a given constant. For $1 < x$ define

$$g(x) := \frac{\psi(x) - x - Cx^\alpha}{x}$$

so that

$$f(s) := \int_1^\infty \frac{g(x)}{x^s}\,dx = \int_1^\infty \frac{\psi(x) - x - Cx^\alpha}{x^{s+1}}\,dx = -\frac{\zeta'(s)}{s\zeta(s)} - \frac{1}{s-1} - \frac{C}{s-\alpha}.$$

Now let σ_0 be the abscissa of convergence of the integral for $f(s)$. The integral represents a holomorphic function on the half plane $\sigma > \sigma_0$. In addition, since $\zeta'(s)/\zeta(s)$ has a pole at an imaginary zeta zero with real part in $(\alpha, 1)$ which is not cancelled by any other term in the derived expression for $f(s)$, the half plane $\sigma > \sigma_0$ can contain no zero of $\zeta(s)$. This implies $\sigma_0 \ge \theta > \alpha$.

On the other hand, because $\zeta(s)$ is non-zero on the positive real axis $(0, \infty)$, and there is a function $h(s)$ which is holomorphic on the positive real axis with

$$\frac{\zeta'(s)}{\zeta(s)} = -\frac{1}{s-1} + h(s) \quad \Longrightarrow \quad f(s) = -\frac{1+h(s)}{s} - \frac{C}{s-\alpha},$$

$f(s)$ (or its continuation) has no singularity on the real axis for $\sigma > \alpha$. Because $\alpha < \sigma_0$, σ_0 is not a singularity of $f(s)$. By Theorem 4.12 we must have that $g(x)$ changes sign infinitely often as $x \to \infty$ so $\psi(x) - x > Cx^\alpha$ and $\psi(x) - x < -Cx^\alpha$, each for an infinite number of values of x with limit value infinity. To complete the proof, replace α by $\theta - \epsilon$. □

Now we can prove a principal result of this chapter, giving an explicit equivalence to RH.

Theorem 4.9 (Schoenfeld criterion) *The Riemann hypothesis is equivalent to the inequality*

$$|\psi(x) - x| \le \frac{\sqrt{x}\log^2 x}{8\pi}, \quad 74 \le x.$$

Proof If RH is true the result follows from Theorem 4.6. If RH is false then $\theta > \frac{1}{2}$. By Theorem 4.8 we can choose $\epsilon > 0$ so that $\theta - \epsilon > \frac{1}{2}$ and, letting the constant $C = 1$ we have

$$|\psi(x) - x| \ge x^{1/2}x^{\theta-\epsilon-1/2}$$

for an infinite sequence of values x with limit plus infinity. But for x sufficiently large we have $x^{\theta-\epsilon-1/2} > \log^2 x/(8\pi)$, so the inequality fails (an infinite number of times). Therefore the inequality is equivalent to RH. □

4.3 Oscillation Theorems

It is useful to note for what follows that, when the Riemann hypothesis fails, there exist sequences of real numbers or integers with limit infinity which contradict the estimates given by the Riemann hypothesis in different ways. To describe this, the notation $f(x) = \Omega_\pm(g(x))$ is used. Here $f(x)$ and $g(x)$ are real-valued and $g(x) > 0$ for all $x > 0$. This notation means that for some positive constant C and an infinite increasing real sequence $(x_n)_{n\in\mathbb{N}}$ with limit value infinity, we have $f(x_n) \geq Cg(x_n)$, and for some other such sequence $(y_n)_{n\in\mathbb{N}}$ we have $f(y_n) \leq -Cg(y_n)$.

The material in this section is based mainly on the account of Grosswald [70]. As given above, in 1901 von Koch proved, using the Riemann hypothesis, that $\pi(x) = \mathrm{Li}(x) + O(\sqrt{x}\log x)$. This was used by E. Schmidt in 1903 [152] to prove the following:

Theorem 4.10 (Schmidt) *Let* $\theta := \sup\{\Re s : \zeta(s) = 0\}$ *and*

$$f(x) := \sum_{m=1}^{\infty} \frac{\pi(x^{1/m})}{m}.$$

Then for all $\lambda < \theta$ *we have* $f(x) = \Omega_\pm(x^\lambda)$ *where the implied constant* C *may be chosen arbitrarily large. In addition there are real sequences* $x_n, y_n \to \infty$ *such that*

$$f(x_n) - \mathrm{Li}(x_n) > \frac{\sqrt{x_n}}{29\log x_n} \quad \text{and} \quad f(y_n) - \mathrm{Li}(y_n) < -\frac{\sqrt{y_n}}{29\log y_n}.$$

Then in 1905 Landau developed a new method to attack these "oscillation" problems. As usual, we define $\theta(x) := \sum_{p\leq x} \log p$.

Theorem 4.11 (Landau) *As* $x \to \infty$ *we have* $\theta(x) = \Omega_\pm(x)$.

The principal tool in this work is the theorem of Landau given in his 1905 paper [104], but also in [80, theorem H] and [166, chapter II.1, theorems 6 and 7]. The proof is included because the result is fundamental to this work, and used in many places.

Theorem 4.12 (Landau) *Let* $s \in \mathbb{C}$ *and let* $f(x) : [1,\infty) \to \mathbb{R}$ *be measurable and bounded on all bounded intervals. Suppose that*

$$F(s) := \int_1^\infty f(x)\frac{dx}{x^s}$$

has a finite abscissa of convergence σ_c *so* $F(s) \in \mathbb{C}$ *if* $\Re s > \sigma_c$.

(a) *If there exists an* $a \in \mathbb{R}$ *such that* $f(x)$ *is non-negative or non-positive for* $x \geq a$, *then the integral* $F(s)$ *for* $\sigma = \Re s > \sigma_c$ *has a singularity at* $s = \sigma_c$ *and* $F(s)$ *converges in a half plane such that it is holomorphic for* $\sigma > \sigma_c$ *but not in any half plane* $\sigma > \sigma_c - \epsilon$ *for any* $\epsilon > 0$.

(*b*) *If* $F(s)$ *is holomorphic at* $s = \sigma_c$, *then* $f(x)$ *changes sign at all points in an infinite set* x_n *with* $x_n \to \infty$. *We also have for every* $\epsilon > 0$

$$f(x) = \Omega_\pm(x^{\sigma_c - \epsilon}).$$

Proof (*a*) Suppose, to derive a contradiction, there exists an open connected set Ω containing the half plane $\Re s > \sigma_c$ and the point $s = \sigma_c$ so the continuation of $F(s)$ is holomorphic in Ω. Let $\alpha > \sigma_c$ and let $\delta > \alpha - \sigma_c$ be such that the open ball $\{s : |s - \alpha| < \delta\} \subset \Omega$.

Since $F(s)$ is holomorphic in Ω, for $|s - \alpha| < \delta$ we can express it as a convergent Taylor series,

$$F(s) = \sum_{n=0}^{\infty} \frac{(s-\alpha)^n}{n!} F^{(n)}(\alpha).$$

Suppose that $f(x) \geq 0$ for $x \geq a$. (If $f(x) \leq 0$ replace $f(x)$ by $-f(x)$.) Therefore if $b > a$ and $n \geq 0$ we can write

$$(-1)^n F^{(n)}(\alpha) = \int_1^{\infty} \frac{f(x) \log^n x}{x^\alpha} dx \geq \int_1^b \frac{f(x) \log^n x}{x^\alpha} dx.$$

Therefore if $s \in \mathbb{R}$ satisfies $\alpha - \delta < s \leq \alpha$, $F(s)$ is real also and we can write

$$F(s) \geq \sum_{n=0}^{\infty} \frac{(\alpha - s)^n}{n!} \int_1^b \frac{f(x) \log^n x}{x^\alpha} dx = \int_1^b \sum_{n=0}^{\infty} \frac{(\alpha - s)^n}{n!} \frac{f(x) \log^n x}{x^\alpha} dx.$$

To justify the interchange of the sum and the integral note that for $1 \leq x \leq b$ we have

$$\left| \frac{(s-\alpha)^n}{n!} \frac{f(x) \log^n x}{x^\alpha} \right| \leq M \frac{\delta^n \log^n b}{n!}$$

where $M := \max\{|f(x) x^{-\alpha}| : 1 \leq x \leq b\}$, so the series is uniformly convergent. Therefore

$$F(s) \geq \int_1^b \frac{f(x)}{x^\alpha} e^{(\alpha - s)\log x} dx = \int_1^b \frac{f(x)}{x^s} dx.$$

Because for real $s \geq a$ the integrand is positive, this integral is increasing with b for $b \geq a$, and bounded above by $F(s)$. Hence the integral is convergent at all points of the interval $s \in (\alpha - \delta, \alpha]$. But this is impossible since the interval properly contains $s = \sigma_c$. Therefore the function defined by the integral has a singularity at $s = \sigma_c$.

(*b*) Assume, to derive a contradiction, that there exists a constant $M > 0$ such that $f(x) \leq M x^{\sigma_c - \epsilon}$ for all $x \geq 1$. Then if we define

$$g(u) := M e^{(\sigma_c - \epsilon)u} - f(e^u)$$

we get $g(u) \geq 0$ for $u \geq 0$, and setting $x = e^u$

$$F(s) - \frac{M}{s - \sigma_c + \epsilon} = \int_1^\infty (f(x) - Mx^{\sigma_c - \epsilon}) \frac{dx}{x^{s+1}} = -\int_0^\infty g(u)e^{-su}\, du.$$

Because the integral

$$\int_1^\infty x^{\sigma_c - \epsilon - s - 1}\, dx$$

is absolutely convergent when $\Re s = \sigma > \sigma_c - \epsilon$, the abscissa of convergence of the integral of $g(u)e^{-su}$ is still σ_c. By part (a), the point $s = \sigma_c$ is a singularity of $F(s) - M/(s - \sigma_c + \epsilon)$ so $F(s)$ cannot be holomorphic at $s = \sigma_c$. This contradiction completes the proof. The proof for $f(x) \leq -Mx^{\sigma_c - \epsilon}$ is similar. □

Then in 1914 Littlewood [109] improved Theorem 4.11 by showing the following:

Theorem 4.13 *As $x \to \infty$ we have*

$$\psi(x) = x + \Omega_\pm(\sqrt{x}\operatorname{logloglog} x), \tag{4.2}$$

where the same formula is true if we replace $\psi(x)$ by $\theta(x)$.

Proof **(1)** The proof is in eight steps. First we note that if RH is false then the theorem follows from Theorem 4.8. Therefore in the rest of this proof we can assume RH is true.

(2) To begin the proof proper, it follows from the proof of Dirichlet's lemma, Lemma 5.32, that if we define $||x||$ to be the distance from x to the nearest integer, given fixed real numbers $\alpha_1, \ldots, \alpha_K$ and a positive integer N, there exists an integer n with $1 \leq n \leq N^K$ such that for $1 \leq j \leq K$ we have $||\alpha_j n|| \leq 1/N$.

(3) Let $\rho_j = \frac{1}{2} + i\gamma_j$ be the jth positive imaginary zeta zero with increasing γ_j. Let N be a sufficiently large positive integer and define $\alpha_j = \gamma_j \log N/(2\pi)$, where $0 < \gamma_j \leq T := N \log N$ (see Section 2.2), so the number of the γ_j under consideration is $K \asymp T \log T$. By Step (2) there is an integer n with $1 \leq n \leq N^K$ such that for each j with $0 < \gamma_j \leq T$ we have

$$\left\| \frac{n\gamma_j \log N}{2\pi} \right\| \leq \frac{1}{N}.$$

Because for all real x we have $|\sin(\pi x)| \leq \pi ||x||$ we can write

$$|\sin(2\pi\alpha) - \sin(2\pi\beta)| = |2\sin(\pi(\alpha - \beta))\cos(\pi(\alpha + \beta))| \leq 2\pi ||\alpha - \beta||.$$

Using these two bounds, if $x = N^n \exp(1/N)$ we have

$$\begin{aligned} |\sin(\gamma_j \log x) - \sin(\gamma_j/N)| &\leq 2\pi \left\| \frac{\gamma_j(\log x - 1/N)}{2\pi} \right\| \\ &= 2\pi \left\| \frac{n\gamma_j \log N}{2\pi} \right\| \leq \frac{2\pi}{N}. \end{aligned}$$

Similarly, if $x = N^n \exp(-1/N)$ we have

$$|\sin(\gamma_j \log x) + \sin(\gamma_j/N)| \le 2\pi \left\| \frac{\gamma_j(\log x + 1/N)}{2\pi} \right\| = 2\pi \left\| \frac{n\gamma_j \log N}{2\pi} \right\| \le \frac{2\pi}{N}.$$

(4) Next we **claim** that for $1/(2x) \le \delta \le \frac{1}{2}$ and $\delta = 1/N$ we have an expression for the average value of $\psi(x) - x$ over the interval $[x/e^\delta, xe^\delta]$, namely, uniformly for $x \ge 4$,

$$\begin{aligned} A(\delta) &:= \frac{1}{x\exp(\delta) - x\exp(-\delta)} \int_{x\exp(-\delta)}^{x\exp(\delta)} (\psi(t) - t)\,dt \\ &= -2\sqrt{x} \sum_{j=1}^{\infty} \frac{\sin(\gamma_j/N)}{\gamma_j/N} \frac{\sin(\gamma_j \log x)}{\gamma_j} + O(\sqrt{x}). \end{aligned} \tag{4.3}$$

To see this we first apply von Mangoldt's theorem, Theorem 2.2, to get, for some constant $c \in \mathbb{R}$,

$$\int_0^x (\psi(t) - t)\,dt = -\sum_\rho \frac{x^{\rho+1}}{\rho(\rho+1)} - cx + O(1),$$

where the sum is over all of the non-trivial zeta zeros. This implies

$$A(\delta) = -\frac{\delta}{\sinh(\delta)} \sum_\rho \frac{e^{\delta(\rho+1)} - e^{-\delta(\rho+1)}}{2\delta\rho(\rho+1)} x^\rho + O(1). \tag{4.4}$$

We have for $\rho = \frac{1}{2} + i\gamma$ as $\delta \to 0+$,

$$e^{\delta(\rho+1)} = e^{\delta i\gamma} + O(\delta) \quad \text{and} \quad e^{-\delta(\rho+1)} = e^{-\delta i\gamma} + O(\delta),$$

and (see Table 3.1)

$$\sum_\rho \frac{1}{\rho(\rho+1)} < \sum_\rho \frac{1}{\gamma^2} < \infty.$$

In addition we have $|x^\rho| = \sqrt{x}$ and, as $\delta \to 0$,

$$\frac{\delta}{\sinh(\delta)} = 1 + O(\delta^2) \ll 1.$$

Using these estimates in the summation of (4.4) gives

$$\begin{aligned} \sum_\rho \frac{e^{\delta(\rho+1)} - e^{-\delta(\rho+1)}}{2\delta\rho(\rho+1)} x^\rho &= ix^{1/2} \sum_\rho \frac{\sin(\gamma\delta)}{\delta} \frac{x^{i\gamma}}{\rho(\rho+1)} + O\left(x^{1/2} \sum_\rho \frac{1}{\rho(\rho+1)} \right) \\ &= ix^{1/2} \sum_\rho \frac{\sin(\gamma\delta)}{\delta} \frac{x^{i\gamma}}{\rho(\rho+1)} + O(x^{1/2}). \end{aligned}$$

Next using

$$\int_T^\infty \frac{d(t\log t)}{t^2} = \frac{2+\log T}{T} \ll \frac{\log T}{T},$$

we have

$$\begin{aligned}\sum_\rho \frac{\sin(\gamma\delta)}{\delta}\frac{x^{i\gamma}}{\rho(\rho+1)} &\ll \sum_{\gamma\geq 1}\left|\frac{\sin(\gamma\delta)}{\delta}\frac{x^{i\gamma}}{\rho(\rho+1)}\right| \\ &\ll \sum_{1\leq\gamma\leq 1/\delta}\frac{\sin(\gamma\delta)}{\gamma\delta}\frac{\gamma}{\gamma^2}+\frac{1}{\delta}\sum_{1/\delta<\gamma}\frac{|\sin(\gamma\delta)|}{\gamma^2} \\ &\ll \sum_{1\leq\gamma\leq 1/\delta}\frac{1}{\gamma}+\frac{1}{\delta}\sum_{1/\delta<\gamma}\frac{1}{\gamma^2} \\ &\ll \log^2(1/\delta)+\log(1/\delta) \ll \log^2(1/\delta).\end{aligned}$$

If we substitute $1+O(\delta^2)$ for $\delta/\sinh(\delta)$ in the right-hand side of (4.4) then we get

$$A(\delta) = -ix^{1/2}\sum_\rho \frac{\sin(\gamma\delta)}{\delta}\frac{x^{i\gamma}}{\rho(\rho+1)} + O(x^{1/2}).$$

We have $\sin(\gamma\delta)/\delta \leq |\gamma|$ and by RH, $1/\rho = 1/(i\gamma)+O(1/\gamma^2)$ and $1/(\rho+1) = 1/(i\gamma)+O(1/\gamma^2)$. Therefore

$$\begin{aligned}A(\delta) &= -\sqrt{x}\sum_\rho \frac{\sin(\gamma\delta)}{\gamma\delta}\frac{x^{i\gamma}}{i\gamma}+O(\sqrt{x}) \\ &= -2\sqrt{x}\sum_\rho \frac{\sin(\gamma\delta)}{\gamma\delta}\frac{x^{i\gamma}-x^{-i\gamma}}{i\gamma}+O(\sqrt{x}) \\ &= -2\sqrt{x}\sum_{\gamma>0}\frac{\sin(\gamma\delta)}{\gamma\delta}\frac{\sin(\gamma\log x)}{\gamma}+O(\sqrt{x}),\end{aligned}$$

which completes the proof of the claim.

(5) Next, using the trivial bound for $\sin(x)$, Lemma 4.1 and the estimate again

$$\int_T^\infty \frac{d(t\log t)}{t^2} \ll \frac{\log T}{T},$$

we have

$$\sum_{N\log N<\gamma_j}\frac{\sin(\gamma_j/N)}{\gamma_j/N}\frac{\sin(\gamma_j\log x)}{\gamma_j} \ll N\sum_{N\log N<\gamma_j}\frac{1}{\gamma_j^2} \ll \frac{N\log(N\log N)}{N\log N} \ll 1.$$

Therefore for $x = N^n \exp(1/N)$, $\delta = 1/N$ and using the result of Step (3) (this is the key to the proof) and Step (5), we see $A(\delta)$ can be expressed as

$$A(\delta) = -2x^{\frac{1}{2}} \sum_{j=1}^{\infty} \frac{\sin(\gamma_j/N)}{\gamma_j/N} \cdot \frac{\sin(\gamma_j \log x)}{\gamma_j} + O(\sqrt{x}) \tag{4.5}$$

$$\begin{aligned}
&= -2x^{\frac{1}{2}} \sum_{\gamma_j \le N\log N} \frac{\sin(\gamma_j/N)}{\gamma_j/N} \cdot \frac{\sin(\gamma_j \log x)}{\gamma_j} \\
&\quad + O\left(\sqrt{x} \sum_{\gamma_j > N\log N} \frac{\sin(\gamma_j/N)}{\gamma_j/N} \cdot \frac{\sin(\gamma_j \log x)}{\gamma_j} \right) + O(\sqrt{x}) \\
&= -2\frac{x^{\frac{1}{2}}}{N} \sum_{\gamma_j \le N\log N} \left(\frac{\sin(\gamma_j/N)}{\gamma_j/N} \right)^2 + (\sqrt{x}) \\
&\ll -2\frac{\sqrt{x}}{N} \int_1^{N\log N} \left(\frac{\sin(t/N)}{t/N} \right)^2 \log t \, dt + O(\sqrt{x}) \\
&\ll -2\frac{x^{\frac{1}{2}}}{N} N\log N + O(\sqrt{x}) \\
&= -2x^{\frac{1}{2}} \log N + O(\sqrt{x}).
\end{aligned} \tag{4.6}$$

Similarly for $x = N^n \exp(-1/N)$ we have

$$-2x^{1/2} \sum_{j=1}^{\infty} \frac{\sin(\gamma_j/N)}{\gamma_j/N} \frac{\sin(\gamma_j \log x)}{\gamma_j} \gg +2x^{1/2} \log N + O(\sqrt{x}). \tag{4.7}$$

(6) Because $x \le N^{N^K} e^{1/N}$, $T = N\log N$ and $K \asymp T\log T \asymp N\log^2 N$ we get

$$\operatorname{loglog} x \ll N \log^3 N.$$

Taking logarithms for a constant $C > 0$ we get from this

$$\log N + \log C + 3 \operatorname{loglog} N = (1 + o(1)) \log(N) \ge \operatorname{logloglog} x,$$

which implies

$$\log N \ge (1 + o(1)) \operatorname{logloglog} x.$$

(7) Finally, by (4.5) and (4.7), and the bound established in Step (6), letting $\delta \to 0+$ at a point of continuity of $\psi(x)$ we can write

$$\psi(x) = x + \Omega_{\pm}(\sqrt{x} \operatorname{logloglog} x).$$

The same is true at points of discontinuity, since the jump is $O(\log x)$.

(8) Considering the expansion of $\psi(x)$ in terms of $\theta(x^{1/j})$, with $1 \le j \le \log x/\log 2$, and the bound $\theta(x) \ll x$, we can write

$$\psi(x) = \theta(x) + \theta(\sqrt{x}) + O(x^{1/3} \log x) \quad \Longrightarrow \quad 0 \le \psi(x) - \theta(x) \ll \sqrt{x}.$$

Thus $\psi(x) = \theta(x) + O(\sqrt{x})$, and substituting this in (4.2) completes the proof. □

This theorem can be shown to be equivalent to

$$\pi(x) = \mathrm{Li}(x) + \Omega_{\pm}\left(\frac{\sqrt{x}\,\mathrm{logloglog}\,x}{\log x}\right).$$

The reader might compare this with theorem of Walfisz, giving the best-known error in the prime number theorem.

Littlewood (Figure 4.1) gives an explicit form for the estimates of Theorem 4.13. There are sequences of positive real numbers $x_n \to \infty$ and $y_n \to \infty$ such that for all $n \in \mathbb{N}$ we have

$$\pi(x_n) > \mathrm{Li}(x_n) + \frac{\sqrt{x_n}\,\mathrm{logloglog}\,x_n}{3\log x_n}\pi(x_n)$$

and

$$\pi(y_n) < \mathrm{Li}(y_n) + \frac{\sqrt{y_n}\,\mathrm{logloglog}\,y_n}{3\log y_n}.$$

It follows from these inequalities that there is an $x > 0$ such that $\pi(x) > \mathrm{li}(x)$. The **Skews number** is the smallest natural number S for which this inequality is true. This number is very celebrated, its proved upper bound being at one time the largest number appearing in a mathematical proof. Indeed Skews proved in 1955 [157] that

$$S < 10^{10^{10^{964}}}.$$

Figure 4.1 J. E. Littlewood (1885–1977).

There has been a great deal of work reducing the size of the known upper bound and finding lower bounds for S. For example Zegowitz [183] showed in 2010 that $S < \exp(727.951347)$ and Kotnik [97] in 2008 that $10^{14} < S$.

For this section, and indeed all of this chapter, the reader would do well to consider the article by Don Zagier [182].

4.4 Errors in Arithmetic Sums

Recall the definition of $\mu(n)$: $\mu(1) = 1$ and, if n is not squarefree, $\mu(n) = 0$; otherwise $\mu(n)$ is $(-1)^{\omega(n)}$ where $\omega(n)$ is the number of distinct primes dividing n. Then the prime number theorem is equivalent to the statement [6]

$$\lim_{N\to\infty} \frac{\mu(1)+\mu(2)+\cdots+\mu(N)}{N} = 0 \quad \text{or} \quad \sum_{n\le x} \mu(n) = o(x).$$

Now we will see below one of the most closely considered equivalences to RH, due originally to Littlewood (see Littlewood [109] or Edwards [51, section 12.1]), namely that RH is equivalent to the estimate that for all $\epsilon > 0$

$$\sum_{n\le x} \mu(n) \ll x^{1/2+\epsilon}.$$

First though two estimates are needed:

Lemma 4.14 *[51, p. 54] Let $a > 0$, $T > 0$ and $0 < x < 1$. Then*

$$\left| \frac{1}{2\pi i} \int_{a-iT}^{a+iT} \frac{x^s}{s}\, ds \right| \le \frac{x^a}{\pi T |\log x|}.$$

Proof Let $a < A$ and apply Cauchy's residue theorem to the function x^s/s on the rectangle $[a,A]\times[-T,T]$, on which the function is holomorphic. Thus the contour integral of the function around the boundary of the rectangle is zero so

$$\frac{1}{2\pi i}\int_{a-iT}^{a+iT} \frac{x^s}{s}\, ds = -\frac{1}{2\pi i}\int_{A-iT}^{A+iT} \frac{x^s}{s}\, ds - \frac{1}{2\pi i}\int_{a-iT}^{A-iT} \frac{x^s}{s}\, ds - \frac{1}{2\pi i}\int_{A+iT}^{a+iT} \frac{x^s}{s}\, ds.$$

Therefore, since the modulus of the integrand is bounded by x^A/A on the right-hand vertical section of the contour and, if $\sigma = \Re s$, by x^σ/T on the horizontal sections, we have

$$\left| \frac{1}{2\pi i}\int_{a-iT}^{a+iT} \frac{x^s}{s}\, ds \right| \le \frac{1}{2\pi}\frac{2Tx^A}{A} + 2\frac{1}{2\pi}\int_a^A \frac{x^\sigma}{T}\, d\sigma \le \frac{Tx^A}{\pi A} + \frac{(x^a - x^A)}{\pi T|\log x|}.$$

Letting $A \to \infty$ the lemma follows. □

Lemma 4.15 *[51, p. 55] Let $a > 0$, $T > 0$ and $x > 1$. Then*

$$\left|\frac{1}{2\pi i}\int_{a-iT}^{a+iT}\frac{x^s}{s}\,ds - 1\right| \le \frac{x^a}{\pi T|\log x|}.$$

Proof The proof is similar to that of Lemma 4.14, except this time integrate around $[-A,a]\times[-T,T]$ and evaluate the residue of the function x^s/s at $s = 0$. □

Theorem 4.16 (Littlewood criterion) *[108] The Riemann hypothesis is equivalent to the estimate that for all $\epsilon > 0$ we have*

$$\sum_{n\le x}\mu(n) \ll x^{1/2+\epsilon}, \qquad x \to \infty,$$

where the implied constant depends on ϵ.

Proof **(1)** First assume the given condition, i.e. for all $\epsilon > 0$ we have

$$\sum_{n\le x}\mu(n) \ll x^{1/2+\epsilon}, \qquad x \to \infty.$$

This part of the proof will show the condition implies the Riemann hypothesis is true. Now

$$\frac{1}{\zeta(s)} = \sum_{n=1}^{\infty}\frac{\mu(n)}{n^s}, \qquad \Re s > 1.$$

Using Abel summation [6, theorem 4.2] with $M(x) := \sum_{n\le x}\mu(n)$,

$$\sum_{n=1}^{x}\frac{\mu(n)}{n^s} = \frac{M(x)}{x^s} + s\int_1^x \frac{M(t)}{t^{s+1}}\,dt.$$

Therefore, letting $x\to\infty$, for $\Re s > 1$ we have

$$\frac{1}{\zeta(s)} = s\int_1^{\infty}\frac{M(x)}{x^{s+1}}\,dx.$$

If $M(x)$ grew less rapidly than $x^{1/2+\epsilon}$ for every $\epsilon > 0$ then the integral representation for $1/\zeta(s)$ would converge for $\Re s > \frac{1}{2} + \epsilon$ to an analytic function. Therefore $1/\zeta(s)$ would not have a pole in this half plane, so $\zeta(s)$ would fail to have a zero. Since this is so for every $\epsilon > 0$, $\zeta(s)$ would have no zero to the right of the critical line, so the Riemann hypothesis would be true.

(2) Now assume RH. Let $T > 0$ and $x > 0$, considered large, $x \notin \mathbb{N}$ and

$$\Delta(x,T) := \left|M(x) - \frac{1}{2\pi i}\int_{2-iT}^{2+iT}\frac{x^s}{\zeta(s)}\frac{ds}{s}\right|$$

$$= \left| M(x) - \frac{1}{2\pi i} \int_{2-iT}^{2+iT} \sum_{n=1}^{\infty} \mu(n) \left(\frac{x}{n}\right)^s \frac{ds}{s} \right|$$
$$= \left| \sum_{n<x} \mu(n) - \frac{1}{2\pi i} \int_{2-iT}^{2+iT} \sum_{n=1}^{\infty} \mu(n) \left(\frac{x}{n}\right)^s \frac{ds}{s} \right|$$
$$= \left| \sum_{n<x} \mu(n) - \frac{1}{2\pi i} \int_{2-iT}^{2+iT} \sum_{n<x} \mu(n) \left(\frac{x}{n}\right)^s \frac{ds}{s} \right.$$
$$\left. - \frac{1}{2\pi i} \int_{2-iT}^{2+iT} \sum_{n>x} \mu(n) \left(\frac{x}{n}\right)^s \frac{ds}{s} \right|.$$

Therefore, using Lemma 4.14, Lemma 4.15, $|\mu(n)| \leq 1$, for values of x restricted to being half an integer so that $|\log(x/n)| \geq 2/n$ for $x/2 \leq n < 2x$ and $|\log(x/n)| \geq \log 2$ for $x/2 > n$ or $n > 2x$, this gives

$$\begin{aligned}
\Delta(x,T) &\leq \sum_{n<x} \frac{(x/n)^2}{\pi T |\log(x/n)|} + \sum_{n>x} \frac{(x/n)^2}{\pi T |\log(x/n)|} \\
&\leq \sum_{x/2>n \text{ or } n>2x} \frac{(x/n)^2}{\pi T |\log(x/n)|} + \sum_{x/2 \leq n < 2x} \frac{(x/n)^2}{\pi T |\log(x/n)|} \\
&\leq \frac{x^2}{\pi T} \left(\frac{1}{\log 2} \sum_{n=1}^{\infty} \frac{1}{n^2} + \sum_{x/2<n\leq 2x} \frac{1}{n} \right) \\
&\leq \frac{x^2}{\pi T} \left(\frac{\pi^2}{6 \log 2} + (\log(2x) - \log(x/2)) \right) \\
&\ll \frac{x^2}{T}.
\end{aligned}$$

Now let $\epsilon > 0$ be given and integrate $x^s/(\zeta(s)s)$ around the rectangle

$$[\tfrac{1}{2} + \epsilon - iT,\ 2 - iT] \times \left[\tfrac{1}{2} + \epsilon - iT,\ \tfrac{1}{2} + \epsilon + iT\right]$$

to derive an alternative expression for the integral approximation to $M(x)$ using the approximation to $1/|\zeta(s)| \ll_\delta t^\delta$ given, since we are assuming RH, in item (8) of Section 2.2:

$$\begin{aligned}
\frac{1}{2\pi i} \int_{2-iT}^{2+iT} \frac{x^s}{\zeta(s)} \frac{ds}{s} &= \frac{1}{2\pi i} \int_{2-iT}^{1/2+\epsilon-iT} \frac{x^s}{\zeta(s)} \frac{ds}{s} + \frac{1}{2\pi i} \int_{1/2+\epsilon-iT}^{1/2+\epsilon+iT} \frac{x^s}{\zeta(s)} \frac{ds}{s} \\
&\quad + \frac{1}{2\pi i} \int_{1/2+\epsilon+iT}^{2+iT} \frac{x^s}{\zeta(s)} \frac{ds}{s} \\
&\leq 2K_\delta T^{\delta-1} x^2 + \left| \frac{1}{2\pi i} \int_{1/2+\epsilon-iT_0}^{1/2+\epsilon+iT_0} \frac{x^s}{\zeta(s)} \frac{ds}{s} \right| \\
&\quad + \frac{2}{2\pi} \int_{T_0}^{T} x^{1/2+\epsilon} K_\delta t^\delta \frac{dt}{t} \\
&\ll x^{1/2+\epsilon+2\delta} + \frac{x^{1/2+\epsilon} K_\delta T^\delta}{\pi\delta}.
\end{aligned}$$

If $T = x^2$ then $M(x)$ will be less than some constant (dependent on δ) times $x^{1/2+\epsilon+2\delta}$ for all large x of the form of an integer plus one half. But since $|\mu(n)| \le 1$, $M(x)$ can change by at most one between these values of x, so replacing ϵ by $\epsilon/3$ and δ by $2\epsilon/3$ we get $M(x) \ll_\epsilon x^{1/2+\epsilon}$, which completes the proof of Liouville's theorem. □

There is an interesting story concerning the bounds for $M(x) := \sum_{n\le x}\mu(n)$ [6, p. 91]. What became known as Mertens' conjecture was the statement

$$|M(x)| \le \sqrt{x}, \quad x \ge 1.$$

That this might be so came from computational evidence (see for example Figure 4.2). However it was shown in 1985, by Odlyzko and te Riele [129], that this conjecture was false, and failed for an infinite number of values x with limit value infinity. The best-known unconditional result is that of Walfisz [177] who showed in 1963 that

$$M(x) \ll x\exp\left(A\frac{\log^{3/5} x}{\log\log^{1/5} x}\right).$$

It follows from Theorem 4.16 directly that if the squarefree integers less than or equal to N are considered, and $D(N)$ denotes the absolute value of the difference between the number of those divisible by an even number of primes and the number of those divisible by an odd number of primes, then we have the following:

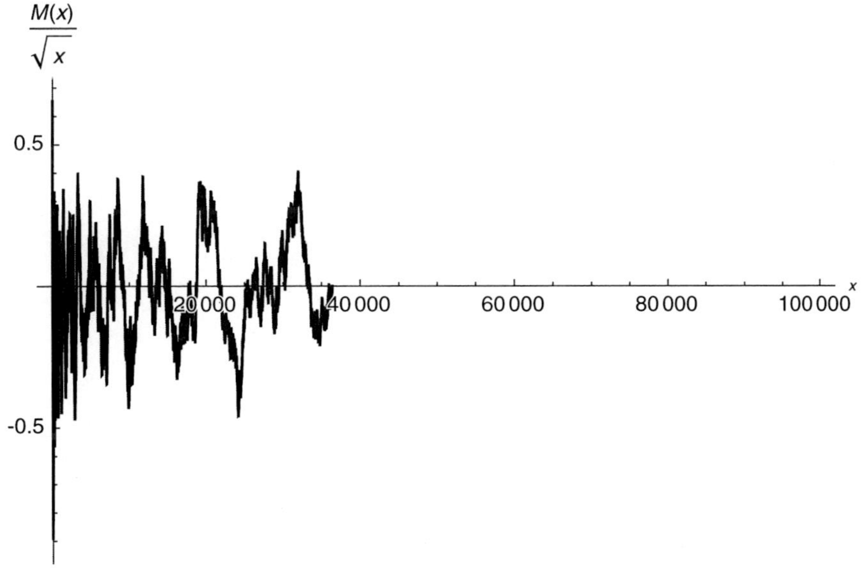

Figure 4.2 The function $M(x)/\sqrt{x}$ for $1 \le x \le 10^5$.

Corollary 4.17 (Parity criterion) *The Riemann hypothesis is equivalent to the statement, as $N \to \infty$,*

$$D(N) \ll N^{1/2+\epsilon}.$$

Another equivalence to RH, sometimes attributed to Landau, but an easy consequence of the Littlewood criterion, is an expression in terms of Liouville's function. Let $n \in \mathbb{N}$ and as usual $\Omega(n)$ is the number of primes dividing n including multiplicity, and let $\lambda(n) = (-1)^{\Omega(n)}$ be Liouville's completely multiplicative function. Then the prime number theorem is equivalent to the statement

$$\lim_{N\to\infty} \frac{\lambda(1) + \lambda(2) + \cdots + \lambda(N)}{N} = 0,$$

or in other words

$$L(x) := \sum_{n \le x} \lambda(n) = o(x), \quad x \to \infty.$$

Theorem 4.18 (Liouville criterion) *The Riemann hypothesis is equivalent to the estimate that for all $\epsilon > 0$ we have*

$$\lim_{N\to\infty} \frac{\lambda(1) + \lambda(2) + \cdots + \lambda(N)}{N^{1/2+\epsilon}} = 0.$$

Proof Checking at prime powers and using the multiplicativity of λ and μ we have for $n \in \mathbb{N}$

$$\lambda(n) = \sum_{d^2 | n} \mu\left(\frac{n}{d^2}\right).$$

Then

$$L(x) = \sum_{1 \le n \le x} \sum_{d^2 | n} \mu\left(\frac{n}{d^2}\right) = \sum_{d \le \sqrt{x}} \sum_{e \le x/d^2} \mu(e) = \sum_{d \le \sqrt{x}} M\left(\frac{x}{d^2}\right).$$

Therefore if $M(x) \ll_\epsilon x^{1/2+\epsilon}$ then it follows that $L(x) \ll_\epsilon x^{1/2+\epsilon}$. By Möbius inversion we also have

$$M(x) = \sum_{d=1}^{\sqrt{x}} \mu(d) L\left(\frac{x}{d^2}\right).$$

From this it follows that if $L(x) \ll_\epsilon x^{1/2+\epsilon}$ then $M(x) \ll_\epsilon x^{1/2+\epsilon}$. The criterion then follows from the Littlewood criterion, Theorem 4.16. □

4.5 Unsolved Problems

(1) Find an $\epsilon > 0$ for which as $x \to \infty$ we have $M(x) \ll x^{1-\epsilon}$.
(2) Find explicit forms of the equivalence for the extended or generalized Riemann hypothesis for Dirichlet L-functions, first for a single non-trivial character and then for a class of characters [13, p. 56].
(3) Use Trudgian's form for $B(T)$ in Section 2.2, and determine whether Theorem 4.5 could be significantly improved.
(4) Examine the proof of Littlewood's theorem, Theorem 4.13, and determine whether it could be improved. For example maybe using

$$\psi(x) = x + \Omega_{\pm}(\sqrt{x}\log\log x).$$

5
Euler's Totient Function

5.1 Introduction

This chapter is based on the work of Jean-Louis Nicolas, J. Barkley Rosser and Lowell Schoenfeld, as published in a variety of articles, but most especially in [126, 127, 145].

J. Barkley Rosser (1907–1989) was a logician and student of Alonzo Church. Together they proved the "Church–Rosser theorem" of the lambda calculus. He also undertook research in prime number theory where "Rosser's rule" is named after him. Less is known about Lowell Schoenfeld, who worked in the main in number theory and complex analysis. His joint 1962 paper with Rosser, giving explicit estimates for many arithmetical functions, has 358 citations registered by MathSciNet. Parts of that work are detailed in Chapter 3.

Jean-Louis Nicolas received his Ph.D. in 1968 with a thesis entitled "The order of the maximal element of the group S_n and permutations of highly composite numbers" supervised by Charles Pisot. His work on the symmetric group is outlined in Chapter 10. His first student was Guy Robin, who graduated in 1983, some of whose work is described in Chapter 7.

The equivalences to RH, which are the subject of this chapter, are expressed in terms of the well-known arithmetical function $\varphi(n)$. This was defined by Euler in 1763 as "the multitude of numbers less than n which have no common divisor with it". He proved that

$$\frac{\varphi(n)}{n} = \prod_{p|n} \left(1 - \frac{1}{p}\right),$$

where the product is over primes which divide n, and proved his theorem $a^{\varphi(n)} \equiv 1 \bmod n$ when $(a,n) = 1$. It was Gauss who introduced the now standard notation $\varphi(n)$ in his 1801 "Disquisitiones Arithmeticae", and J. J. Sylvester in 1879 who called the function a "totient".

Figure 5.1 Jean-Louis Nicolas.

Some idea of the behaviour of $\varphi(n)$ for large n can be obtained from the average order

$$\varphi(1) + \varphi(2) + \cdots + \varphi(n) \sim \frac{3n^2}{\pi^2}$$

and the spread of values. These are such that $\{n/\varphi(n) : n \in \mathbb{N}\}$ is dense in $(1, \infty)$,

$$\liminf_{n\to\infty} \frac{\varphi(n)}{n} = 0 \quad \text{and} \quad \limsup_{n\to\infty} \frac{\varphi(n)}{n} = 1.$$

In addition we have for all $\epsilon > 0$

$$\lim_{n\to\infty} \frac{\varphi(n)}{n^{1-\epsilon}} = \infty.$$

Finally, in this brief survey of some of the properties of $\varphi(n)$, we have

$$\liminf_{n\to\infty} \frac{\varphi(n) \log\log n}{n} = e^{-\gamma}$$

but for $n \geq 2$,

$$\frac{n}{\varphi(n)} < e^{\gamma} \log\log n + \frac{3}{\log\log n},$$

whereas $e^\gamma \log\log n < n/\varphi(n)$ for an infinite number of integers n. These properties are treated in this chapter.

Background material on $\varphi(n)$, including many of these results, can be found in the standard texts by Hardy and Wright [75] and De Koninck and Luca [95, chapter 8], and in the "Handbook of Number Theory", volume I [150, chapter I]; see also Erdős [54]. Historical material is in Dickson [46, volume I, chapter V].

Here is a summary of the content of the chapter. We begin with estimates relating to $\varphi(n)$ developed by Rosser and Schoenfeld. For example these include the proof (Lemma 5.12) that for all $x \geq 286$ we have

$$e^\gamma \log x\left(1 - \frac{1}{2\log^2 x}\right) < \prod_{p\leq x} \frac{p}{p-1} < e^\gamma \log x\left(1 + \frac{1}{2\log^2 x}\right),$$

where the lower bound also holds for $1 < x < 286$. We also call on explicit estimates from Chapters 3 and 4. These explicit estimates provide an essential and important underpinning for the RH equivalences.

The first main result, of Rosser and Schoenfeld, Theorem 5.25, comes after quite a large number of lemmas.

Theorem 5.25 *Without any hypotheses, for all $n \geq 3$ except for the ninth primorial (see below) $N_9 = 2\cdot 3\cdots 23$, we have*

$$\frac{n}{\varphi(n)} < e^\gamma \log\log n + \frac{5}{2\log\log n}. \tag{5.1}$$

The second main result, that of Nicolas, Theorem 5.29, is an essential tool, used in this chapter and in Chapter 7, when we assume RH is false. Here we use the function

$$f(x) := e^\gamma \log\theta(x) \prod_{p\leq x}\left(1 - \frac{1}{p}\right),$$

where as usual $\theta(x) := \sum_{p\leq x} \log p$.

Theorem 5.29 *Assuming RH is false, there exists a real number b with $0 < b < \frac{1}{2}$ such that as $x \to \infty$*

$$f(x) = \Omega_\pm\left(\frac{1}{x^b}\right). \tag{5.2}$$

We then build to proving what we call **Nicolas' first criterion**, which is Theorem 5.31, from 1983. It is very similar in structure to that of Robin's theorem, Theorem 7.11, but is expressed in terms of primorials. Let N_k be the kth **primorial**, i.e. the product of the first k prime numbers in increasing order of size.

Theorem 5.31 (Nicolas' first criterion) *If RH is **true** then for all $k \geq 1$ we have*

$$e^{\gamma} \log\log N_k < \frac{N_k}{\varphi(N_k)}. \tag{5.3}$$

*If however RH is **false** then this inequality is true for an infinite number of k and false for an infinite number of k.*

There has been, apparently, little discussion of expanding the set of solutions to the inequality (5.3) beyond the primorials, even with RH. The author suspects that much more is true than Theorem 5.31. For example, up to 10^9, every positive integer which is a multiple of a primorial and is less than the following primorial satisfies the given inequality. There has also been little discussion on which primorials qualify in the different situations when RH fails.

Nicolas' second theorem, Theorem 5.34, is much more recent and includes four equivalences to RH in terms of $\varphi(n)$. We will state just the first in this introduction.

Theorem 5.34 *Let*

$$c(n) := \left(\frac{n}{\varphi(n)} - e^{\gamma} \log\log n \right) \sqrt{\log n}.$$

Then RH is equivalent to

$$\limsup_{n \to \infty} c(n) = e^{\gamma}(2 + \beta),$$

where $\beta = 2 + \gamma - \log 4\pi$.

The equivalence, Corollary 5.35, is easy to prove, but shows that Theorem 5.34 is consistent with other results, such as those of Chapter 7. It says that RH is equivalent to the inequality

$$\frac{n}{\varphi(n)} < e^{\gamma} \log\log n + \frac{e^{\gamma}(2 + \beta)}{\sqrt{\log n}}$$

for $n \geq N_{120569}$. We call this form **Nicolas' second criterion**.

Proving these theorems requires a set of lemmas which we prove in several sections below. Estimates for prime products and sums are in Section 5.2. Preliminary results with RH true, mostly dating from 1983, are in Section 5.3. The further results with RH true, given in Section 5.4, correspond roughly to Nicolas' more recent work.

Nicolas' first theorem follows in Section 5.6. In this section we include a note regarding what would result if RH fails. In that case an infinite number of primorials for which the inequality (5.3) were true would each be followed by a primorial for which it was false. Finally we have in Section 5.7 the proof of Nicolas' second theorem.

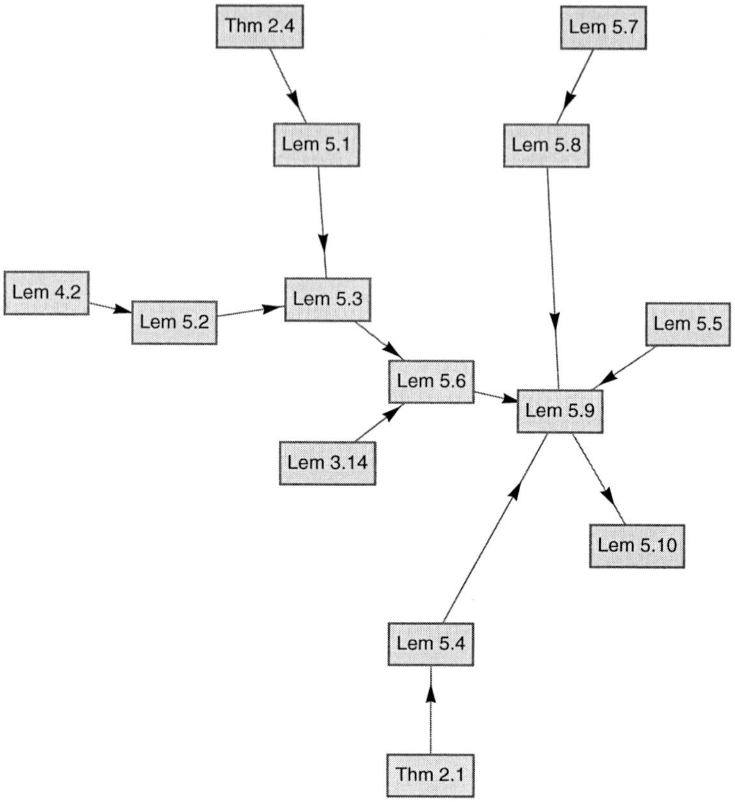

Figure 5.2 Some relationships between results in Chapter 5.

Navigating through the results in this chapter may at times be challenging. To aid the reader, the relationships between some of the lemmas, theorems and corollaries are set out in the directed graphs in Figures 5.2, 5.3 and 5.7.

The following RHpack (see Appendix B) functions relate to the material in this chapter: NOverPhi, NicolasInequalityQ, Computef, NicolasTwoFailures, ComputeH, ComputeDeltas, FHalf, PlotEulerPhiRatio, PlotInequality, CheckCNk, ComputeC and PlotCNk.

5.2 Estimates for Euler's Function $\varphi(n)$

We commence with a definition of the function $V(x)$, which forms an essential part of the statements and proofs of results in this part of the development of ideas. Note that in the definition of $V(x)$, the sum is over

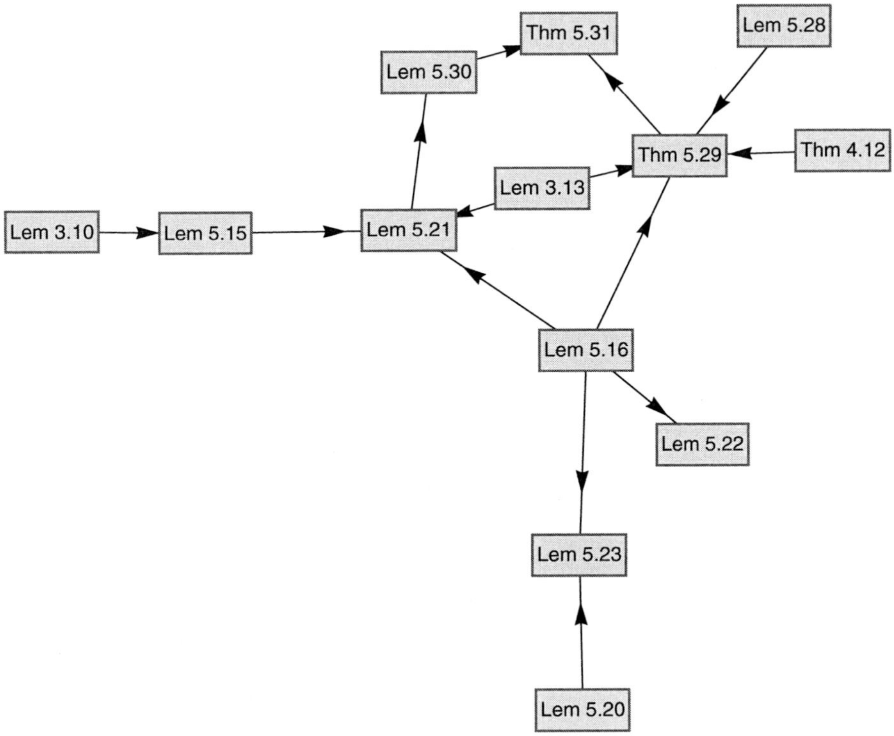

Figure 5.3 Some relationships between results in Chapter 5.

all imaginary zeros $\rho = \beta + i\gamma$ of $\zeta(s)$:

$$V(x) := \sum_{\rho} \frac{x^{\beta-1}}{\gamma^2}.$$

Considering zeta zeros $\rho = \beta + i\gamma$, a critical parameter is denoted H, where for all $0 < \gamma \leq H, \beta = \Re\rho = \frac{1}{2}$, i.e. H is the height below which all zeta zeros with positive imaginary part have been shown to be on the critical line. A variety of values for H were used in the papers of Rosser and Schoenfeld. Even in the 1962 paper both γ_{57} and γ_{25000} were used. We will assume that $\Re\rho = \frac{1}{2}$ up to $H := 5.4 \times 10^8 < \gamma_{15\times 10^8}$ (see Table 2.2), rather than the best-known available value of H, which is larger. A variety of values for the parameter R were also used in their papers. First we recall the description of a zero-free region in the critical strip implied by Theorem 2.4: if $\zeta(\rho) = 0$ with $\rho = \beta + i\gamma$ and $H < \gamma$ then $\beta < 1 - 1/(R \log \gamma)$.

This leads to the definition of the principal function in this part of the development, $\phi(x,\gamma,R)$, which we write as $\phi(\gamma)$:

$$\phi(\gamma) := \frac{\exp\left(-\dfrac{\log x}{R\log\gamma}\right)}{\gamma^2}.$$

We begin by deriving a preliminary upper bound for $V(x)$ in terms of the $\phi(\gamma)$ values. This will be made more explicit later.

Lemma 5.1 *For $x > 1$ we have*

$$V(x) \le \frac{1}{\sqrt{x}}\sum_{\rho}\frac{1}{\gamma^2} + \sum_{H<\gamma}\phi(\gamma).$$

Proof Splitting the sum in the definition of $V(x)$ we get

$$V(x) = \sum_{\beta\le 1/2}\frac{x^{\beta-1}}{\gamma^2} + \sum_{\beta>1/2}\frac{x^{\beta-1}}{\gamma^2}.$$

For the first sum

$$\sum_{\beta\le 1/2}\frac{x^{\beta-1}}{\gamma^2} \le \frac{1}{\sqrt{x}}\sum_{\beta\le 1/2}\frac{1}{\gamma^2} \le \frac{1}{\sqrt{x}}\sum_{\rho}\frac{1}{\gamma^2},$$

and for the second

$$\sum_{\beta>1/2}\frac{x^{\beta-1}}{\gamma^2} = 2\sum_{\beta>1/2,\ \gamma>0}\frac{x^{\beta-1}}{\gamma^2}.$$

In addition, by Theorem 2.4, for $\gamma > H$ we have

$$\frac{x^{\beta-1}}{\gamma^2} < \frac{\exp\left(-\dfrac{\log x}{R\log\gamma}\right)}{\gamma^2} = \phi(\gamma),$$

and so

$$\sum_{\beta>1/2}\frac{x^{\beta-1}}{\gamma^2} < 2\sum_{\beta>1/2,\ \gamma>0}\phi(\gamma) \le \sum_{\gamma>H}\phi(\gamma).$$

Putting these bounds together completes the proof. □

To see how this and the next few lemmas work together, we have the network flow diagram given in Figure 5.2.

Recall from Section 2.2 Backlund's result, and its improvements, for the difference between the number $N(T)$ of zeta zeros with positive imaginary part up to height T and its approximation $F(T)$, namely $B(T) := a_1\log T + a_2\log\log T + a_3$. Also recall the definition $\phi(\gamma) = \exp(-\log x/(R\log\gamma))/\gamma^2$.

We are now able to bound sums of the $\phi(\gamma)$ in terms of two supplementary functions which depend on a_1, a_2 and a_3 using the generalized form of Abel summation.

Lemma 5.2 *[145, theorem 27] If $2 \le h$ then*

$$\sum_{h<\gamma} \phi(\gamma) < 2P(h)e^{-\log x/(R\log h)} + Q(h)\int_h^\infty \phi(t)\log\frac{t}{2\pi}\,dt$$

where

$$P(h) = \frac{a_1\log h + a_2\log\log h + a_3}{h^2}$$

and

$$Q(h) = \frac{1}{2\pi} + \frac{a_1\log h + a_2}{h\log h\log(h/(2\pi))}.$$

Proof By Lemma 4.2 and Backlund's bound $B(T)$ we can write

$$\sum_{h<\gamma}\phi(\gamma) < \int_h^\infty\left(\frac{1}{2\pi}\log\left(\frac{t}{2\pi}\right)+\left(a_1+\frac{a_2}{\log h}\right)\frac{1}{t}\right)\phi(t)\,dt + 2B(h)\phi(h).$$

But for $h \le t$ we have

$$\frac{h}{t} \le \frac{\log(t/(2\pi))}{\log(h/(2\pi))} \quad \text{and} \quad \frac{1}{t\log(t/(2\pi))} \le \frac{1}{h\log(h/(2\pi))}.$$

Therefore

$$\begin{aligned}\sum_{h<\gamma}\phi(\gamma) &< \left(\frac{1}{2\pi}+\frac{a_1\log h + a_2}{h\log h\log(h/(2\pi))}\right)\int_h^\infty \phi(t)\log\left(\frac{t}{2\pi}\right)dt + 2B(h)\phi(h)\\ &= 2P(h)e^{-\log x/(R\log h)} + Q(h)\int_h^\infty \phi(t)\log\left(\frac{t}{2\pi}\right)dt.\end{aligned}$$

This completes the proof. □

For $m \ge 1$ recall the definition $k_m(h) := \sum_{\rho,\gamma>h} 1/\gamma^{m+1}$. These functions received quite a lot of attention in Section 3.2 – see for example Table 3.1. We next derive an upper bound for $V(x)$ in terms of $k_1(0)$ and $k_1(H)$.

Lemma 5.3 *[145, lemma 6] For $x \ge 1$ and for $H \le h$ we have*

$$V(x) < \frac{k_1(0)}{\sqrt{x}} + \frac{2B(h)}{h^2} + Q(H)\frac{1+\log(h/(2\pi))}{h} + \frac{k_1(H)}{x^{1/(R\log h)}}.$$

Proof First note that

$$\int_h^\infty \frac{x^{-1/(R\log t)}\log(t/(2\pi))}{t^2}\,dt \le \int_h^\infty \frac{\log(t/(2\pi))}{t^2}\,dt = \frac{1+\log(h/(2\pi))}{h}.$$

Using this integral together with Lemmas 5.1 and 5.2 we get:

$$
\begin{aligned}
V(x) &= \sum_\rho \frac{x^{\beta-1}}{\rho^2} \le \frac{1}{\sqrt{x}}\sum_\rho \frac{1}{\gamma^2} + \sum_{H<\gamma}\phi(\gamma) \\
&< \frac{k_1(0)}{\sqrt{x}} + \sum_{H<\gamma\le h}\phi(\gamma) + \sum_{h<\gamma}\phi(\gamma) \le \frac{k_1(0)}{\sqrt{x}} + \sum_{H<\gamma\le h}\frac{x^{-1/(R\log h)}}{\gamma^2} + \sum_{h<\gamma}\frac{1}{\gamma^2} \\
&< \frac{k_1(0)}{\sqrt{x}} + \frac{2B(h)}{h^2} + Q(H)\frac{1+\log(h/(2\pi))}{h} + \frac{k_1(H)}{x^{1/(R\log h)}}.
\end{aligned}
$$

This completes the derivation. □

Next we link $\psi(x)$ with $V(x)$. This takes the form of upper bounds for two infinite integrals. Von Mangoldt's first theorem plays a crucial role here.

Lemma 5.4 *[145, lemmas 7 and 8] For all $x > 1$ we have the upper bounds*

$$
(1)\ I_1 := \left|\int_x^\infty \frac{(\psi(t)-t)}{t^2}\,dt\right| < V(x) + \frac{\log 2\pi}{x} + \frac{1-\log 2}{x^3}
$$

and

$$
\begin{aligned}
(2)\ I_2 &:= \left|\int_x^\infty \frac{(\psi(t)-t)(1+\log t)}{t^2\log^2 t}\,dt\right| \\
&< \frac{2+\log x}{\log^2 x}\left(V(x) + \frac{\log 2\pi}{x} + \frac{1-\log 2}{x^3}\right),
\end{aligned}
$$

where $\log 2\pi < 1.84$ *and* $1-\log 2 < 0.31$.

Proof **(1)** Let ρ be a complex zeta zero and note that

$$
\int_x^\infty t^{\rho-2}\,dt = -\frac{x^{\rho-1}}{\rho-1}.
$$

Using von Mangoldt's theorem, Theorem 2.1, and observing that $|\gamma| \le |\rho|$ and $|\gamma| \le |1-\rho|$ gives

$$
\begin{aligned}
I_1 = \left|\int_x^\infty \frac{\psi(t)-t}{t^2}\,dt\right| &\le \sum_\rho \frac{x^{\beta-1}}{|\rho(\rho-1)|} + \int_x^\infty \frac{\log 2\pi - \frac{1}{2}\log(1-1/t^2)}{t^2}\,dt \\
&\le V(x) + \frac{\log 2\pi}{x} + \sum_{n=1}^\infty \frac{x^{-2n-1}}{2n(2n+1)} \\
&\le V(x) + \frac{\log 2\pi}{x} + \sum_{n=1}^\infty \frac{x^{-3}}{2n(2n+1)} \\
&< V(x) + \frac{\log 2\pi}{x} + \frac{1-\log 2}{x^3},
\end{aligned}
$$

which gives the bound for the integral.

(2) For the second integral, integrating twice by parts we get

$$\int_x^\infty \frac{t^{\rho-2}(1+\log t)}{\log^2 t}\,dt = -\frac{x^{\rho-1}}{(\rho-1)\log x} - \frac{\rho}{(\rho-1)^2}\left(\frac{x^{\rho-1}}{\log^2 x} - 2\int_x^\infty \frac{t^{\rho-2}}{\log^3 t}\,dt\right).$$

Also

$$\begin{aligned}\left|\frac{x^{\rho-1}}{\log^2 x} - 2\int_x^\infty \frac{t^{\rho-2}}{\log^3 t}\,dt\right| &\le \frac{x^{\beta-1}}{\log^2 x} + 2\int_x^\infty \frac{t^{\beta-2}}{\log^3 t}\,dt \\ &< \frac{x^{\beta-1}}{\log^2 x} + 2x^{\beta-1}\int_x^\infty \frac{1}{t\log^3 t}\,dt = \frac{2x^{\beta-1}}{\log^2 x}.\end{aligned}$$

Therefore

$$\begin{aligned}\left|\frac{1}{\rho}\int_x^\infty \frac{t^{\rho-2}(1+\log t)}{\log^2 t}\,dt\right| &< \frac{1}{|\rho(\rho-1)|}\frac{x^{\beta-1}}{\log x} + \frac{1}{|\rho-1|^2}\frac{2x^{\beta-1}}{\log^2 x} \\ &< \frac{2+\log x}{\log^2 x}\frac{x^{\beta-1}}{\gamma^2}.\end{aligned}$$

Summing we get

$$\sum_\rho \left|\frac{1}{\rho}\int_x^\infty \frac{t^{\rho-2}(1+\log t)}{\log^2 t}\,dt\right| < \frac{2+\log x}{\log^2 x}V(x).$$

Now we apply von Mangoldt's theorem, Theorem 2.1, to get

$$I_2 = \left|\int_x^\infty \frac{(\psi(t)-t)(1+\log t)}{t^2\log^2 t}\,dt\right| < \frac{2+\log x}{\log^2 x}V(x) + I_3(x),$$

where

$$\begin{aligned}I_3(x) &= \int_x^\infty \frac{(2\log 2\pi - \log(1-1/t^2))(1+\log t)}{2t^2\log^2 t}\,dt \\ &< \frac{1+\log x}{2\log^2 x}\int_x^\infty \left(2\log 2\pi - \log\left(1-\frac{1}{t^2}\right)\right)\frac{dt}{t^2} \\ &< \frac{2+\log x}{\log^2 x}\left(\frac{\log 2\pi}{x} + \frac{1-\log 2}{x^3}\right).\end{aligned}$$

This completes the proof. □

Next we have an easy-to-derive upper bound for an infinite integral which appears in the proof of the previous lemma.

Lemma 5.5 *[145, lemma 9] For $0 \le \alpha < n$ and $x > 1$ we have*

$$\int_x^\infty \frac{t^\alpha(1+n\log t)}{t^{n+1}\log^2 t}\,dt \le \frac{n}{n-\alpha}\frac{1}{x^{n-\alpha}\log x}.$$

Proof Rewrite the form of the integral and then use integration by parts:

$$\begin{aligned}\int_x^\infty \frac{t^\alpha(1+n\log t)}{t^{n+1}\log^2 t}\,dt &= -\int_x^\infty t^\alpha \frac{d}{dt}\left(\frac{1}{t^n\log t}\right)dt\\ &= \frac{1}{x^{n-\alpha}\log x} + \alpha\int_x^\infty \frac{t^\alpha}{t^{n+1}\log t}\,dt\\ &\le \frac{1}{x^{n-\alpha}\log x} + \frac{\alpha}{\log x}\int_x^\infty t^{\alpha-n-1}\,dt\\ &= \frac{n}{n-\alpha}\,\frac{1}{x^{n-\alpha}\log x},\end{aligned}$$

which completes the proof. □

The next step uses three functions to estimate $\theta(x)$:

$$V(x) := \sum_\rho \frac{x^{\beta-1}}{\gamma^2},$$

$$L(x) := \frac{2+\log x}{\log x}\left(V(x) + \frac{1.84}{x} + \frac{0.31}{x^3}\right) + \frac{2.04}{x^{1/2}} + \frac{4.5}{x^{2/3}} + \frac{1.02}{x-1}$$

and

$$M(x,\alpha) := \log x \log\left(1 + \frac{1}{2\log^2 x}\right) - \frac{\alpha}{\log x}.$$

We have already employed and found bounds for the function $V(x)$. The function $L(x)$ arises below as part of a term in an upper bound, but is too difficult to work with. We will find ranges of variables and values of a parameter α for which $L(x) < M(x,\alpha)$, so we can replace $L(x)$ by $M(x,\alpha)$ in those ranges. The reader is invited to investigate whether $L(x)$ and $M(x,\alpha)$ are of comparable orders. If not, this step could be improved.

Lemma 5.6 *[145, lemmas 11 and 12, theorems 29 and 31] We have the following bounds:*

(1) $10^8 \le x \le e^{600} \implies L(x) < M(x, 0.31)$,
(2) $e^{375} \le x \implies L(x) < M(x, 0.47)$,
(3) $1451 \le x \le e^{375} \implies x\left(1 - \frac{0.31}{\log x}\right) < \theta(x) < x\left(1 + \frac{0.31}{\log x}\right)$,
(4) $569 \le x \implies x\left(1 - \frac{0.47}{\log x}\right) < \theta(x) < x\left(1 + \frac{0.47}{\log x}\right)$.

Proof First we will derive a useful sufficient criterion. Let $H \le u \le v$ and $0 < \alpha < \frac{1}{2}$ be real numbers with

$$\log u \ge \kappa(\alpha) := \sqrt{\frac{\frac{1}{2}+\alpha}{1-2\alpha}} \quad \text{and} \quad L(u) < M(v,\alpha),$$

and let $u \le x \le v$. Then, since $L(x)$ and $M(x,\alpha)$ are decreasing as x increases, we have

$$L(x) \le L(u) < M(v,\alpha) \le M(x,\alpha) \quad \Longrightarrow \quad L(x) < M(x,\alpha).$$

(1) In this case $u = 10^8,\ v = e^{600}$. Let $h \to \infty$ in Lemma 5.3 and set $\alpha = 0.31$ to get $\kappa(0.31) < 1.460$ to obtain

$$L(u) = L(10^8) < 3.0 \times 10^{-4} < M(e^{600}, 0.31) = M(v, 0.31)$$
$$\Longrightarrow \quad L(x) < M(x, 0.31) \quad \text{for } 10^8 \le x \le e^{600},$$

proving part (1).

(2) Now set $h = e^{25}$ in Lemma 5.3 to get

$$V(x) < \frac{k_1(0)}{\sqrt{x}} + \frac{k_1(H)}{e^{1/(25R)}} + \frac{1}{10^{10}}.$$

Note also that $\kappa(0.47) < 4.021$ so

$$L(e^{375}) < 2.9 \times 10^{-5} < M(e^{10^3}, 0.47),$$
$$L(e^{10^3}) < 7.0 \times 10^{-6} < M(e^{4\times 10^3}, 0.47)$$

and

$$L(e^{4\times 10^3}) < 9.0 \times 10^{-9} < M(e^{3\times 10^6}, 0.47).$$

Then let in turn

$$[u, v] = [\exp(375), \exp(10^3)],\ [\exp(10^3), \exp(4 \times 10^3)] \text{ and}$$
$$[\exp(4 \times 10^3), \exp(3 \times 10^6)],$$

to show $L(x) < M(x, 0.47)$ for $\exp(375) \le x \le \exp(3 \times 10^6)$.

Finally, for $x \ge \exp(3 \times 10^6)$ set

$$\log h = \frac{\log x}{R(\log\log x - \log 400)}$$

and again use the definitions of $L(x)$ and $M(x,\alpha)$, the bound $k_1(0) < 0.0463$ and Lemma 5.3 to show

$$L(x) < \frac{0.028}{\log x} < M(x, 0.47).$$

(3) and **(4)** Parts (3) and (4) are proved in Lemma 3.14(b) and (d). □

Next we have a simple application of Abel summation to obtain an estimate for an infinite sum of primes to negative powers.

Lemma 5.7 *Let $\alpha > 1$. Then as $x \to \infty$ we have*

$$\sum_{p \geq x} \frac{1}{p^\alpha} = \frac{1}{(\alpha-1)x^{\alpha-1}\log x} + O\left(\frac{1}{x^{\alpha-1}\log^2 x}\right),$$

where the sum is over primes.

Proof Use Abel's identity [6, theorem 4.2], $\pi(x) = x/(\log x) + O(x/(\log x)^2)$ and integration by parts to derive

$$\begin{aligned}
\sum_{p \geq x} \frac{1}{p^\alpha} &= -\frac{\pi(x)}{x^\alpha} + \alpha \int_x^\infty \frac{\pi(t)}{t^{\alpha+1}}\,dt \\
&= -\frac{1}{x^\alpha}\left(\frac{x}{\log x} + O\left(\frac{x}{\log^2 x}\right)\right) + \alpha \int_x^\infty \frac{dt}{t^\alpha \log t} + O\left(\int_x^\infty \frac{dt}{t^\alpha \log^2 t}\right) \\
&= -\frac{1}{x^{\alpha-1}\log x} + \frac{\alpha}{\alpha-1}\left[-\frac{1}{t^{\alpha-1}\log t}\right]_x^\infty + O\left(\int_x^\infty \frac{dt}{t^\alpha \log^2 t}\right) \\
&\quad + O\left(\frac{1}{x^{\alpha-1}\log^2 x}\right).
\end{aligned}$$

Therefore

$$\begin{aligned}
\sum_{p \geq x} \frac{1}{p^\alpha} &= \frac{1}{(\alpha-1)x^{\alpha-1}\log x} + O\left(\frac{1}{x^{\alpha-1}\log^2 x}\right) + O\left(\frac{1}{\log^2 x}\int_x^\infty \frac{dt}{t^\alpha}\right) \\
&= \frac{1}{(\alpha-1)x^{\alpha-1}\log x} + O\left(\frac{1}{x^{\alpha-1}\log^2 x}\right).
\end{aligned}$$

This completes the proof. □

Next we derive a useful upper bound for another type of prime sum. The constant 1.02 could be improved, but we do not need this.

Lemma 5.8 *[145, p. 87] We have the upper bound for a sum over primes valid for $1 < x$:*

$$\sum_{n \geq 2} \frac{1}{n} \sum_{p > x} \frac{1}{p^n} < \frac{1.02}{(x-1)\log x},$$

so if

$$S := -\sum_{p > x}\left(\log\left(1 - \frac{1}{p}\right) + \frac{1}{p}\right),$$

we have $0 \geq S > -1.02/((x-1)\log x)$.

Proof A re-examination of the proof of Lemma 5.7, using $f(x) := 1/x^n$, gives the upper bound for $n \geq 2$:

$$\sum_{p > x} \frac{1}{p^n} < -\frac{\theta(x)}{x^n \log x} + \int_x^\infty \frac{\theta(t)}{t}\,\frac{1 + n\log t}{t^n \log^2 t}\,dt.$$

By Lemma 3.10 we have $\theta(t) < 1.02t$ for all $t > 0$ so therefore using Lemma 5.5 we get

$$\sum_{p>x} \frac{1}{p^n} < 1.02 \int_x^\infty \frac{1+n\log t}{t^n \log^2 t}\,dt \le \frac{1.02n}{(n-1)x^{n-1}\log x} \le \frac{1.02n}{x^{n-1}\log x},$$

and so

$$\sum_{n\ge 2}\frac{1}{n}\sum_{p>x}\frac{1}{p^n} < \frac{1.02}{x\log x}\sum_{n\ge 2}\frac{1}{x^{n-2}} = \frac{1.02}{(x-1)\log x}$$

and the proof is complete. □

With these preliminaries in place, we now finesse the error in an explicit form of Mertens' estimate for the sum of prime reciprocals.

Lemma 5.9 *[145, lemma 13] There is an absolute constant B such that for all $x \ge 10^8$ we have*

$$\left|\sum_{p\le x}\frac{1}{p} - \log\log x - B\right| < \log\left(1+\frac{1}{2\log^2 x}\right) - \frac{1.02}{(x-1)\log x}.$$

Proof **(1)** By Abel's identity [6, theorem 4.2], with $f(t) = 1/(t\log t)$ and $a_n = \log n$ for prime n and 0 otherwise, we have

$$\begin{aligned}\sum_{p\le x}\frac{1}{p} &= \frac{\theta(x)}{x\log x} - \int_2^x \theta(t)\frac{d}{dt}\left(\frac{1}{t\log t}\right)dt \\ &= \frac{\theta(x)-x}{x\log x} + \int_2^x \frac{dt}{t\log t} + B + \int_x^\infty (\theta(t)-t)\frac{d}{dt}\left(\frac{1}{t\log t}\right)dt,\end{aligned}$$

where

$$B := \frac{1}{\log 2} - \log\log 2 - \int_2^\infty (\theta(t)-t)\frac{d}{dt}\left(\frac{1}{t\log t}\right)dt \in \mathbb{R}.$$

It follows that

$$\sum_{p\le x}\frac{1}{p} - \log\log x - B = \frac{\theta(x)-x}{x\log x} + \int_x^\infty \frac{(\theta(t)-t)(1+\log t)}{t^2\log^2 t}\,dt.$$

(2) Next by Lemma 5.5 with $n = 1$ we get

$$\alpha = \frac{1}{2} \quad\Longrightarrow\quad \int_x^\infty \frac{1+\log t}{t^{3/2}\log^2 t}\,dt \le \frac{2}{\sqrt{x}\log x}$$

and

$$\alpha = \frac{1}{3} \quad\Longrightarrow\quad \int_x^\infty \frac{1+\log t}{t^{5/3}\log^2 t}\,dt \le \frac{3}{2x^{2/3}\log x}.$$

Using these bounds and Lemmas 5.4 and 3.10 we get

$$\begin{aligned}\left|\sum_{p\le x}\frac{1}{p}-\log\log x-B\right| &\le \frac{|\theta(x)-x|}{x\log x}+\left|\int_x^\infty \frac{(\psi(t)-t)(1+\log t)}{t^2\log^2 t}\,dt\right| \\ &\quad+\int_x^\infty \frac{(\psi(t)-\theta(t))(1+\log t)}{t^2\log^2 t}\,dt \\ &< \frac{|\theta(x)-x|}{x\log x}+\left(\frac{2+\log x}{\log^2 x}\right)\left(V(x)+\frac{1.84}{x}+\frac{0.31}{x^3}\right) \\ &\quad+\int_x^\infty \frac{(1.02t^{1/2}+3t^{1/3})(1+\log t)}{t^2\log^2 t}\,dt.\end{aligned}$$

Therefore, using the definition of $L(x)$, namely,

$$L(x):=\frac{2+\log x}{\log x}\left(V(x)+\frac{1.84}{x}+\frac{0.31}{x^3}\right)+\frac{2.04}{x^{1/2}}+\frac{4.5}{x^{2/3}}+\frac{1.02}{x-1},$$

we get:

$$\left|\sum_{p\le x}\frac{1}{p}-\log\log x-B\right| < \frac{|\theta(x)-x|}{x\log x}+\frac{L(x)}{\log x}-\frac{1.02}{(x-1)\log x}.$$

(3) To complete the proof we split the range. Firstly let $10^8\le x\le e^{375}$. By Lemma 5.6 parts (1) and (3) we have in this range

$$|\theta(x)-x|<\frac{0.31x}{\log x}\quad\text{and}\quad L(x)<M(x,0.31).$$

Using the formula derived in Step (2), and the definition of $M(x,\alpha)$, namely,

$$M(x,\alpha):=\log x\log\left(1+\frac{1}{2\log^2 x}\right)-\frac{\alpha}{\log x},$$

we therefore have

$$\begin{aligned}\left|\sum_{p\le x}\frac{1}{p}-\log\log x-B\right| &< \frac{0.31}{\log^2 x}+\frac{M(x,0.31)}{\log x}-\frac{1.02}{(x-1)\log x} \\ &= \log\left(1+\frac{1}{2\log^2 x}\right)-\frac{1.02}{(x-1)\log x}.\end{aligned}$$

(4) Now let $e^{375}\le x$. Using Steps (2) and (4) of Lemma 5.6, we have in this range

$$|\theta(x)-x|<\frac{0.47x}{\log x}\quad\text{and}\quad L(x)<M(x,0.47),$$

and the result follows as in Step (3). □

We now have a useful estimate which can easily be derived from Lemma 5.9.

Lemma 5.10 *[145, theorem 5] For all $x \geq 286$ we have*

$$\log\log x + B - \frac{1}{2\log^2 x} < \sum_{p \leq x} \frac{1}{p} < \log\log x + B + \frac{1}{2\log^2 x},$$

where the lower bound also holds for $1 < x < 286$.

Proof Use the upper bound $\log(1 + x) < x$ for $0 < x < 1$ and then remove the absolute value to derive the result directly from Lemma 5.9 in the range $x \geq 10^8$. For $2 \leq x \leq 10^8$ a computation verifies the result – see Section 3.3. □

We now give two estimates for partial Euler products which apply in an explicit finite range. The reader might compare [145, theorem 23, p. 73].

Lemma 5.11 *For $2 \leq x \leq 10^8$ we have*

$$\prod_{p \leq x} \left(1 - \frac{1}{p}\right)^{-1} \leq e^{\gamma} \log x \left(1 + \frac{2}{\sqrt{x} \log x}\right)$$

and

$$\left| \sum_{p \leq x} \frac{1}{p} - \log\log x - B \right| < \frac{1}{2\log^2 x}.$$

Proof These follow from a computation at integers – see Section 3.3. □

The following explicit form of Mertens' theorem [115] plays an essential role in the development. See for example [75, chapter 22, theorems 427 and 428] or [6]. Recall that the constant B is normally defined by [95, p. 56]

$$B := \gamma + \sum_{p} \left(\log\left(1 - \frac{1}{p}\right) + \frac{1}{p} \right),$$

where as usual γ is Euler's constant.

Lemma 5.12 *[145, theorem 8] For all $x \geq 286$ we have*

$$e^{\gamma} \log x \left(1 - \frac{1}{2\log^2 x}\right) < \prod_{p \leq x} \frac{p}{p-1} < e^{\gamma} \log x \left(1 + \frac{1}{2\log^2 x}\right),$$

where the lower bound also holds for $1 < x < 286$.

Proof First

$$\sum_{p \leq x} \log\left(\frac{p}{p-1}\right) = \sum_{p \leq x} -\log\left(1 - \frac{1}{p}\right)$$

$$= \sum_{p \leq x} \frac{1}{p} + \sum_{n \geq 2} \frac{1}{n} \sum_{p \leq x} \frac{1}{p^n}.$$

Now we can write, from the definition of B,

$$B = \gamma + \sum_p \left(\log\left(1 - \frac{1}{p}\right) + \frac{1}{p}\right) = \gamma - \sum_{n\geq 2} \frac{1}{n} \sum_{p>x} \frac{1}{p^n}.$$

From Lemma 5.9 we have

$$\begin{aligned} B + \log\log x - \log\left(1 + \frac{1}{2\log^2 x}\right) \\ < \sum_{p\leq x} \frac{1}{p} < B + \log\log x + \log\left(1 + \frac{1}{2\log^2 x}\right) - \frac{1.02}{(x-1)\log x}. \end{aligned}$$

Using these bounds and replacing B with the expression we have derived gives

$$\begin{aligned} \gamma + \log\log x - \log\left(1 + \frac{1}{2\log^2 x}\right) < \sum_{p\leq x} -\log\left(1 - \frac{1}{p}\right) \\ < \gamma + \log\log x + \log\left(1 + \frac{1}{2\log^2 x}\right). \end{aligned}$$

Then because $\log(1-x) < -\log(1+x)$ for $0 < x < 1$ we get

$$\begin{aligned} \gamma + \log\log x + \log\left(1 - \frac{1}{2\log^2 x}\right) < \sum_{p\leq x} -\log\left(1 - \frac{1}{p}\right) \\ < \gamma + \log\log x + \log\left(1 + \frac{1}{2\log^2 x}\right). \end{aligned}$$

Finally, take exponentials to complete the proof. □

5.3 Preliminary Results With RH True

Since we are deriving equivalences of RH, one main part of any proof must assume RH is true, and the other main part considers properties which apply if it is assumed to be false. In order that these two vastly different situations not be confused, every lemma or theorem which uses one or other of these will include a reference to the particular assumption in the statement. If no such reference appears, then the reader can safely assume no such assumption is needed.

We use the standard definitions for $\theta(x)$ and $\psi(x)$. For $x > 0$,

$$\theta(x) := \sum_{p\leq x} \log p, \quad \psi(x) := \sum_{n\geq 1,\ p^n\leq x} \log p, \tag{5.4}$$

where p is a prime number.

Let $S(x) := \theta(x) - x$ and $T(x) := \sqrt{x}\log^2 x/(8\pi)$. Recall the result of Theorem 4.6 – an estimate for $S(x)$, assuming RH, which is used many times in what follows: for $x \geq 599$ we have $|S(x)| \leq T(x)$.

The function $f(x)$, which we will now define, is fundamental in this work. It comes from slightly varying Mertens' theorem, an explicit form of which has just been proved in Lemma 5.12:

$$1 - \frac{1}{2\log^2 x} < \left(e^\gamma \log x \prod_{p\leq x}\left(1 - \frac{1}{p}\right)\right)^{-1} < 1 + \frac{1}{2\log^2 x}, \qquad x > 1.$$

Now for $x \geq 2$ let

$$f(x) := e^\gamma \log \theta(x) \prod_{p\leq x}\left(1 - \frac{1}{p}\right). \tag{5.5}$$

Estimating $f(x)$ is a primary goal. We begin with an upper bound on a finite domain.

Lemma 5.13 *Assume RH is true. For $2 \leq x \leq 10^8$ we have $f(x) < 1$.*

Proof By Lemma 3.10, for $1 < x \leq 10^8$ we have $\theta(x) < x$ and by a computation in this range (see Section 3.3)

$$e^\gamma \log x < \prod_{p\leq x} \frac{p}{p-1}.$$

Therefore for $2 \leq x < 10^8$ we get, using Lemma 5.11,

$$f(x) = e^\gamma \log \theta(x) \prod_{p\leq x} \frac{p-1}{p} < e^\gamma \log x \prod_{p\leq x} \frac{p-1}{p} < 1.$$

□

Treating the infinite range $x > 10^8$ is much more difficult and will take us on quite a journey.

To begin, for $x > 1$ define $R(x) := \psi(x) - x$ and define also an associated integral expression which is also fundamental to this work:

$$J(x) := \int_x^\infty \frac{R(t)}{t^2}\left(\frac{1}{\log t} + \frac{1}{\log^2 t}\right) dt. \tag{5.6}$$

We need a simple lemma:

Lemma 5.14 *Let $a, b > 0$ be real and let $g(t) := \log^a t/t^b$. Then $g(t)$ is positive and decreasing for $t > e^{a/b}$ and* $\max\{g(t) : t \geq 1\} = (a/(eb))^a$.

Proof Since

$$\frac{g'(t)}{g(t)} = \frac{a - b\log t}{t\log t},$$

the proof follows with some simple calculus. □

To show how Lemma 5.15, and some of the results which follow, fit into the overall scheme of the proof, we have another network flow diagram (Figure 5.3).

We have defined $S(x) := \theta(x) - x$. Analogous to $J(x)$ let

$$K(x) := \int_x^\infty \frac{S(t)}{t^2}\left(\frac{1}{\log t} + \frac{1}{\log^2 t}\right) dt. \tag{5.7}$$

Note that the prime number theorem gives $S(t) = O(t/\log t)$ and ensures the convergence of the integral. Then we have an estimate for $f(x)$ for $x > 1$:

Lemma 5.15 *[126, proposition 1] For $x > 1$*

$$K(x) - \frac{S(x)^2}{x^2 \log x} \leq \log f(x) \leq K(x) + \frac{1}{2(x-1)}. \tag{5.8}$$

Proof First use Abel summation [6, theorem 4.2] and $\theta(x) = x + S(x)$ to derive an expression, in terms of $S(x)$, for the partial sum over primes p of prime reciprocals:

$$\begin{aligned}\sum_{p\leq x} \frac{1}{p} &= \frac{\theta(x)}{x\log x} + \int_2^x \frac{\theta(t)(1+\log t)}{t^2 \log^2 t}\, dt \\ &= \frac{S(x)}{x\log x} + \log\log x - \log\log 2 + \frac{1}{\log 2} + \int_2^x \frac{S(t)(1+\log t)}{t^2\log^2 t}\, dt,\end{aligned}$$

so

$$\begin{aligned}\sum_{p\leq x} \frac{1}{p} &= \frac{S(x)}{x\log x} + \log\log x + B - \int_x^\infty \frac{S(t)(1+\log t)}{t^2\log^2 t}\, dt \\ &= \frac{S(x)}{x\log x} + \log\log x + B - K(x),\end{aligned}$$

with

$$B := -\log\log 2 + \frac{1}{\log 2} + \int_2^\infty \frac{S(t)(1+\log t)}{t^2 \log^2 t}\, dt.$$

Thus

$$K(x) = \sum_{p\leq x}\left(-\frac{1}{p}\right) + \frac{S(x)}{x\log x} + \log\log x + B. \tag{5.9}$$

Note also the given expression for B can be compared with the classical formula given by Lemma 5.9:

$$\sum_{p\leq x}\frac{1}{p} = \log\log x + B + O\left(\frac{1}{\log x}\right),$$

where in this case

$$B = \gamma + \sum_{p} \left(\log\left(1 - \frac{1}{p}\right) + \frac{1}{p} \right).$$

Now we are able to derive an expression for $\log f(x)$ where the reader will recall the definition $f(x) := e^{\gamma} \log\theta(x) \prod_{p \le x}(1 - 1/p)$ (see Figure 5.4). Then

$$\begin{aligned} \log f(x) &= \log\log\theta(x) + \gamma + \sum_{p \le x} \log\left(1 - \frac{1}{p}\right) \\ &= \left(\log\log\theta(x) + \sum_{p \le x} \left(-\frac{1}{p}\right) + B \right) + \left(\sum_{p \le x} \left(\log\left(1 - \frac{1}{p}\right) + \frac{1}{p} \right) + \gamma - B \right). \end{aligned}$$

Thus

$$\log f(x) = U(x) + u(x),$$

where, using the classical expression for B, we have

$$u(x) = \sum_{p > x} \left(-\log\left(1 - \frac{1}{p}\right) - \frac{1}{p} \right).$$

Therefore

$$U(x) := \log\log\theta(x) + \sum_{p \le x} \left(-\frac{1}{p}\right) + B$$

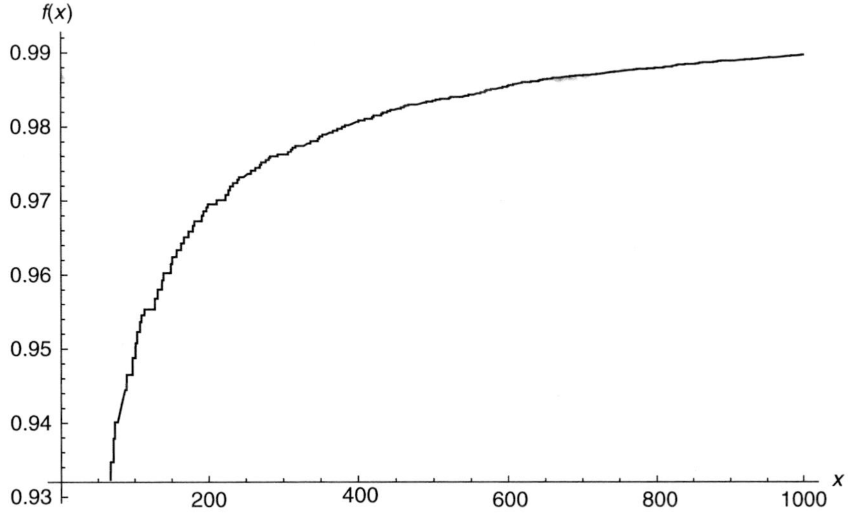

Figure 5.4 The function $f(x)$.

and

$$0 < u(x) < \sum_{p>x} \frac{1}{2p^2}\left(1 + \frac{1}{p} + \frac{1}{p^2} + \cdots\right) \leq \sum_{n>x} \frac{1}{2n(n-1)} \leq \frac{1}{2(x-1)}.$$

Note that by Lemma 3.10 (*d*), for $x \geq 121$ we have $\theta(x) \geq 4x/5$. Note also that for $t > 1$ the function

$$\frac{d^2}{dt^2}(\log\log t) = -\frac{1+\log t}{t^2 \log^2 t}$$

is increasing. A direct computation gives, for $x \geq 121$ and $t := 4x/5$,

$$-\frac{1+\log t}{t^2 \log^2 t} = -\frac{25}{16} \frac{\log(4x/5)+1}{x^2 \log^2(4x/5)} \geq -\frac{2}{x^2 \log x}.$$

Now, using this bound, apply Taylor's theorem to the function $\log\log x$ for $\theta(x) = x + S(x)$, expanded about x, to get for $x \geq 121$

$$\log\log x + \frac{S(x)}{x \log x} - \frac{S(x)^2}{x^2 \log x} \leq \log\log \theta(x) \leq \log\log x + \frac{S(x)}{x \log x}.$$

Then, to derive the upper bound of the lemma, use (5.9) to get

$$\begin{aligned}
\log f(x) &= U(x) + u(x) < U(x) + \frac{1}{2(x-1)} \\
&\leq \log\log \theta(x) + \sum_{p \leq x}\left(-\frac{1}{p}\right) + B + \frac{1}{2(x-1)} \\
&\leq \log\log x + \frac{S(x)}{x \log x} + \sum_{p \leq x}\left(-\frac{1}{p}\right) + B + + \frac{1}{2(x-1)} \\
&= K(x) + \frac{1}{2(x-1)}.
\end{aligned}$$

The derivation of the lower bound is similar using the lower bound from the application of Taylor's theorem. □

Next define for complex z with $\Re z < 1$ and real $x > 1$ the integral

$$F_z(x) := \int_x^\infty \frac{t^z}{t^2}\left(\frac{1}{\log t} + \frac{1}{\log^2 t}\right) dt. \tag{5.10}$$

This very useful function has some easy-to-derive properties set out in the next lemma. Again, integration by parts is the workhorse! There is an illustrative plot of $F_{1/2}(x)$ later in Figure 5.6.

Lemma 5.16 *[127, lemma 2.2] For $x > 1$ and $\Re z < 1$,*

$$F_z(x) = \frac{x^{z-1}}{(1-z)\log x} + r_z(x) \quad \textit{where} \quad r_z(x) := -\int_x^\infty \frac{z t^{z-2}}{(1-z)\log^2 t}\, dt. \tag{5.11}$$

If $\Re z = 1/2$ *then*

$$|r_z(x)| \le \frac{1}{|1-z|\sqrt{x}\log^2 x}\left(1 + \frac{4}{\log x}\right). \tag{5.12}$$

If $z = 1/2$ *then*

$$\frac{2}{\sqrt{x}\log x} - \frac{2}{\sqrt{x}\log^2 x} \le F_{1/2}(x) \le \frac{2}{\sqrt{x}\log x} - \frac{2}{\sqrt{x}\log^2 x} + \frac{8}{\sqrt{x}\log^3 x}. \tag{5.13}$$

Finally, if $z = 1/3$ *we have*

$$0 \le F_{1/3}(x) \le \frac{3}{2x^{2/3}\log x}. \tag{5.14}$$

Proof Split the integral and integrate the first term by parts to derive

$$\begin{aligned} F_z(x) &= \int_x^\infty t^{z-2}\left(\frac{1}{\log t} + \frac{1}{\log^2 t}\right) dt \\ &= \int_x^\infty \frac{t^{z-2}}{\log t}\, dt + \int_x^\infty \frac{t^{z-2}}{\log^2 t}\, dt \\ &= \frac{x^{z-1}}{(1-z)\log x} + \int_x^\infty \frac{t^{z-2}}{(z-1)\log^2 t}\, dt + \int_x^\infty \frac{t^{z-2}}{\log^2 t}\, dt \\ &= \frac{x^{z-1}}{(1-z)\log x} + \int_x^\infty \frac{-zt^{z-2}}{(1-z)\log^2 t}\, dt \\ &= \frac{x^{z-1}}{(1-z)\log x} + r_z(x). \end{aligned}$$

Integrating by parts again we have

$$r_z(x) = -\frac{z}{1-z}\left(\frac{x^{z-1}}{(1-z)\log^2 x} + \int_x^\infty \frac{2t^{z-2}}{(z-1)\log^3 t}\, dt\right) dt. \tag{5.15}$$

If $\Re z = 1/2$ then $1 - z = \bar{z}$ so (5.12) follows from

$$\begin{aligned} |r_z(x)| &\le \frac{1}{|1-z|\sqrt{x}\log^2 x} + \frac{2}{|1-z|\log^3 x}\int_x^\infty \frac{1}{t^{3/2}}\, dt \\ &= \frac{1}{|1-z|\sqrt{x}\log^2 x}\left(1 + \frac{4}{\log x}\right). \end{aligned}$$

Let $z = 1/2$ in (5.11) and (5.15) and use the upper bound for $|r_z(x)|$ to get (5.13). Finally, to get (5.14) use (5.11), noting that $r_{1/3}(x) < 0$. □

The following lemma involves rather intricate sums and estimates. It is a crucial ingredient for Lemma 5.18. Recall the definition $T(x) := \sqrt{x}\log^2 x/(8\pi)$.

Lemma 5.17 *[127, lemma 2.3] Let $\kappa = \kappa(x) := \lfloor \log x / \log 2 \rfloor$. For $x \geq 16$ define*

$$H(x) := 1 + \sum_{n=4}^{\kappa} x^{1/n - 1/3}$$

(see Figure 5.5). For $x \geq 4$ and $n \geq 2$ let

$$l_n(x) := \frac{T(x^{1/n})}{x^{1/3}} = \frac{\log^2 x}{8\pi n^2 x^{1/3 - 1/(2n)}}.$$

Finally, let

$$L(x) := \sum_{n=2}^{\kappa} l_n(x).$$

Then

(a) for $j \geq 9$ and $x \geq 2^j$ we have $H(x) \leq H(2^j)$, and
(b) for $j \geq 35$ and $x \geq 2^j$ we have $L(x) \leq L(2^j)$.

Proof (*a*) The function $H(x)$ is continuous and decreasing on the interval $[2^j, 2^{j+1})$ for all $j \geq 1$. Therefore to prove (*a*) we need only verify $H(2^j) \geq H(2^{j+1})$ for all $j \geq 9$. If $j \geq 20$ then

$$\begin{aligned} H(2^j) - H(2^{j+1}) &= \sum_{k=4}^{j} 2^{j(1/k - 1/3)}(1 - 2^{1/k - 1/3}) - 2^{(j+1)(1/(j+1) - 1/3)} \\ &\geq 2^{j(1/4 - 1/3)}(1 - 2^{1/4 - 1/3}) - 2^{(j+1)(1/(j+1) - 1/3)} \end{aligned}$$

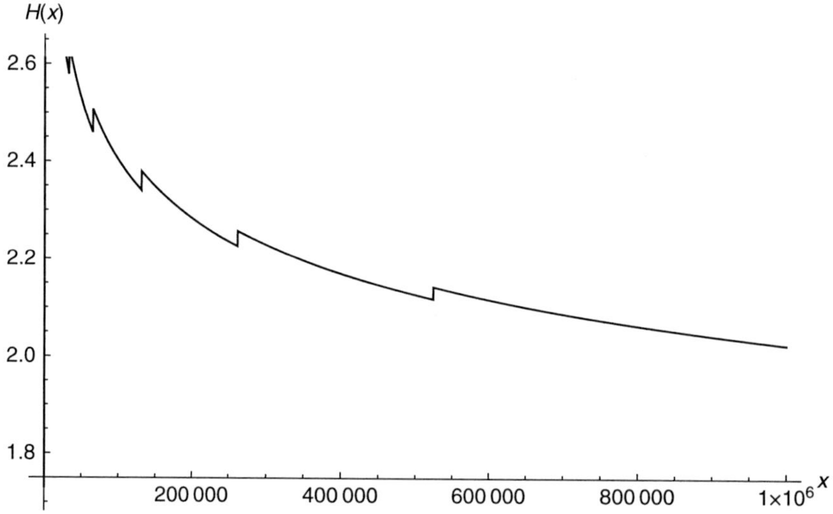

Figure 5.5 The function $H(x)$.

$$= 2^{-j/3}((1-2^{-1/12})2^{j/4} - 2^{2/3})$$
$$\geq 2^{-j/3}((1-2^{-1/12})2^{20/4} - 2^{2/3}) > 0.$$

For $9 \leq j \leq 19$, the inequality $H(2^j) \geq H(2^{j+1})$ can be verified numerically.

(*b*) If $j \geq 35$ then $2^j \geq e^{24}$. By Lemma 5.14, for each k the function $l_k(x)$ is decreasing for $x \geq 2^j$ so $L(x)$ is decreasing on $[2^j, 2^{j+1})$. Then

$$\begin{aligned} L(2^j) - L(2^{j+1}) &= \sum_{k=2}^{j}(l_k(2^j) - l_k(2^{j+1})) - l_{j+1}(2^{j+1}) \\ &\geq l_2(2^j) - l_2(2^{j+1}) - l_{j+1}(2^{j+1}) \\ &= \frac{\log^2 2}{32\pi} 2^{-j/3}[2^{j/4}(j^2 - 2^{-1/12}(j+1)^2) - 4 \times 2^{1/6}] \\ &> 0 \end{aligned}$$

for $j \geq 35$ since the term inside the outer square brackets is positive at $j = 35$ and increasing for $j \geq 35$. □

The next lemma gives upper and lower bounds for $\psi(x) - \theta(x)$ assuming RH. The reader might compare these with Lemma 3.13 where RH is not assumed.

Lemma 5.18 *[127, lemma 2.4] Assume RH is true. Then* (*a*) *for* $x \geq 121$ *we have*

$$\sqrt{x} \leq \psi(x) - \theta(x), \tag{5.16}$$

(*b*) *for* $x \geq 1$ *we have*

$$(\psi(x) - \theta(x) - \sqrt{x})x^{-1/3} \leq \frac{4}{3}. \tag{5.17}$$

Proof (*a*) We prove this part in two ranges of x values.

(1) If $11^2 \leq x < 599^3$ the result may be verified numerically as follows. Let q_i be the increasing sequence of squarefull powers of primes p which are less than 599^3, so $q_1 = 4$, $q_2 = 8$, $q_3 = 9$, $q_4 = 16$, $q_5 = 25$, ... On each interval $[q_i, q_{i+1})$ the function $\psi(x) - \theta(x)$ is non-negative and constant and $(\psi(x) - \theta(x))/\sqrt{x}$ is decreasing.

Then for each i with $5^2 \leq q_i < q_{i+1} \leq 599^3$ (so $5 \leq i \leq 1921$), compute recursively $\delta_i := (\psi(q_i) - \theta(q_i)/\sqrt{q_{i+1}}$ using

$$\Delta_1 = 2\log 2,$$
$$\delta_1 = \frac{\Delta_1}{\sqrt{q_2}},$$

and, for $i \geq 1$, $q_i = p_i^n$ for some $n \geq 2$ defines p_i,

$$\Delta_{i+1} = \Delta_i + \log(p_{i+1}),$$
$$\delta_i = \frac{\Delta_i}{\sqrt{q_{i+1}}},$$

to get

$$\min_{11\le i\le 1921} \delta_i = \delta_{1886} > 1.03795 > 1,$$

while $q_5 = 11^2$ and $\delta_4 < 1$. (See the RHpack function ComputeDeltas in Appendix B.) Therefore (a) is true when $11^2 \le x < 599^3$.

(2) Now assume $x \ge 599^3$. By Theorem 4.6, with $S(x) = \theta(x) - x$, since we are assuming RH, we have

$$|S(x)| \le T(x) := (\sqrt{x}\log^2 x)/(8\pi), \tag{5.18}$$

so

$$\psi(x) - \theta(x) \ge \theta(x^{1/2}) + \theta(x^{1/3}) \ge x^{1/2} + x^{1/3} - T(x^{1/2}) - T(x^{1/3}).$$

By Lemma 5.14 for $x \ge 599^2$ we get

$$\frac{T(x^{1/2})}{x^{1/3}} + \frac{T(x^{1/3})}{x^{1/3}} = \frac{1}{8\pi}\left(\frac{\log^2 x}{4x^{1/12}} + \frac{\log^2 x}{9x^{1/6}}\right) \le \frac{20}{\pi e^2} < 1.$$

Therefore

$$\psi(x) - \theta(x) \ge \sqrt{x} + \left(1 - \frac{20}{\pi e^2}\right)x^{1/3} \ge \sqrt{x}.$$

(b) To derive the stated inequality, consider three ranges for values of x.

(1) First let $1 \le x < 2^{32}$. The largest q_i smaller than 2^{32} is $q_{6947} = 65\,521^2$. On each of the intervals $[q_i, q_{i+1})$ in this range we find explicitly that the function

$$G(x) := \frac{\psi(x) - \theta(x) - \sqrt{x}}{x^{1/3}}$$

is decreasing and that $G(x) \le G(q_{103}) < 1.3330 < 4/3$.

(2) Now let $2^{32} \le x < 64 \times 10^{22}$. Here we use [48, table 6.6] which gives $\theta(x) < x$ for $x \le 8 \times 10^{11}$ (compare Lemma 3.10). Then, if as before we set $\kappa := \lfloor \log x/\log 2 \rfloor$, we get

$$\psi(x) - \theta(x) = \sum_{n=2}^{\kappa} \theta(x^{1/n}) \le \sum_{n=2}^{\kappa} x^{1/n}$$

so by Lemma 5.17

$$G(x) \le H(x) \le H(2^{32}) < 1.32 < \frac{4}{3}.$$

(3) Finally let $x \ge 64 \times 10^{22} \ge 2^{79}$. Use $\theta(x) \le x + T(x)$ to write

$$\psi(x) - \theta(x) = \sum_{n=2}^{\kappa} \theta(x^{1/n}) \le \sum_{n=2}^{\kappa} (x^{1/n} + T(x^{1/n})),$$

so using the notation and result of Lemma 5.17 we get

$$\begin{aligned}
G(x) &= \frac{\psi(x) - \theta(x) - \sqrt{x}}{x^{1/3}} \\
&\leq \sum_{n=2}^{K} \left(x^{1/n-1/3} + \frac{T(x^{1/n})}{x^{1/3}} \right) - x^{1/2-1/3} \\
&= 1 + \sum_{n=4}^{K} x^{1/n-1/3} + \sum_{n=2}^{K} \frac{T(x^{1/n})}{x^{1/3}} \\
&= H(x) + L(x) \\
&\leq H(2^{79}) + L(2^{79}) < 1.33 < \frac{4}{3}.
\end{aligned}$$

This completes the proof of the lemma. □

The reader might wish to compare the lower bound of Lemma 5.18 with the upper bound from Theorem 4.6, both of which assume RH:

$$\sqrt{x} \leq \psi(x) - \theta(x) \leq \frac{\sqrt{x}\log^2 x}{4\pi}, \qquad x \geq 599,$$

or even, without assuming RH,

$$\psi(x) - \theta(x) < 1.021\sqrt{x}, \qquad x > 0,$$

which is Lemma 3.13 (c).

Next recall the definitions $R(x) := \psi(x) - x$ and $S(x) = \theta(x) - x$ and the related integrals:

$$\begin{aligned}
J(x) &:= \int_x^\infty \frac{R(t)}{t^2}\left(\frac{1}{\log t} + \frac{1}{\log^2 t}\right) dt, \\
K(x) &:= \int_x^\infty \frac{S(t)}{t^2}\left(\frac{1}{\log t} + \frac{1}{\log^2 t}\right) dt, \qquad (5.19) \\
F_z(x) &:= \int_x^\infty \frac{t^z}{t^2}\left(\frac{1}{\log t} + \frac{1}{\log^2 t}\right) dt.
\end{aligned}$$

Substituting the bounds for $\psi(x) - \theta(x)$ from Lemma 5.18 in the definitions of $J(x)$ and $K(x)$, the following bounds for $J(x) - K(x)$ can be derived directly:

Lemma 5.19 *[127, corollary 2.1] Assume RH is true. Then for $x \geq 121$ we get*

$$F_{1/2}(x) \leq J(x) - K(x) \leq F_{1/2}(x) + \tfrac{4}{3}F_{1/3}(x).$$

This lemma enables us to estimate $K(x)$ by first estimating $J(x)$, a task which begins with Lemma 5.20 below, and which employs zeta zeros.

To plot the graph of $F_{1/2}(x)$ in Figure 5.6, we used the representation

$$F_{1/2}(x) = \frac{2}{\sqrt{x}\log x} - \mathrm{Ei}\left(-\frac{\log x}{2}\right).$$

See the RHpack function FHalf in Appendix B.

We now express $J(x)$ in terms of the values of $F_\rho(x)$ with an error which is small for large x.

Lemma 5.20 *[126, proposition 2] Assume the Riemann hypothesis is true. There is a function $J_1(x)$ such that*

$$J(x) = -\sum_\rho \frac{1}{\rho} F_\rho(x) - J_1(x),$$

where $J_1(x)$ satisfies $0 \le J_1(x) \le \log(2\pi)/(x\log x)$, and where the sum over ρ is over all of the imaginary zeros of $\zeta(s)$ with increasing absolute value of the imaginary part.

Proof Firstly, there is a real positive number A such that

$$A := \sum_\rho \frac{1}{|\rho(\rho+1)|}.$$

(The reader can explore properties of this and similar sums in [51].)

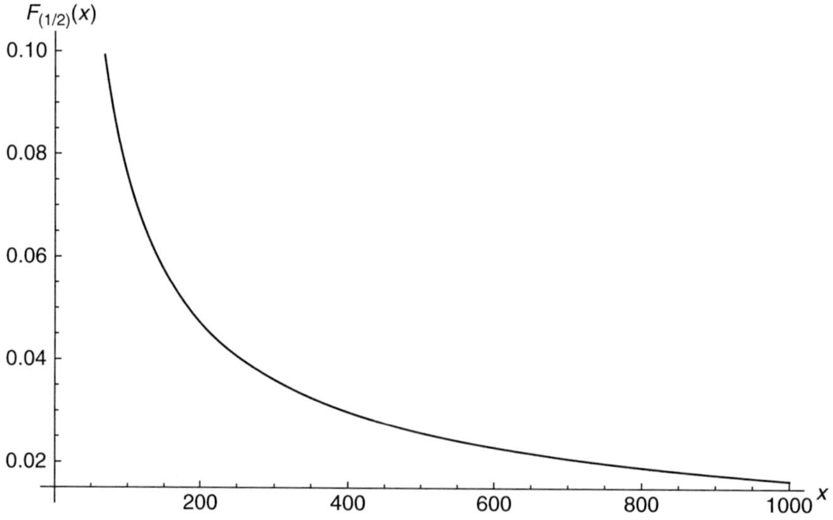

Figure 5.6 The function $F_{1/2}(x)$.

Next define $g(t) := (1/\log t + 1/\log^2 t)/t^2$, $t > 1$, so

$$g'(t) = -\frac{1}{t^3}\left(\frac{2}{\log t} + \frac{3}{\log^2 t} + \frac{2}{\log^3 t}\right).$$

The series $\sum_\rho g'(t)t^{\rho+1}/(\rho(\rho+1))$ is absolutely convergent for all $t > 1$, and by the Riemann hypothesis the partial sums are all bounded above by $A|g'(t)|t^{3/2}$, for some constant $A > 0$. For $x > 1$ this bound is integrable on $[x, \infty)$. By the Weierstrass M-test we can integrate term by term and write

$$\int_x^\infty \left(\sum_\rho g'(t)\frac{t^{\rho+1}}{\rho(\rho+1)}\right) dt = \sum_\rho \int_x^\infty g'(t)\frac{t^{\rho+1}}{\rho(\rho+1)}\, dt. \tag{5.20}$$

Next, in the explicit formula of von Mangoldt's theorem, Theorem 2.2, namely,

$$\psi_1(x) := \int_0^x \psi(t)\, dt = \frac{x^2}{2} - \sum_\rho \frac{x^{\rho+1}}{\rho(\rho+1)} - \sum_{n=1}^\infty \frac{x^{-2n+1}}{2n(2n-1)} - \frac{\zeta'(0)}{\zeta(0)}x + \frac{\zeta'(-1)}{\zeta(-1)},$$

which is valid for all $x > 1$, set

$$g_1(x) := x\frac{\zeta'(0)}{\zeta(0)} - \frac{\zeta'(-1)}{\zeta(-1)} + \sum_{n=1}^\infty \frac{x^{1-2n}}{2n(2n-1)}.$$

For $x > 1$ using the value for $\zeta'(0)/\zeta(0)$ derived in Lemma 2.5, i.e. $\zeta'(0)/\zeta(0) = \log(2\pi)$, we get

$$g_1'(x) = \log(2\pi) + \frac{1}{2}\log\left(1 - \frac{1}{x^2}\right),$$

and so for $x \geq 2$ we have

$$0 \leq g_1'(x) < \log(2\pi). \tag{5.21}$$

Now consider the left-hand side of (5.20). First use integration by parts, noting that $\psi_1(t) = O(t^2)$ and setting

$$J_1(x) := \int_x^\infty g(t)g_1'(t)\, dt, \tag{5.22}$$

to obtain

$$\begin{aligned}\int_x^\infty g'(t)\sum_\rho \frac{t^{\rho+1}}{\rho(\rho+1)}\, dt &= \int_x^\infty g'(t)(\tfrac{1}{2}t^2 - \psi_1(t) - g_1(t))\, dt\\ &= g(x)(\psi_1(x) - \tfrac{1}{2}x^2 + g_1(x)) + J(x) + J_1(x),\end{aligned}$$

where $J(x)$ has been defined as

$$\int_x^\infty (\psi(t) - t)g(t)\, dt = \int_x^\infty \frac{\psi(t) - t}{t^2}\left(\frac{1}{\log t} + \frac{1}{\log^2 t}\right) dt.$$

The right-hand side of (5.20), using integration by parts again, is

$$\sum_\rho \left(-g(x)\frac{x^{\rho+1}}{\rho(\rho+1)} - \int_x^\infty g(t)\frac{t^\rho}{\rho}\,dt \right) = g(x)(\psi_1(x) - \tfrac{1}{2}x^2 + g_1(x)) - \sum_\rho \frac{F_\rho(x)}{\rho}.$$

These derivations show that (5.20) can be written

$$J(x) = -\sum_\rho \frac{F_\rho(x)}{\rho} - J_1(x).$$

By (5.21) and (5.22) for $x \geq 2$ we have

$$0 \leq J_1(x) \leq \log(2\pi) \int_x^\infty d\left(\frac{-1}{t\log t}\right) = \frac{\log(2\pi)}{x\log x}.$$

This completes the proof of the lemma. □

Lemma 5.21 *[126, proposition 2] For $x \geq 121$ we have*

$$\log f(x) \leq J(x) - 0.98F_{1/2}(x) + \frac{1}{2(x-1)} \tag{5.23}$$

and for $x \geq 3000$ we have

$$\frac{0.9}{\sqrt{x}\log x} \leq 0.98F_{1/2}(x) - \frac{1}{2(x-1)}. \tag{5.24}$$

Proof By Lemma 3.13 (d), for $x \geq 121$ we have $\theta(x) \leq \psi(x) - 0.98\sqrt{x}$. This shows $S(x) \leq R(x) - 0.98\sqrt{x}$ so by Lemma 5.15, in the range $x \geq 121$, we have

$$\log f(x) \leq J(x) - 0.98F_{1/2}(x) + \frac{1}{2(x-1)}.$$

Next by Lemma 5.16 we have

$$F_{1/2}(x) = \frac{2}{\sqrt{x}\log x} + r_{1/2}(x),$$

and for $x \geq 3000 > e^8$ we have $|r_{1/2}(x)| \leq 1/(\sqrt{x}\log x)$. Therefore $F_{1/2}(x) \geq 1/(\sqrt{x}\log x)$ and thus, since $1/(2x-2) < 0.08/(\sqrt{x}\log x)$,

$$\frac{0.9}{\sqrt{x}\log x} \leq 0.98F_{1/2}(x) - \frac{1}{2(x-1)},$$

which completes the proof. □

Now we are able to bound the sum in the statement of Lemma 5.20 which depends on non-trivial zeta zeros:

Lemma 5.22 *[126, proposition 2] Assume RH so $\Re\rho = \frac{1}{2}$. For $x \geq e^4$ we have*

$$\left|\sum_{\rho} \frac{F_\rho(x)}{\rho}\right| \leq \frac{0.08}{\sqrt{x}\log x}.$$

Proof By Lemma 5.16 we have

$$\begin{aligned}-\sum_{\rho} \frac{F_\rho(x)}{\rho} &= \sum_{\rho} \frac{x^{\rho-1}}{\rho(\rho-1)\log x} - \sum_{\rho} \frac{r_\rho(x)}{\rho} \\ &= \frac{1}{\sqrt{x}\log x} \sum_{\rho} \frac{x^{i\Im\rho}}{\rho(\rho-1)} - \sum_{\rho} \frac{r_\rho(x)}{\rho}.\end{aligned}$$

The first series is convergent. Note that $\rho(1-\rho) > 0$ and recall that by Lemma 2.10,

$$\beta := \sum_{\rho} \frac{1}{\rho(1-\rho)} = 2 + \gamma - \log(4\pi) \leq 0.047.$$

Then using the estimate for $r_\rho(x)$ given by inequality (5.12) of Lemma 5.16 we get for $x \geq e^4$

$$\left|\sum_{\rho} \frac{F_\rho(x)}{\rho}\right| \leq \frac{0.047 \times 1.5}{\sqrt{x}\log x} \leq \frac{0.08}{\sqrt{x}\log x},$$

which completes the proof of the lemma. □

5.4 Further Results With RH True

Now we bound $J(x)$ in terms of a function $W(x)$, which will play a central role in the proof of Nicolas' second theorem. Define

$$W(x) := \sum_{\rho} \frac{x^{i\Im\rho}}{\rho(1-\rho)},$$

and recall the definitions $R(t) := \psi(t) - t$ and

$$J(x) := \int_x^\infty \frac{R(t)}{t^2}\left(\frac{1}{\log t} + \frac{1}{\log^2 t}\right) dt.$$

Lemma 5.23 *[127, lemma 2.5] Assume RH is true. For $x > 1$ we have*

$$J(x) = -\frac{W(x)}{\sqrt{x}\log x} - J_1(x) - J_2(x),$$

where $J_1(x)$ and $J_2(x)$ are functions which satisfy

$$0 < J_1(x) \leq \frac{\log(2\pi)}{x\log x} \quad \textit{and} \quad |J_2(x)| \leq \frac{\beta}{\sqrt{x}\log^2 x}\left(1 + \frac{4}{\log x}\right).$$

Proof First, by Lemma 5.20, for $x \geq 2$ we have

$$J(x) = -\sum_{\rho} \frac{1}{\rho} F_{\rho}(x) - J_1(x),$$

where $J_1(x)$ is a function which satisfies $0 \leq J_1(x) \leq \log(2\pi)/(x \log x)$ – so is positive and very small for large x.

Then, by (5.11) of Lemma 5.16 we can write $F_{\rho}(x) = x^{\rho-1}/((1-\rho)\log x) + r_{\rho}(x)$, so if we define $J_2(x) := \sum_{\rho} r_{\rho}(x)/\rho$ the result follows using the bound for $|r_{\rho}(x)|$ which is given by the estimate (5.12). □

In the next step, which provides the fundamental lemma for this chapter, we bound $f(x)$ in terms of $W(x)$:

Lemma 5.24 *Assume RH is true and let $x \geq 10^9$. Then*

$$-\frac{2+W(x)}{\sqrt{x}\log x} + \frac{0.055}{\sqrt{x}\log^2 x} \leq \log f(x) \leq -\frac{2+W(x)}{\sqrt{x}\log x} + \frac{2.062}{\sqrt{x}\log^2 x} \tag{5.25}$$

and

$$\frac{2+W(x)}{\sqrt{x}\log x} - \frac{2.062}{\sqrt{x}\log^2 x} \leq \frac{1}{f(x)} - 1 \leq \frac{2+W(x)}{\sqrt{x}\log x} - \frac{0.055}{\sqrt{x}\log^2 x}. \tag{5.26}$$

Proof Let $x \geq 599$. Use (5.8), (5.13), (5.14) and (5.18) and Lemmas 5.19 and 5.23 to derive the lower bound

$$\begin{aligned}
\log f(x) &\geq J(x) - F_{1/2}(x) - \frac{4}{3}F_{1/3}(x) - \frac{\log^3 x}{64\pi^2 x} \\
&= -\frac{W(x)}{\sqrt{x}\log x} - J_1(x) - J_2(x) - F_{1/2}(x) - \frac{4}{3}F_{1/3}(x) - \frac{\log^3 x}{64\pi^2 x} \\
&\geq -\frac{W(x)+2}{\sqrt{x}\log x} + \frac{2-\beta}{\sqrt{x}\log^2 x} - \frac{8+4\beta}{\sqrt{x}\log^3 x} - \frac{\log(2\pi)}{x\log x} \\
&\quad - \frac{2}{x^{2/3}\log x} - \frac{\log^3 x}{64\pi^2 x}
\end{aligned}$$

and similarly the upper bound

$$\log f(x) \leq -\frac{W(x)+2}{\sqrt{x}\log x} + \frac{2+\beta}{\sqrt{x}\log^2 x} + \frac{4\beta}{\sqrt{x}\log^3 x} + \frac{1}{2(x-1)}. \tag{5.27}$$

Now let $x_0 = 10^9$. Because $x \geq x_0$ these bounds imply

$$\begin{aligned}
\log f(x) \geq -\frac{W(x)+2}{\sqrt{x}\log x} + \frac{1}{\sqrt{x}\log^2 x}\Bigg(2 - \beta - \frac{8+4\beta}{\log x_0} \\
- \frac{\log(2\pi)\log x_0}{\sqrt{x_0}} - \frac{2\log x_0}{x_0^{1/6}} - \frac{\log^5 x_0}{64\pi^2\sqrt{x_0}}\Bigg)
\end{aligned}$$

and

$$\log f(x) \leq -\frac{W(x)+2}{\sqrt{x}\log x} + \frac{1}{\sqrt{x}\log^2 x}\left(2+\beta+\frac{4\beta}{\log x_0}+\frac{\sqrt{x_0}\log^2 x_0}{2(x_0-1)}\right),$$

from which (5.25) follows on substituting for x_0.

Now let $v := -\log f(x)$ and $v_0 := (2+\beta)/(\sqrt{x_0}\log x_0)$. For all real v we have $e^v - 1 \geq v$ so, because

$$\frac{1}{f(x)} - 1 = e^v - 1 \geq v = -\log f(x),$$

the upper bound of (5.25) implies the lower bound of (5.26). Then by (5.5) and (5.25) we can write

$$v \leq \frac{W(x)+2}{\sqrt{x}\log x} \leq \frac{2+\beta}{\sqrt{x}\log x} \leq v_0 < 3.13 \times 10^{-6}.$$

Then, to obtain the upper bound notice that

$$e^v - 1 - v \leq \frac{e^{v_0}v^2}{2} \leq \frac{e^{v_0}(2+\beta)^2}{2x\log^2 x} \leq \frac{e^{v_0}(2+\beta)^2}{2\sqrt{x_0}\sqrt{x}\log^2 x} < \frac{6.63\times 10^{-5}}{\sqrt{x}\log^2 x}.$$

Then

$$\frac{1}{f(x)} - 1 = e^v - 1 \leq v + \frac{6.63\times 10^{-5}}{\sqrt{x}\log^2 x} \leq -\log f(x) + \frac{6.63\times 10^{-5}}{\sqrt{x}\log^2 x}.$$

This complctes the proof. □

Now we need three estimates, first given by Rosser and Schoenfeld and proved here in Lemmas 3.14, 5.12 and 3.13, respectively:

(1) For $563 \leq x$ we have $x(1 - 1/(2\log x)) < \theta(x)$ and for $1 < x$ we have $\theta(x) < x(1 + 1/(2\log x))$.

(2) For $286 \leq x$ we have

$$\prod_{p\leq x}\frac{p}{p-1} < e^\gamma \log x\left(1+\frac{1}{2\log^2 x}\right).$$

(3) For $1423 \leq x \leq 10^8$ we have $x - 2\sqrt{x} < \theta(x)$.

The next theorem should be compared with an analogous upper bound for the ratio $\sigma(n)/n$ given in Chapter 7, which is also unconditional. It might also be compared with the unconditional lower bound given in Chapter 6, Lemma 6.8, where $n > 1$ is restricted to being superabundant, i.e. for all $1 \leq m < n$ we have

$$\frac{\sigma(m)}{m} < \frac{\sigma(n)}{n}.$$

To show how it leads on to other results, we have a network flow diagram, Figure 5.7. The proof is in six steps.

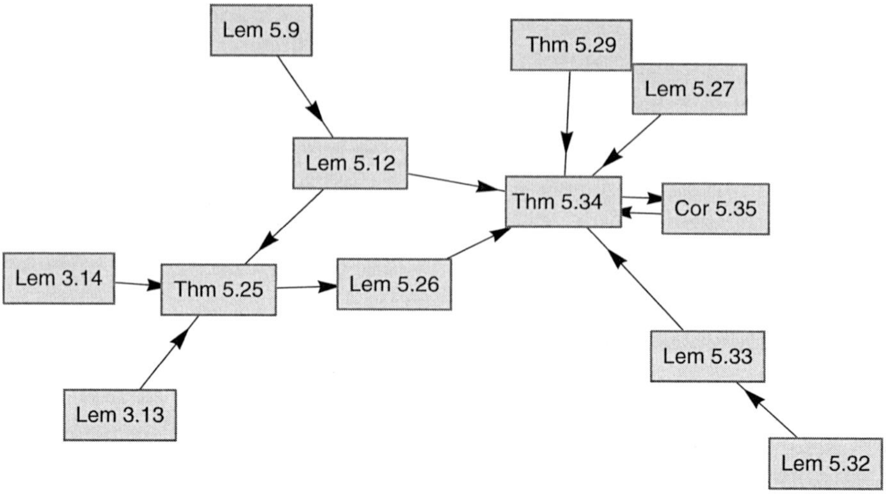

Figure 5.7 Some relationships between results in Chapter 5.

Theorem 5.25 *[145, theorem 15] For all $n \geq 3$, except $n = N_9 = 2 \cdot 3 \cdot 5 \cdot 7 \cdot 11 \cdot 13 \cdot 17 \cdot 19 \cdot 23$ we have (with or without RH)*

$$\frac{n}{\varphi(n)} < e^{\gamma} \log\log n + \frac{5}{2 \log\log n} =: g(n).$$

Proof **(1)** As usual let p_j be the jth prime with $p_1 = 2$ and suppose that the integers n and m are such that $n < p_1 \cdots p_{m+1}$. Let q_j for $1 \leq j \leq r$ be all of the distinct primes dividing n ordered so that $q_1 < q_2 < \cdots < q_r$. Thus for all j we have $p_j \leq q_j$. Then

$$p_1 \cdots p_r \leq q_1 \cdots q_r \leq n < p_1 \cdots p_{m+1},$$

so we must have $r \leq m$. Therefore

$$\frac{n}{\varphi(n)} = \prod_{j=1}^{r} \frac{q_j}{q_j - 1} \leq \prod_{j=1}^{r} \frac{p_j}{p_j - 1} \leq \prod_{j=1}^{m} \frac{p_j}{p_j - 1} = \frac{N_m}{\varphi(N_m)}.$$

(2) Next we verify the theorem for $\log n \leq 294$: indeed, if N_m is the mth primorial, and $g(n)$ the right-hand side of the inequality of this theorem, which is increasing for n in this range, then by Step (1), if $N_m \leq n < N_{m+1}$, and we have verified the result of the theorem at each primorial in the range $9 < m \leq M$, then for these values of m we would have

$$\frac{n}{\varphi(n)} \leq \prod_{p \leq p_m} \frac{p}{p-1} < g(N_m) \leq g(n).$$

Now checking the theorem at primorials is easily accomplished up to $M = 10^6$ giving at least $\log n \leq 1.5 \times 10^7$.

(3) Now let $x \geq 5$ and suppose $\log n < \theta(x)$. Choose m so $p_{m+1} \leq x < p_{m+2}$. If p_{m+1} was 2 or 3 then p_{m+2} would be 3 or 5 so we would have the bound $x < 5$, which is false. Hence $5 \leq p_{m+1}$, and thus $p_m \leq p_{m+1} - 2 \leq x - 2$. By Step (1) we therefore have

$$\frac{n}{\varphi(n)} \leq \prod_{p \leq x-2} \frac{p}{p-1}.$$

(4) The next step is quite clever. Let $n > 1$ be an integer and the real number $y \geq 2$ be chosen to satisfy three conditions:

$$288 \leq \log n + y,$$
$$\log n < \theta(\log n + y),$$
$$y - 2 \leq \frac{0.9 \log n}{\log\log n}.$$

Then, by Lemma 5.12, replacing x by $\log n + y$ and using Step (3), since $\log n < \theta(\log n + y)$ and $(y-2)/\log n \leq 0.9/\log\log n$ we get

$$\begin{aligned}\frac{n}{e^\gamma \varphi(n)} &< \log(\log n + y - 2) + \frac{0.5}{\log(\log n + y - 2)} \\ &\leq \log\log n + \log\left(1 + \frac{y-2}{\log n}\right) + \frac{0.5}{\log\log n} \\ &\leq \log\log n + \frac{1.4}{\log\log n}.\end{aligned}$$

Therefore

$$\frac{n}{\varphi(n)} < e^\gamma \log\log n + \frac{5}{2\log\log n}.$$

(5) Now assume $255 \leq \log n \leq 1340$ and set $y = 2 + 2\sqrt{1 + \log n}$. This implies $288 \leq \log n + y \leq 1420$.

Therefore using [145, theorem 19] (or compare with Lemma 3.13(a)), which in particular shows that for $0 < x \leq 1420.9$ we have $x - 2\sqrt{x} < \theta(x)$, we get

$$\log n = \log n + y - 2\sqrt{\log n + y} < \theta(\log n + y).$$

The other two conditions in Step (4) are readily found to hold, so in the given range the theorem follows by the result of Step (4).

(6) Finally assume $1340 \leq \log n$. Choose y as $0.9 \log n/\log\log n$ and use Lemma 3.14(e). With these choices for $\log n$ and y each of the hypotheses listed in (4) are easily checked. This completes the proof. □

Recall $N_k := p_1 \cdot p_2 \cdots p_k$ is the kth primorial, with p_i the ith prime and $p_1 = 2$. Also recall the definition given in the introduction to this chapter:

$$c(n) := \left(\frac{n}{\varphi(n)} - e^\gamma \log\log n\right)\sqrt{\log n}. \tag{5.28}$$

The following lemma shows that $c(n)$ is very well behaved. In particular it has a local maximum on the right-hand side at every primorial, and that if $\omega(n)$ is small so is $c(n)$.

Lemma 5.26 *[127, lemma 3.1] Let $n \geq 2$ and $k \geq 1$ with $n,k \in \mathbb{N}$. Fix k and let n be such that either $k = \omega(n)$ or $N_k \leq n < N_{k+1}$. Then in each of these cases we have $c(n) \leq c(N_k)$.*

Proof If $k = \omega(n)$ then $N_k \leq n$ and $\omega(n) \leq k$. If $\omega(n) > k$ then $n \geq N_{k+1}$, so $N_k \leq n < N_{k+1}$ implies $N_k \leq n$ and $\omega(n) \leq k$ also.

Write $n = P_1^{\alpha_1} \cdots P_j^{\alpha_j}$ where $P_1 < P_2 < \cdots < P_j$ are primes and $\alpha_i \geq 1$ integers. Then each $P_i \geq p_i$, where p_i is the ith prime, so

$$\frac{n}{\varphi(n)} = \prod_{i=1}^{j} \frac{1}{1-1/P_i} \leq \prod_{i=1}^{j} \frac{1}{1-1/p_i} \leq \prod_{i=1}^{k} \frac{1}{1-1/p_i} = \frac{N_k}{\varphi(N_k)}$$

and therefore

$$c(n) \leq \left(\frac{N_k}{\varphi(N_k)} - e^{\gamma} \log\log n \right) \sqrt{\log n} =: h(n), \tag{5.29}$$

where for fixed k the expression on the right-hand side of (5.29) defines a function $h(n)$. Interpolate h on $[N_k, N_{k+1})$ with a differentiable real function $h(x)$ so

$$\begin{aligned} h'(n) &= \frac{1}{2n\sqrt{\log n}} \left(\frac{N_k}{\varphi(N_k)} - e^{\gamma} \log\log n - 2e^{\gamma} \right) \\ &\leq \frac{1}{2n\sqrt{\log n}} \left(\frac{N_k}{\varphi(N_k)} - e^{\gamma} \log\log N_k - 2e^{\gamma} \right). \end{aligned}$$

We **claim** the term in parentheses in the upper bound is negative. For $k = 1,2$ this can be tested numerically. For $k \geq 3$ it is also negative because $N_3 = 30$, so $\log\log N_k \geq \log\log 30$ and we can use the bound given in Theorem 5.25, namely for all $n \geq 3$, without any special assumptions,

$$\frac{n}{\varphi(n)} \leq e^{\gamma} \log\log n + \frac{2.50637}{\log\log n},$$

where the slightly increased numerator $2.50637 > 2.5$ has been introduced to accommodate the exception N_9. Therefore $h(n) \leq h(N_k) = c(N_k)$ so by (5.29), $c(n) \leq c(N_k)$ as claimed. □

In Figure 5.8 there is plotted $c(N_3)$ and $c(n)$ for $30 = N_3 \leq n < N_4 = 210$. For $x \geq 1$ recall the definition

$$W(x) = \sum_{\rho} \frac{x^{i\Im \rho}}{\rho(1-\rho)},$$

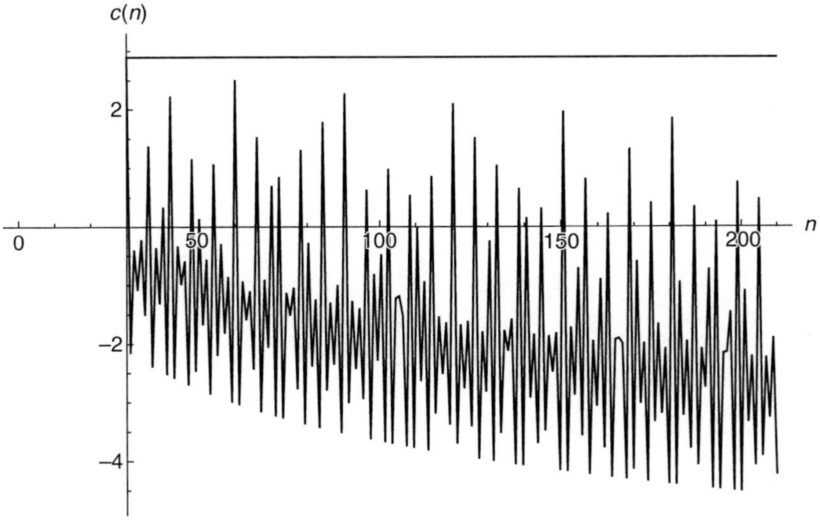

Figure 5.8 The relation $c(n) \le c(N_k)$ for $k = 3$.

which for real x is real, since the zeros of $\zeta(s)$ occur in complex conjugate pairs.

Lemma 5.27 *[127, proposition 3.1] Assume RH is true and that $10^9 \le p_k \le x < p_{k+1}$ where p_i is the ith prime. Then for $c(N_k)$ we have the upper bound*

$$c(N_k) \le e^\gamma(2 + W(x)) - \frac{0.07}{\log x} \le e^\gamma(2 + \beta) - \frac{0.07}{\log x}$$

and the lower bound

$$e^\gamma(2 - \beta) - \frac{3.7}{\log x} \le e^\gamma(2 + W(x)) - \frac{3.7}{\log x} \le c(N_k).$$

Proof Firstly, the definitions of $f(x)$ and $c(n)$ ((5.5) and (5.28) respectively) and the identity $\log N_k = \theta(p_k) = \theta(x)$ imply

$$c(N_k) = e^\gamma \sqrt{\theta(x)} \log \theta(x) \left(\frac{1}{f(x)} - 1 \right). \tag{5.30}$$

Equation (5.18) and the bound $\theta(x) \ge 4x/5$, $x \ge 121$, from Lemma 3.10(d), give

$$\left| \sqrt{\theta(x)} \log \theta(x) - \sqrt{x} \log x \right| = \left| \int_x^{\theta(x)} \frac{2 + \log t}{2\sqrt{t}}\, dt \right|$$
$$\le |\theta(x) - x| \left(\frac{\log(4x/5) + 2}{2\sqrt{4x/5}} \right)$$

$$\leq \frac{\sqrt{5}}{4} T(x) \left(\frac{2 + \log x}{\sqrt{x}} \right)$$
$$= \frac{\sqrt{5}}{32\pi} \log^2 x (\log x + 2).$$

Therefore, using $x_0 = 10^9$ again:

$$\left| \frac{\sqrt{\theta(x)} \log \theta(x)}{\sqrt{x} \log x} - 1 \right| \leq \frac{\sqrt{5} \log^2 x (2 + \log x)}{32\pi \sqrt{x} \log x}$$
$$\leq \frac{\sqrt{5} \log^2 x_0 (2 + \log x_0)}{32\pi \sqrt{x_0} \log x} \leq \frac{6.9 \times 10^{-3}}{\log x}.$$

Now use Lemma 5.24 and (5.30), (5.26) and (5.5), recall that by Lemma 2.10

$$\beta := \sum_{\rho} \frac{1}{\rho(1-\rho)} = 2 + \gamma - \log(4\pi),$$

and observe $|W(x)| \leq \beta$, to derive

$$c(N_k) \leq e^{\gamma} \left(2 + W(x) - \frac{0.055}{\log x} \right) \left(1 + \frac{0.0069}{\log x} \right)$$
$$\leq e^{\gamma} (2 + W(x)) - \frac{e^{\gamma}}{\log x} (0.055 - 0.0069(2 + \beta))$$
$$\leq e^{\gamma} (2 + W(x)) - \frac{0.07}{\log x},$$

to complete the proof of the upper bound. To derive the lower bound, use the lower bound from (5.26) of Lemma 5.24. □

5.5 Preliminary Results With RH False

In this section we assume RH is false and derive, in Theorem 5.29, a consequence for the asymptotic values of $\log f(x)$. Logarithmic differentiation of the Euler product representation for $\zeta(s)$ leads to the following lemma:

Lemma 5.28 *For* $\Re s > 1$,

$$-\frac{\zeta'(s)}{\zeta(s)} = s \int_2^{\infty} \frac{\psi(t)}{t^{s+1}} \, dt.$$

Proof Since for $\Re s > 1$ we have $\zeta(s) = \prod_p (1 - 1/p)^{-1}$,

$$\log \zeta(s) = -\sum_p \log \left(1 - \frac{1}{p^s} \right) = \sum_p \sum_{m \geq 1} \frac{1}{m p^{sm}}.$$

Therefore, using integration by parts and uniform convergence for $1 < \sigma_0 \leq \Re s$:

$$-\frac{\zeta'(s)}{\zeta(s)} = \sum_p \sum_{m\geq 1} \frac{\log p}{p^{sm}} = \sum_{n\geq 2} \frac{\Lambda(n)}{n^s}$$
$$= \lim_{x\to\infty} \int_1^x \frac{1}{t^s}\, d\psi(t) = \lim_{x\to\infty} s \int_1^x \frac{\psi(t)}{t^{s+1}}\, dt = s\int_2^\infty \frac{\psi(t)}{t^{s+1}}\, dt,$$

which completes the derivation. □

Recall the definition: if $g(x)$ is a real-valued function defined on some domain (x_0, ∞) and $h(x)$ is a positive real-valued function on the same domain, we say $g(x) = \Omega_+(h(x))$ as $x \to \infty$ if there is a constant $M > 0$ and a real sequence $(x_i)_{i\in\mathbb{N}}$ with $x_i \to \infty$ such that $g(x_i) \geq Mh(x_i)$. We say $g(x) = \Omega_-(h(x))$ if $-g(x) = \Omega_+(h(x))$. Finally $g(x) = \Omega_\pm(h(x))$ means both of these properties are true for $g(x)$ with respect to $h(x)$.

To prove the following theorem of Nicolas, the method developed by Landau to obtain "Ω-theorems" is employed; see Theorem 4.12. (The reader might compare [80, chapter V].) Also recall the definitions:

$$f(x) := e^\gamma \log\theta(x) \prod_{p\leq x}\left(1 - \frac{1}{p}\right),$$
$$\psi(x) := \sum_{p^n\leq x} \log p,$$
$$R(x) := \psi(x) - x,$$
$$J(x) := \int_x^\infty \frac{R(t)}{t^2}\left(\frac{1}{\log t} + \frac{1}{\log^2 t}\right) dt.$$

The following theorem provides the key element for Nicolas' theorems, Theorems 5.31 and 5.34, when RH is false, and, looking ahead, also for the proof of Robin's theorem, Theorem 7.11, in Chapter 7.

Theorem 5.29 *If the Riemann hypothesis is* ***false*** *then there exists a real b with $0 < b < 1/2$ such that, as $x \to \infty$,*

$$\log f(x) = \Omega_\pm\left(\frac{1}{x^b}\right).$$

Proof The proof is in seven steps.

(1) First apply Theorem 4.12 with $h(x) = J(x)$ for $x \geq 2$ and $h(x) = 0$ for $1 \leq x \leq 2$. For $\Re s > 1$ set

$$G(s) := \int_2^\infty \frac{J(x)}{x^s}\, dx = \int_2^\infty \frac{1}{x^s}\left(\int_x^\infty \frac{R(t)}{t^2}\left(\frac{1}{\log t} + \frac{1}{\log^2 t}\right) dt\right) dx.$$

The prime number theorem gives $|R(t)| = O(t/\log t)$. Hence, the double integral is absolutely convergent. Therefore we can use Fubini's theorem to change the order of integration and write

$$G(s) = \int_2^\infty \left(\frac{R(t)}{t^2}\left(\frac{1}{\log t} + \frac{1}{\log^2 t}\right)\int_2^t \frac{dx}{x^s}\right) dt,$$

which gives

$$G(s) = \frac{1}{s-1}\left(2^{1-s}J(2) - \int_2^\infty \frac{R(t)}{t^{1+s}}\left(\frac{1}{\log t} + \frac{1}{\log^2 t}\right) dt\right). \tag{5.31}$$

Note that the term on the right-hand side inside the outer parentheses is holomorphic for $\Re s > 1$ and vanishes at $s = 1$.

(2) Next, for $\Re s > 1$ we can write

$$\int_2^\infty \frac{R(t)}{t^{s+1}} dt = \int_2^\infty \frac{\psi(t)}{t^{s+1}} dt - \int_1^\infty \frac{dt}{t^s} + \int_1^2 \frac{dt}{t^s}.$$

The third integral on the right-hand side is an entire function $(1-2^{1-s})/(s-1)$ which we call $E_1(s)$.

Therefore for $\Re s > 1$ we have, by Lemma 5.28,

$$\int_2^\infty \frac{R(t)}{t^{s+1}} dt = -\frac{1}{s}\frac{\zeta'(s)}{\zeta(s)} - \frac{1}{s-1} + E_1(s). \tag{5.32}$$

We know from the relationship [6, theorem 13.11], applied with $0 < s < 1$,

$$(1-2^{1-s})\zeta(s) = \sum_{n=1}^\infty \frac{(-1)^{n+1}}{n^s} = \left(1 - \frac{1}{2^s}\right) + \left(\frac{1}{3^s} - \frac{1}{4^s}\right) + \cdots > 0,$$

that, since $1 - 2^{1-s} < 0$ and $\zeta(0) = -\frac{1}{2}$, $\zeta(s)$ has no zero in $[0, \infty)$, so therefore there is a real number $\delta > 0$ such that there are no zeros of $\zeta(s)$ in a simply connected set

$$\Theta := \{s : \Re s > 1\} \cup \{s : 0 < \Re s \leq 1,\ |\Im s| < \delta\}.$$

Calculations of zeta zeros show we can take $\delta = 14$.

Equation (5.32) also shows that if σ_0 is the abscissa of convergence of the integral on the left, then $\sigma_0 \leq 1$.

(3) Now note that $\zeta(s) = 1/(s-1) + E_0(s)$, where E_0 is entire, implies that the combination of the first and second terms on the right-hand side of (5.32) represents a function which is holomorphic in a neighbourhood of $s = 1$ in Θ, and therefore has, on Θ, a holomorphic antiderivative $G_1(s)$ and a second antiderivative $G_2(s)$. That is to say, there are complex numbers λ and μ such that

$$\int_2^\infty \frac{R(t)}{t^{s+1}} \frac{1}{\log t}\, dt = -G_1(s) + \lambda,$$
$$\int_2^\infty \frac{R(t)}{t^{s+1}} \frac{1}{\log^2 t}\, dt = G_2(s) + \lambda s + \mu.$$

By (5.31) we have

$$G(s) = \frac{1}{s-1}(G_1(s) - G_2(s) + E_2(s)), \tag{5.33}$$

where $E_2(s)$ is entire. The expression in parentheses in (5.33) is a function holomorphic on Θ which by Step (1) vanishes at $s = 1$. Therefore the right-hand side of (5.33) is holomorphic on Θ.

In the neighbourhood of a zero ρ of $\zeta(s)$ of multiplicity m we have

$$G_1''(s) \sim -\frac{d}{ds}\left(\frac{1}{s}\frac{\zeta'(s)}{\zeta(s)}\right) \sim \frac{1}{\rho}\frac{m}{(s-\rho)^2},$$
$$G_2''(s) \sim -\frac{1}{s}\frac{\zeta'(s)}{\zeta(s)} \sim -\frac{1}{\rho}\frac{m}{s-\rho},$$

so $G_1''(s) - G_2''(s)$ would have a pole of order 2 at $s = \rho$ and $\Re\rho \geq \frac{1}{2}$. This shows that $\sigma_0 \geq \frac{1}{2}$ so $\sigma_0 \in \Theta$ and $G(s)$ is holomorphic at σ_0.

If $R(x)$ had the same sign for x sufficiently large, by Theorem 4.12, $G(s)$ would have a singularity at σ_0, but is holomorphic at that point. Therefore $R(x)$ changes sign infinitely often as $x \to \infty$.

(4) Now let θ be a fixed real number satisfying $0 < \theta < 1/2$. Suppose there exists a zero ρ of $\zeta(s)$ with $\Re\rho = \beta > 1 - b$ where b is a real number which satisfies 1 $0 < b < 1/2$. We will show that $J(x) - x^{-b}$ and $J(x) + x^{-b}$ do not keep a constant sign as $x \to \infty$. To see this first calculate the Mellin transform

$$\int_2^\infty \frac{J(x) - x^{-b}}{x^s}\, dx = G(s) - \frac{1}{s-1+b} + E_3(s), \quad \Re s > 1,$$

where $E_3(s) := \int_1^2 x^{-b-s}\, dx$ is an entire function. The second term extends to a holomorphic function on $\Theta_b := \Theta \cap \{s : \Re s > 1 - b\}$.

If $J(x) - x^{-b}$ had a constant sign as $x \to \infty$, by Theorem 4.12 the abscissa of convergence of the Mellin transform would be less than or equal to $1 - b \leq \beta$. This is impossible since, by Step (4), $G(s)$ has a singularity ρ with real part β. Then adopt the same procedure for $J(x) + x^{-b}$ to show that it too does not ultimately have constant sign.

(5) Recall the definition

$$K(x) = \int_x^\infty \frac{S(t)}{t^2}\left(\frac{1}{\log t} + \frac{1}{\log^2} t\right) dt$$

where $S(t) = \theta(t) - t$. By Lemma 3.13 (*e*), for $x > 0$ we have

$$\psi(x) - \theta(x) < 1.42620\sqrt{x} < 3\sqrt{x}/2,$$

which gives directly, using $\theta(x) \le \psi(x)$ for the upper bound:

$$J(x) - \tfrac{3}{2}F_{1/2}(x) \le K(x) \le J(x).$$

The estimates we have for $F_{1/2}(x)$ given in Lemma 5.16, together with the sign changes exhibited by $J(x) \pm x^{-b}$, then show that $K(x) = \Omega_\pm(x^{-b})$ for all b with $1 - \theta < b < 1/2$.

(6) We next study the function $y(x) := K(x) - x^{-b}$. It is differentiable for all x not equal to a prime number, and at such a point

$$y'(x) = -\frac{S(x)}{x^2}\left(\frac{1}{\log x} + \frac{1}{\log^2 x}\right) + \frac{b}{x^{b+1}}.$$

This derivative can change sign at $x = p$, p a prime, and considering the equation

$$(x - a)(\log x + 1) + bx^{1-b}\log^2 x = 0,$$

with $a = S(x)$, which has only one root for $a > 1$ in $[1, \infty)$, shows that $y'(x)$ vanishes at most once in any interval (p, p'), where p and p' are consecutive primes with $p \ge 3$. To see this note that the derivative of

$$x - a + \frac{bx^{1-b}\log^2 x}{1 + \log x}$$

is

$$1 + \frac{bx^{-b}\log x((1-b)\log^2 x + (2-b)\log x + 2)}{(\log x + 1)^2} > 0$$

and use Rolle's theorem.

Therefore the sign changes of $y'(x)$ form a discrete set in $[1, \infty)$. Suppose they constitute the increasing sequence (x_i). This sequence is infinite, since we are assuming the Riemann hypothesis is false, and this implies $S(x) = \Omega_\pm(x^{(\theta+1-b)/2})$; cf. [80, chapter V] or Theorem 4.12.

At a point x_i where $y'(x)$ changes sign, so does

$$S(x) - \frac{bx^{1-b}\log^2 x}{1 + \log x}.$$

This means that this quantity is either 0 or, in case x_i is prime, lies between 0 and $\log x_i$. In all cases we have

$$S(x_i) = O(x_i^{1-b}\log x_i) \quad \text{and thus} \quad \frac{S(x)^2}{x_i^2 \log x_i} = O\left(\frac{\log x_i}{x_i^{2b}}\right).$$

(7) This is the final step. We **claim** there are an infinite number of values of i with $y(x_i) > 0$. Indeed, if the function $y(x)$ was negative or zero at all local extrema, it would be negative everywhere, and this would mean, for all x, $K(x) \leq x^{-b}$. But, as has been shown, $K(x) = \Omega_+(x^{-b'})$ where $b' := (1+b-\theta)/2$. At such a point x_i we have

$$K(x_i) - \frac{S(x_i)^2}{x_i^2 \log x_i} > x_i^{-b} + O(x_i^{-2b} \log x_i),$$

giving, by Lemma 5.15, $\log f(x) = \Omega_+(x^{-b})$. To show that $\log f(x) = \Omega_-(x^{-b})$, also consider $K(x) + x^{-b}$ and use the same approach. This completes the proof. □

5.6 Nicolas' First Theorem

Now that we have the key result of Section 5.5 we are able to prove Nicolas' first theorem. First we derive a good upper bound for the logarithm of $f(x)$ provided x is sufficiently large.

Lemma 5.30 *Assume RH is true. For* $x \geq 3000$ *we have*

$$\log f(x) \leq -\frac{0.8}{\sqrt{x}\log x}.$$

Proof By Lemma 5.21 we have

$$\log f(x) \leq J(x) - 0.98F_{1/2}(x) + \frac{1}{2(x-1)}, \qquad x \geq 121.$$

By the second part of Lemma 5.21, for $x \geq 3000$ we have

$$-0.98F_{1/2}(x) + \frac{1}{2(x-1)} \leq -\frac{0.9}{\sqrt{x}\log x}.$$

By Lemmas 5.13, 5.20 and 5.22 we have

$$J(x) \leq \left|\sum_{\rho} \frac{F_\rho}{\rho}\right| \leq \frac{0.1}{\sqrt{x}\log x}, \qquad x \geq 55,$$

so using these inequalities it follows that

$$\begin{aligned}\log f(x) &\leq J(x) - 0.98F_{1/2}(x) + \frac{1}{2(x-1)} \\ &\leq \frac{0.1}{\sqrt{x}\log x} - \frac{0.9}{\sqrt{x}\log x} \\ &= -\frac{0.8}{\sqrt{x}\log x}\end{aligned}$$

which completes the proof. □

Now Nicolas' theorem is an easy consequence:

Theorem 5.31 (Nicolas first criterion) *If the Riemann hypothesis is* ***true*** *then for all $x \geq 2$ we have $f(x) < 1$ and*

$$e^\gamma \log\log N_k < \frac{N_k}{\varphi(N_k)}, \quad k \geq 1.$$

If the Riemann hypothesis however is ***false*** *then the given inequality is true for an infinite number of k and false for an infinite number of k.*

Proof If RH is true, then by Lemmas 5.13 and 5.30 for all $x \geq 2$, $f(x) < 1$ so we can substitute p_k, the kth prime, for x and use $\log N_k = \theta(p_k)$ to derive the given inequality.

If RH is false then by Theorem 5.29 there is a real sequence $(x_k)_{k\in\mathbb{N}}$ with $x_k \to \infty$ with $\log f(x_k) < 0$ for all $k \in \mathbb{N}$. Let p_{j_k} be the largest prime less than or equal to x_k. By taking a subsequence of the (x_k) we can ensure the p_{j_k} are distinct. Then

$$1 > f(x_k) = f(p_{j_k}) = e^\gamma \log\log N_{j_k} \frac{\varphi(N_{j_k})}{N_{j_k}},$$

so the given inequality is true for this infinite set of indices k. Similarly by using x_k such that $\log f(x_k) > 0$ the same approach can be used to find an infinite set of k such that the inequality is false. □

As pointed out in the introduction to this chapter, little appears to be known about the solutions in general to the inequality

$$e^\gamma \log\log n < \frac{n}{\varphi(n)}$$

where n is not necessarily a primorial. A limited amount of numerical exploration appears to indicate that the proportion of solutions to this Nicolas' reversed inequality slowly declines, so the natural conjecture is that they have density zero. See Figure 5.9 and Section A.6 in the Appendix.

We now briefly introduce an implication for when RH fails, so in this last part of the section let RH be false. Then by Nicolas' first theorem, Theorem 5.31, there exists an infinite sequence of integers $k \to \infty$ such that

$$\frac{N_k}{\varphi(N_k)} \leq e^\gamma \log\log N_k < e^\gamma \log\log N_{k+1} < \frac{N_{k+1}}{\varphi(N_{k+1})}.$$

These inequalities have some delicate consequences. We have the numerical example as shown in Figure 5.10, which is a plot of

$$\Delta_k := \left(\frac{N_{k+1}}{\varphi(N_k)} - \frac{N_k}{\varphi(N_k)}\right) - e^\gamma \log\left(\frac{\log N_{k+1}}{\log N_k}\right).$$

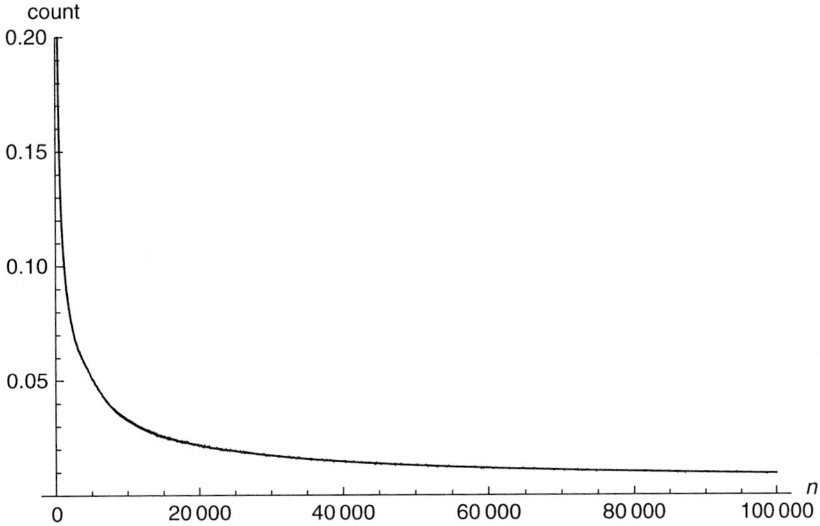

Figure 5.9 Proportion of solutions to $e^\gamma \log\log n < n/\varphi(n)$ for $1 \le n \le 10^5$.

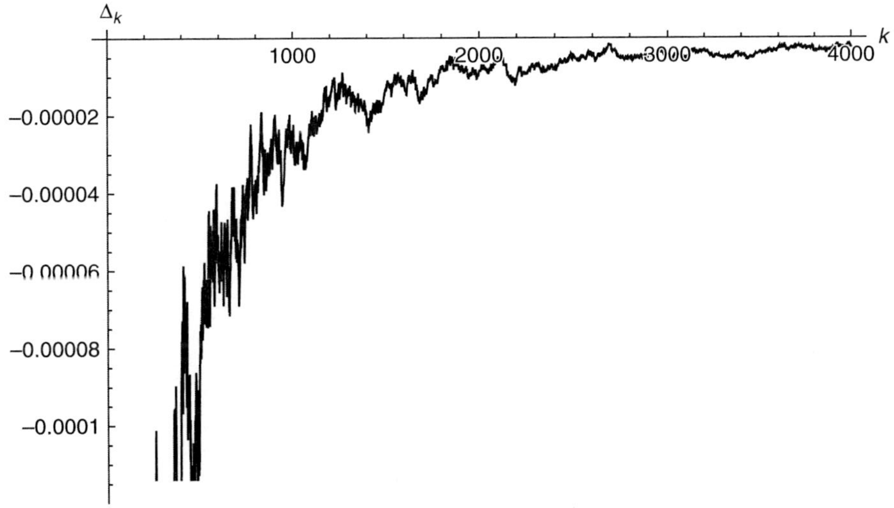

Figure 5.10 The function Δ_k for $1 \le k \le 4000$.

5.7 Nicolas' Second Theorem

Primorials also play an important role in Nicolas' second theorem. All the hard work has been done, especially in Lemma 5.24. Corollary 5.35 shows that the theorem is close in content to Robin's theorem and Nicolas' first theorem. Let, as before, $N_k = 2 \cdot 3 \cdot 5 \cdots p_k$ be the kth primorial. Recall the definitions:

$$c(n) := \left(\frac{n}{\varphi(n)} - e^{\gamma}\,\mathrm{loglog}(n)\right)\sqrt{\log n},$$

$$W(x) := \sum_{\rho} \frac{x^{i\Im\rho}}{\rho(1-\rho)},$$

$$\beta := \sum_{\rho} \frac{1}{\rho(1-\rho)} = 2 + \gamma - \log(4\pi) = 0.046191\ldots,$$

$$f(x) := e^{\gamma} \log\theta(x) \prod_{p\le x}\left(1 - \frac{1}{p}\right).$$

For a plot of values of the sequence $c(n)$ for $n \in [1, 1000]$, see Figure 5.11.

Next we give a form of Dirichlet's approximation theorem [5, section 7.2], which we need to show the supremum of $W(x)$ is β:

Lemma 5.32 *Given $\epsilon > 0$, fixed real numbers $\alpha_1, \ldots, \alpha_N$ and a real $t_0 > 0$, then there exist integers $x_1, \ldots, x_N$ and real $t \ge t_0$ such that*

$$|t\alpha_j - x_j| < \epsilon, \quad 1 \le j \le N.$$

Proof Let $q \in \mathbb{N}$ satisfy $1/q < \epsilon$ and set $a = (\alpha_1, \ldots, a_N) \in \mathbb{R}^N$. Let the real variable u take the $1 + q^N$ values

$$0,\ t_0,\ 2t_0,\ \ldots,\ q^N t_0.$$

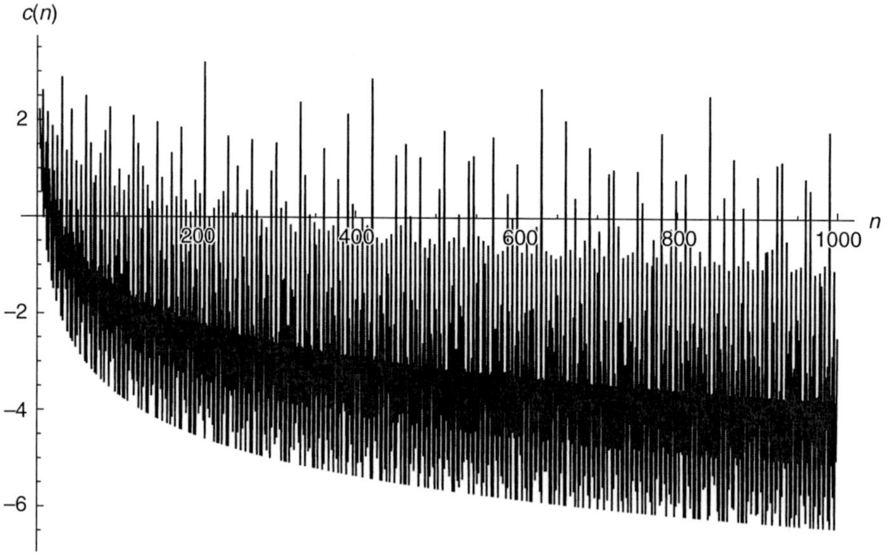

Figure 5.11 The sequence $c(n)$.

Then, if as usual we denote the fractional part of the real variable x by $\{x\}$, the $1+q^N$ points $(\{u\alpha_1\},\ldots,\{u\alpha_N\}) \in [0,1]^N$. This cube can be decomposed into exactly q^N regular subcubes each having all sides of length $1/q$, so at least two of these points are in a particular subcube. If the corresponding values of u are u_1 and u_2 with $u_1 < u_2$, then for $1 \le j \le N$ we have integers n_j and m_j such that

$$\begin{aligned}
|\{u_2\alpha_j\} - \{u_1\alpha_j\}| \le \frac{1}{q} \quad &\Longrightarrow \quad |u_2\alpha_j - n_j - u_1\alpha_j + m_j| \le \frac{1}{q} \\
&\Longrightarrow \quad |(u_2-u_1)\alpha_j - (n_j - m_j)| \le \frac{1}{q} \\
&\Longrightarrow \quad |t\alpha_j - x_j| \le \epsilon.
\end{aligned}$$

Finally observe that each x_j is an integer and $t = u_2 - u_1 \ge t_0$ to complete the proof. □

Lemma 5.33 *Assume RH is true. Then*

$$\limsup_{x\to\infty} W(x) = \beta.$$

Proof Let $\epsilon > 0$ and $t_0 > 0$ be given. Choose N sufficiently large so that

$$\sum_{j=1}^{N} \frac{2}{|\rho_j|^2} > \beta - \epsilon,$$

and note that we can write

$$W(x) = \sum_{\gamma>0} \frac{2\cos(\gamma \log x)}{|\rho|^2}.$$

Let $\alpha_j := \gamma_j/(2\pi)$ where $\rho_1 = \frac{1}{2} + i\gamma_1$ is the first complex zeta zero with positive imaginary part. By Lemma 5.32 there is a $t \ge t_0$ such that for $1 \le j \le N$ we have $|t\alpha_j - x_j| < \epsilon/(2\pi)$. This enables us to write $|t\gamma_j - 2\pi x_j| < \epsilon$ which implies $\cos(t\gamma_j) \ge \cos(\epsilon) \ge 1-\epsilon$. Therefore letting $x = e^t$ we have

$$\begin{aligned}
W(x) &\ge \sum_{j=1}^{N} \frac{2\cos(\gamma_j \log x)}{|\rho_j|^2} - \sum_{|\rho|>|\rho_N|} \frac{1}{|\rho|^2} \\
&\ge \sum_{j=1}^{N} \frac{2(1-\epsilon)}{|\rho_j|^2} - \epsilon \\
&\ge (1-\epsilon)(\beta-\epsilon) - \epsilon.
\end{aligned}$$

Since x may be chosen arbitrarily large and $W(x) \le \beta$ this shows that $\limsup_{x\to\infty} W(x) = \beta$. □

Theorem 5.34 *The hypothesis RH is equivalent to each of the following:*

(1) $\limsup_{n\to\infty} c(n) = e^{\gamma}(2+\beta) = 3.644415\ldots$.
(2) For all $n \geq N_{120569} = 2\cdot 3\cdots 1591\,883$, *we have* $c(n) < e^{\gamma}(2+\beta)$.
(3) For all $n \geq 2$, $c(n) \leq c(N_{66}) = c(2\cdot 3\cdots 317) = 4.0628356921\ldots$.
(4) For all $k \geq 1$, *we have* $c(N_k) \geq c(N_1) = c(2) = 2.208589\ldots$.

Proof First assume RH is true.

(1) By Lemmas 5.33, 5.26 and 5.27 it follows that

$$\begin{aligned} e^{\gamma}(2+\beta) &= \limsup_{x\to\infty} e^{\gamma}(2+W(x)) \leq \limsup_{k\to\infty} c(N_k) \\ &\leq \limsup_{n\to\infty} c(n) \leq \limsup_{k\to\infty} c(N_k) \leq e^{\gamma}(2+\beta), \end{aligned}$$

so (1) follows.

(2) Let, as before, $x_0 = 10^9$ and $k_0 = \pi(x_0) = 50\,847\,534$. Setting $k_1 := 120\,568$ and testing the equation numerically shows that for $k_1 < k \leq k_0$ we have $c(N_k) < e^{\gamma}(2+\beta)$. To do this we used a *Mathematica* procedure based on the following pseudocode where γ is Euler's constant and p_k is the kth prime:

```
Set k0, k1 and the precision d
Evaluate e^γ numerically
Set the number of failures to 0
Set c to e^γ(4 + γ − log(4π)) numerically
Set rk to the numerical product of pk/(pk − 1) from k = 1 to k = k1 − 1
Set LogNk to the numerical sum of log(pk) from k = 1 to k = k1 − 1
For k from k1 to k0 update rk and LogNk and then evaluate
  cNk which is (rk − e^γ log(LogNk)) √LogNk
  Test to see if cNk > c, report any failure, update the number of failures
Return the number of failures
```

See the RHpack function NicolasTwoFailures. Also see Figure 5.12 for a plot of $c(N_k)$ values.

For $k > k_0$ Lemma 5.27 gives $c(N_k) < e^{\gamma}(2+\beta)$ so, by Lemma 5.26, choosing k appropriately, $c(n) \leq c(N_k) < e^{\gamma}(2+\beta)$ so (2) follows.

(3) A *Mathematica* computation similar to that in (2) shows that for $1 \leq k \leq k_0$ we have $c(N_1) \leq c(N_k) \leq c(N_{66})$ and then use Lemma 5.27 to complete the proof that $c(n) \leq c(N_{66})$ for all $n \geq 2$.

(4) By Lemma 5.27, for all $k \geq k_0$ we get

$$c(N_k) \geq e^{\gamma}(2-\beta) - \frac{3.7}{\log x_0} = 3.30\cdots > c(2),$$

so, using the lower bound from the computation in (3), (4) is true also.

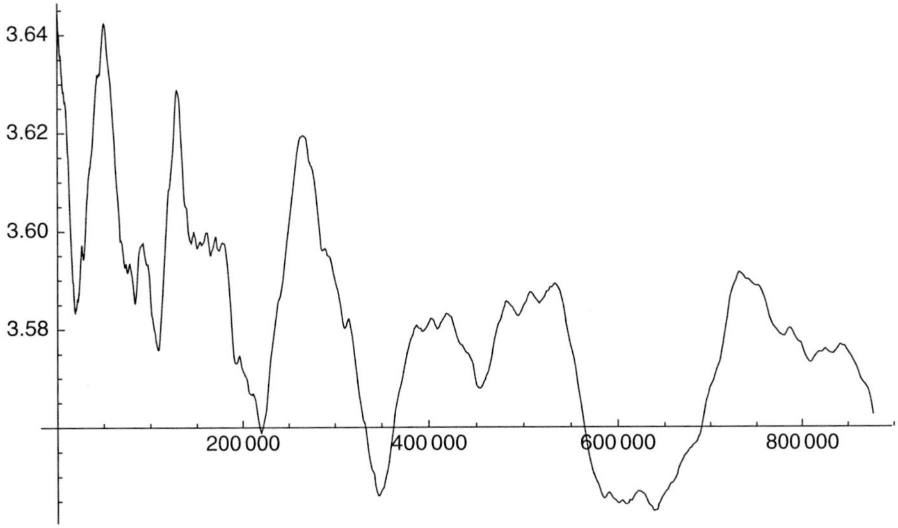

Figure 5.12 The sequence $c(N_k)$ for $120\,568 \le k \le 10^6$.

Now assume RH is false. By the prime number theorem in the form $\theta(x) \sim x$ and using Mertens' theorem (see Lemma 5.12) we have $\lim_{x\to\infty} f(x) = 1$.

Observe that $f(x)$ is locally constant, which goes to the essence of its definition. If p_k is the kth prime, when $p_k \le x < p_{k+1}$, $f(x) = e^{\gamma} \log\log(N_k)\varphi(N_k)/N_k$. Also, when $k \to \infty$, because $\log(1+y)/y \to 1$ as $y \to 0$, if we set $1 + 1/y = f(p_k)$ we get:

$$\log\left(1 + \frac{1}{y}\right) \sim \log f(p_k) \sim \frac{1}{y} = f(p_k) - 1,$$

so $\log f(p_k) \sim f(p_k) - 1$. Therefore, since $\log N_k \sim p_k$ and by Theorem 5.25 and Lemma 5.12 we have

$$\frac{N_k}{\varphi(N_k)} \sim e^{\gamma} \log\log N_k,$$

we can derive

$$\begin{aligned}\log f(p_k) &\sim e^{\gamma} \log\log(N_k)\varphi(N_k)/N_k - 1\\ &= -\frac{\varphi(N_k)c(N_k)}{N_k\sqrt{\log N_k}}\\ &\sim -\frac{c(N_k)}{e^{\gamma}(\log\log N_k)\sqrt{\log N_k}}\\ &\sim -\frac{e^{-\gamma}c(N_k)}{(\log p_k)\sqrt{p_k}}.\end{aligned}$$

Because RH fails, there is a real number b with $0 < b < 1/2$ such that, by Theorem 5.29, we have $\log f(x) = \Omega_{\pm}(1/x^b)$, so in this case, where $f(x)$ is constant between primes, we must have

$$\frac{e^{-\gamma}c(N_k)}{(\log p_k)p_k^{1/2-b}} = \Omega_{\pm}(1),$$

so $\liminf_{n\to\infty} c(n) = -\infty$ and $\limsup_{n\to\infty} c(n) = \infty$. Therefore each of (1), (2), (3) and (4) is false also. This completes the proof of the equivalences. □

Now consider the inequality derived in Theorem 5.25, namely for all $n \geq 3$ except $n = N_9$ we have

$$\frac{n}{\varphi(n)} < e^{\gamma} \log\log n + \frac{5}{2\log\log n}.$$

From Theorem 5.34 part (2), we see that RH is equivalent to a stronger form of this inequality, at least for n sufficiently large. We call this **Nicolas' second criterion**, something of a misnomer since his second theorem is more comprehensive, but is given because of its simplicity and because it takes the same shape as other criteria.

Corollary 5.35 (Nicolas' second criterion) *The Riemann hypothesis is equivalent to the following inequality holding for all $n \geq N_{120569}$:*

$$\frac{n}{\varphi(n)} < e^{\gamma} \log\log n + \frac{e^{\gamma}(2+\beta)}{\sqrt{\log n}}.$$

Note also from Chapter 6, Lemma 6.8, for n a sufficiently large superabundant number with maximum prime divisor p, that, without RH,

$$(1-\epsilon(p))\frac{n}{\varphi(n)} < \frac{\sigma(n)}{n} < \frac{n}{\varphi(n)} < e^{\gamma} \log\log n + \frac{0.6}{\log\log n},$$

where $\epsilon(p) := (1+3/(2\log p))/\log p$ and provided $n \geq N_k$ with $p_k \geq 20000$.

Remark A computation shows the inequality of Corollary 5.35 fails for primorials N_k with k in the following ranges:

$$[9,\ 27220],\ [28168,\ 40969],\ [41368,\ 56605],$$
$$[70250,\ 82904],\ [118349,\ 120568],$$

and succeeds for all other primorials with $1 \leq k \leq 120568$.

5.8 Unsolved Problems

(1) Let $k_1 = 120568$. Find the largest number M such that both inequalities $M < N_{k_1+1}$ and $c(M) \geq e^{\gamma}(2+\beta)$ are true.
(2) Find all the values of $n \in \mathbb{N}$ satisfying $n < N_{k_1+1}$ and $c(n) > e^{\gamma}(2+\beta)$.

(3) Explain the remark following the corollary to Theorem 5.34 in terms of the distribution of primes.
(4) Find a relationship between gaps, or sets of gaps, between primes which is equivalent to the Riemann hypothesis.
(5) Prove that RH implies $e^\gamma \log\log n < n/(\varphi(n)+1)$ for all $n \ge 2$.
(6) Lehmer's conjecture of 1932: if $\varphi(n) \mid n-1$ then necessarily n is prime.
(7) Carmichael's totient function conjecture: for all m, $\#\{n : \varphi(n) = m\} > 1$.
(8) The Duffin–Schaeffer conjecture of 1941: if $f : \mathbb{N} \to (0,\infty)$ then for almost all α (with respect to Lebesgue measure), the inequality

$$\left|\alpha - \frac{p}{q}\right| < \frac{f(q)}{q}$$

has infinitely many solutions in coprime integers p, q with $q > 0$ if and only if the sum

$$\sum_{q=1}^{\infty} f(q)\frac{\varphi(q)}{q} = \infty.$$

6
A Variety of Abundant Numbers

6.1 Introduction

In this chapter we present a study of superabundant and colossally abundant numbers. The material is included because these are the numbers which appear as possible counterexamples to the inequality of Robin, which we will show in Chapter 7 to be equivalent to RH.

Paul Erdős (Figure 6.1) made fundamental contributions to our understanding of these numbers, especially in two joint papers, one with Alaoglu in 1944 [3] and the other with Nicolas in 1975 [56]. Erdős was the most prolific and connected mathematician of all time, in terms of both the number of articles he published and the number of his joint authors. He never held a regular position, but travelled the world, staying for a time with one or other of his joint authors and discussing mathematical problems. These were in a wide range: graph theory, combinatorics, analysis, probability and set theory, as well as number theory.

All of the numbers of interest share the property that, apart from exceptions like 1, they have many divisors, with many of the divisors being large. Here now are some definitions. A positive integer n is said to be **highly composite** if the number-of-divisors function $d(n)$ satisfies, for $1 \leq m < n$, $d(n) > d(m)$. In other words, a highly composite number n has more divisors than any smaller number. It is called **abundant** provided $\sigma(n)/n > 2$. We say it is **highly abundant** if $\sigma(n) > \sigma(m)$ for all $m < n$. It is **superabundant** if $\sigma(n)/n > \sigma(m)/m$ for all $1 \leq m < n$. Some authors denote these numbers as "SA". It is **colossally abundant**, sometimes denoted "CA", if there exists an $\epsilon > 0$, depending on n, such that for all $m > 1$ we have

$$\frac{\sigma(m)}{m^{1+\epsilon}} \leq \frac{\sigma(n)}{n^{1+\epsilon}}.$$

There are relations between these types of numbers. For example all superabundant numbers are highly abundant and all colossally abundant

Figure 6.1 Paul Erdős, 1913–1996.

numbers are superabundant. To see this, suppose that $n > 1$ were colossally abundant but *not* superabundant. Then for some $m < n$ we would have $\sigma(n)/n < \sigma(m)/m$. But this would mean that for some $\epsilon > 0$ we would have

$$\frac{\sigma(n)}{n^{1+\epsilon}} \geq \frac{\sigma(m)}{m^{1+\epsilon}} > \frac{m\sigma(n)}{nm^{1+\epsilon}},$$

which gives $1 > (m/n)^{\epsilon} > 1$, a contradiction. Hence CA $\subset$ SA. Examples of small highly composite, superabundant and colossally abundant numbers are given in Appendix A.

There are a range of counting problems for these types of numbers, some of which have been solved. For example, the set of abundant numbers has an asymptotic density. This was proved independently by Chowla [39], Erdős [52] and Davenport [42]; and see [124, p. 266]. For equivalences to the Riemann hypothesis, it is the superabundant and colossally abundant numbers which appear, as will be seen in Chapters 7, 8 and 9.

A note about the history of the sum-of-divisors function $\sigma(n)$, or, more generally, divisor sum functions $\sigma_s(n)$, where s is a complex number, is in order: they have an ancient past, being known to the Greeks, who were fascinated with perfect numbers. They appear in Euclid's "Elements" VII.22. In Dickson's "History of the Theory of Numbers" volume I [46], there are extensive references in chapters II and X, including references to works by Cardano, Mersenne, Newton, Waring, Descartes and Fermat.

The definition of $\sigma_s(n)$ is as follows:

$$\sigma_s(n) := \sum_{d|n} d^s = \prod_{p|n} \left(\frac{p^{s(\nu_p(n)+1)} - 1}{p^s - 1} \right) = \prod_{p|n} (1 + p^s + p^{2s} + \cdots + p^{\nu_p(n)s}).$$

If $s = 0$ we write $d(n) := \sigma_0(n) = \prod_{p|n}(1 + \nu_p(n))$, and if $s = 1$, the principal focus in this chapter, then

$$\sigma(n) := \sigma_1(n) = \prod_{p|n} \left(\frac{p^{(\nu_p(n)+1)} - 1}{p - 1} \right).$$

Here we are interested in the ratios $\sigma(n)/n$ and, for $\epsilon > 0$, $\sigma(n)/n^{1+\epsilon}$, over initial sequences of natural numbers. If $N_k = 2 \cdot 3 \cdots p_k$ is the kth primorial, then

$$\frac{\sigma(N_k)}{N_k} = \prod_{j=1}^{k} (1 + p_j) \quad \Longrightarrow \quad \limsup_{n \to \infty} \frac{\sigma(n)}{n} = \infty,$$

but since for each prime p, $\sigma(p)/p = 1 + 1/p$, we have $\liminf_{n \to \infty} \sigma(n)/n = 1$. It can be shown that $\{\sigma(n)/n : n \in \mathbb{N}\}$ is, in the usual topology, dense in $(1, \infty)$.

Even though the values of both $\sigma(n)$ and $\sigma(n)/n$ exhibit a high degree of randomness as functions of n, they have very nice Dirichlet series representations. For example

$$\sum_{n=1}^{\infty} \frac{\sigma(n)}{n^s} = \zeta(s)\zeta(s-1), \quad \Re s > 2,$$

and

$$\sum_{n=1}^{\infty} \frac{\sigma_a(n)\sigma_b(n)}{n^s} = \frac{\zeta(s)\zeta(s-a)\zeta(s-b)\zeta(s-a-b)}{\zeta(2s-a-b)}, \quad \Re s > \sigma_0,$$

where

$$\sigma_0 = \max\{\Re(1+a),\ \Re(1+b),\ \Re(1+a+b)\}.$$

For further background material and references for $\sigma(n)$ the reader might consult [6, 75, 150].

The following RHpack (see Appendix B) functions relate to the material in this chapter: HighlyCompositeQ, HardyRamanujanQ, SuperAbundantQ, ColossallyAbundantQ, ColossallyAbundantInteger, ColossallyAbundantF and CriticalEpsilonValues.

6.2 Superabundant Numbers

Recall that a positive integer $n > 1$ is said to be **superabundant** if for all $1 \leq m < n$ we have

$$\frac{\sigma(m)}{m} < \frac{\sigma(n)}{n}.$$

In this section we use the following notation for unique factorizations of integers greater than 1:

$$n = 2^{a_2} 3^{a_3} \cdots p^{a_p},$$

where, for each prime q, $a_q \geq 0$, and p is the maximum prime dividing n, so all of the primes up to p are included in the expression, and "$\cdots$" here means all consecutive primes appear to the same power as the leading and trailing term. In other words, in an explicit prime factorization, $q^j \cdots p^j$ means $q < p$ and all of the consecutive primes between q and p are to be included, each to the power j.

The table in section A.3 gives the 31 smallest superabundant numbers. The database of Kilminster and Noe includes thousands, with the largest at this time being numbered 8436 in the *Online Encyclopedia of Integer Sequences* [158]

https://oeis.org/A004394/b004394.txt

which is:

$$n_{8436} = 2^{14} \cdot 3^8 \cdot 5^5 \cdot 7^4 \cdot 11^3 \cdots 17^3 \cdot 19^2 \cdots 71^2 \cdot 73^1 \cdots 2719^1.$$

When it comes to Riemann hypothesis equivalents, superabundant numbers are not as conspicuous as some other types of integer. However, this may change when the numbers are better understood. To this end we have included some interesting recent results of Nazardonyavi and Yakubovich [122], as well as part of the classical material, now 70 years old, of Erdős and Alaoglu.

First we show that in the prime factorization of any superabundant number all primes up to p appear since their exponents are in non-increasing order.

Lemma 6.1 *If n is superabundant then the exponents are non-increasing, i.e. $a_q \geq a_{q'}$ for all primes q, q' with $q \leq q' \leq p$.*

Proof All primes up to p must appear, since if a prime was missing we could replace a larger prime in n with the missing prime raised to the same power as the larger prime, and thereby obtain an integer $m < n$ with $\sigma(n)/n < \sigma(m)/m$. This is because for primes p and q with $p < q$ we have $\sigma(p^e)/p^e > \sigma(q^e)/q^e$. If the exponents of the prime powers were not non-increasing, first note that for given $e > f > 0$ and $x \in (1, \infty)$ the function

$$g(x) := \frac{1 - x^{-e-1}}{1 - x^{-f-1}}$$

is strictly decreasing because the numerator of the derivative is $ex^f - fx^e + f - e$ and its derivative is 0 at $x = 1$ and is negative. Thus if p, q, f, e are positive integers with p, q prime, $p < q$ and $e > f$, then $p^f q^e > p^e q^f$, but because $g(p) > g(q)$ we have

$$\begin{aligned}\frac{\sigma(p^e q^f)}{p^e q^f} &= \left(\frac{1 - p^{-e-1}}{1 - p^{-1}}\right)\left(\frac{1 - q^{-f-1}}{1 - q^{-1}}\right) \\ &> \left(\frac{1 - p^{-f-1}}{1 - p^{-1}}\right)\left(\frac{1 - q^{-e-1}}{1 - q^{-1}}\right) = \frac{\sigma(p^f q^e)}{p^f q^e},\end{aligned}$$

contradicting n is superabundant. Therefore the prime powers are non-increasing and the proof is complete. □

Lemma 6.2 *[3, theorem 2] For n superabundant and $q < r \le p$ primes we have the inequality*

$$\left| a_r - \left\lfloor a_q \frac{\log q}{\log r} \right\rfloor \right| \le 1.$$

Proof Let $\beta := \lfloor a_q \log q / \log r \rfloor$ and suppose first that $a_r \le \beta - 2$. Define an integer x by the inequalities $q^{x-1} < r < q^x$. If $a_q < x$ then $a_q \le x - 1$ so

$$q^{a_q} \le q^{x-1} < r \le r^{a_r} \le r^{\beta-2} \le q^{a_q} r^{-2} < q^{a_q},$$

which is false. Therefore we must have $a_q \ge x$. This implies nr/q^x is an integer which is less than n. Using the fact that n is superabundant, and cancelling common factors, we can write

$$\frac{\sigma(q^{a_q - x} r^{a_r+1})}{q^{a_q - x} r^{a_r+1}} < \frac{\sigma(q^{a_q} r^{a_r})}{q^{a_q} r^{a_r}},$$

which implies

$$(r-1)q^{a_q+1} + q^x < (q^x - 1)r^{a_r+2} + r.$$

Note that $q^{x-1} \le r - 1$ so $q^x \le q(r-1)$ giving $q^x - 1 < q(r-1)$. Using this we have

$$\begin{aligned}(q^x - 1)r^{a_r+2} + r &\le (q^x - 1)r^\beta + q^x < q(r-1)r^\beta + q^x \\ &\le (r-1)q^{a_q+1} + q^x < (q^x - 1)r^{a_r+2} + r\end{aligned}$$

since $r^\beta \le q^{a_q}$. This contradiction shows $a_r \ge \beta - 1$.

Now suppose that $a_r \geq \beta + 2$ and again define x by $q^{x-1} < r < q^x$ so $x \geq 2$. Then $r > q^{x-1}$ which implies $q^{a_q} r^{a_r} > q^{a_q+x-1} r^{a_r-1}$. Using the fact that n is superabundant, and cancelling common factors, gives

$$q^{a_q+x}(r-1) + q^{x-1} > r^{a_r+1}(q^{x-1}-1) + r > r^{a_r+1}(q^{x-1}-1) + q^{x-1}.$$

This implies $q^x(r-1) > r^2(q^{x-1}-1)$. Since the function $g(r) := r^2/(r-1)$ is increasing with $r \geq 2$, and $h(x) := q^x/(q^{x-1}-1)$ is decreasing with $x \geq 2$, it follows that

$$\frac{r^2}{r-1} < \frac{q^x}{q^{x-1}-1} \leq \frac{q^2}{q-1}.$$

The outermost terms give $r^2/(r-1) < q^2/(q-1)$, which is false since $q < r$ and $g(r)$ is increasing. Therefore we must have also $a_r \leq \beta + 1$, and the proof is complete. □

The example n_{8436} given above had $a_p = 1$. With two exceptions, this is always the case for superabundant numbers.

Lemma 6.3 *The exponent of the maximum prime divisor p of a superabundant number is $a_p = 1$ except for $n = 4$ and $n = 36$ in which cases we have $a_p = 2$.*

Proof Suppose, to get a contradiction, that $a_p > 1$ and let q be the penultimate prime less than p and r the next higher prime to p, so $q < p < r$. Then $m := nr/(pq) < n$ and n is superabundant, so $\sigma(n)/n > \sigma(m)/m$. Because $a_q \geq a_p \geq 2$, this implies

$$1 + \frac{1}{r} < \left(1 + \frac{p-1}{p^{a_p+1}-p}\right)\left(1 + \frac{q-1}{q^{a_q+1}-q}\right) \leq \left(1 + \frac{1}{p^2+p}\right)\left(1 + \frac{1}{q^2+q}\right)$$
$$< \left(1 + \frac{1}{p^2}\right)\left(1 + \frac{1}{q^2}\right).$$

But the order and rank of the three primes implies, by Bertrand's hypothesis, that

$$\frac{p}{2} < q < p < r < 2p \quad \Longrightarrow \quad 1 + \frac{1}{2p} < \left(1 + \frac{1}{p^2}\right)\left(1 + \frac{4}{p^2}\right)$$

and therefore $2 \leq p \leq 7$.

If $(q, p, r) = (3, 5, 7)$ or $(5, 7, 11)$ then

$$1 + \frac{1}{r} > \left(1 + \frac{1}{p^2+p}\right)\left(1 + \frac{1}{q^2+q}\right)$$

so we must have $p \in \{2, 3\}$. If $p = 2$, since 2^2 is superabundant, we are left to consider $n = 2^e$ with $e \geq 3$ so $n > 6$. But

$$2 - \frac{1}{2^e} = \frac{\sigma(2^e)}{2^e} > \frac{\sigma(6)}{6} = 2,$$

which is false. Hence 4 is the only superabundant number which is a power of 2.

Finally if $p = 3$ first note that $2 \cdot 3$, $2^2 \cdot 3$ and $2^2 \cdot 3^2$ are superabundant. If $n = 2^e \cdot 3^3$ with $e \geq 3$ first note that $2^3 \cdot 3^3 = 216 > 120$ so

$$\left(2 - \frac{1}{2^e}\right)\frac{40}{27} = \frac{\sigma(2^e \cdot 3^3)}{2^e \cdot 3^3} < 3 = \frac{\sigma(120)}{120},$$

so no number of the form $2^e \cdot 3^3$, $e \geq 3$, can be superabundant. Therefore 4 and 36 are the only superabundant numbers with $a_p > 1$. This completes the proof. □

The prime power of 2 provides a convenient upper bound for prime power divisors of superabundant numbers. Generalizing this result is a problem included in the list at the end of this chapter.

Lemma 6.4 *If n is superabundant then for all primes $q > 2$ with $q \mid n$ we have $q^{a_q} < 4 \cdot 2^{a_2}$.*

Proof If $a_q \leq \beta := \lfloor a_2 \log 2/\log q \rfloor$ then $q^{a_q} \leq 2^{a_2} < 4 \cdot 2^{a_2}$, so we may assume $a_q > \beta$. With Lemma 6.2 this implies we must have $a_q = \beta + 1$. To get a contradiction, assume also that for some superabundant number n we have $2^{a_2+2} < q^{a_q}$. Note that if we define an integer $x \geq 1$ by $2^x < q < 2^{x+1}$ then we would have the unitary divisor of n, $2^{a_2}q^{\beta+1} = 2^{a_2}q^{a_q} > 2^{a_2+x}q^{\beta}$. Using that n is superabundant and cancelling gives

$$q^{\beta+2}(2^x - 1) + q < 2^{a_2+x+1}(q-1) + 2^x. \tag{6.1}$$

Therefore

$$\begin{aligned} q^{a_q+1}(2^x - 1) &< 2^{a_2+x+1}(q-1) \\ 2^{a_2+2}q(2^x - 1) &< q2^{a_2+x+1} \\ 2^{a_2+2}(2^x - 1) &< 2^{a_2+x+1} \\ 2^{a_2+x+1} &< 2^{a_2+2} \\ 2^x &< 2^1 \text{ which is false.} \end{aligned}$$

This contradiction shows that $q^{a_q} \leq 2^{a_2+2}$ and completes the proof of the lemma. □

It seems that superabundant numbers have many divisors. One way of quantifying this is the result of Caveney:

Lemma 6.5 *[34, proposition 9] Every given integer n greater than 1 divides all but a finite number of superabundant numbers.*

Proof Note that in this proof primes are indexed by their sequential rank so $2 = p_1, \ldots$. Let $n > 1$ be a given integer with factorization

$$n = p_1^{a_1} \cdots p_m^{a_m}$$

with $a_i \geq 0$, $a_m \geq 1$, with p_i the ith prime. Let $S := \{1,2,4,6,24,\ldots\}$ be the set of all superabundant numbers. Let $A := \max(a_i : 1 \leq i \leq m)$ be the maximum of the exponents of n so $n \mid (p_1 \cdots p_m)^A$. Let

$$F := \{s \in S : p_m^A \nmid s\}.$$

Because of Lemma 6.1, if $p_m^A \mid s$ we would have $n \mid s$. Hence the set of superabundant numbers s for which $n \nmid s$ is a subset of those for which p_m^A does *not* divide s, i.e. F.

We **claim** F is a finite set. To see this let $s = p_1^{b_1} \cdots p_l^{b_l}$ be in F so $b_m < A$. By Lemma 6.2 we have

$$|b_m - \lfloor b_1 \log 2/\log p_m \rfloor| \leq 1$$

so b_1 is bounded above, $b_1 \leq B$ say, and therefore, by Lemma 6.4, $p_i^{b_i} < 4 \times 2^B$ is bounded also for $1 \leq i \leq l$, so F is finite and the proof of the lemma is complete. □

Lemma 6.6 *[123, proposition 11] If $n_1 < n_2$ are two successive superabundant numbers then $n_2 \leq 2n_1$.*

Proof Let $n_1 = 2^k \cdots p$ so $2n_1 = 2^{k+1} \cdots p$. Then

$$\frac{\sigma(2n_1)/(2n_1)}{\sigma(n_1)/n_1} = \frac{2^{k+2}-1}{2^{k+2}-2} > 1$$

so $n_2 \leq 2n_1$. □

It follows that there is a superabundant number in each real interval $[x, 2x]$ and that the number of superabundant numbers is thus infinite.

Recall that there is a nice upper bound for the prime power terms in a superabundant number, namely that given in Lemma 6.4, $q^{a_q} < 4 \times 2^{a_2}$. In [3] we see the remark that for $q > 11$ the factor of 4 can be removed. The next lemma gives a convenient lower bound for the exponent:

Lemma 6.7 *[123, lemma 14] If $n = 2^{a_2} \cdots q^{a_q} \cdots p$ is a superabundant number, then for all q with $2 \leq q \leq p$ we have*

$$\lfloor \log p/\log q \rfloor \leq a_q.$$

Proof To get a contradiction set $m := nq^{a_q+1}/p$ and suppose that

$$a_q \leq \left\lfloor \frac{\log p}{\log q} \right\rfloor - 1 \implies q^{a_q+1} < p \implies \frac{m}{n} = \frac{nq^{a_q+1}}{p} < 1.$$

Since n is superabundant we must have

$$1 < \frac{\sigma(n)/n}{\sigma(m)/m} = \frac{1}{1+1/q^{a_q+1}}\left(1+\frac{1}{p}\right) \implies p < q^{a_q+1},$$

which is a contradiction. Therefore $\lfloor \log p/\log q \rfloor \leq a_q$. This completes the proof. □

Note that for all integers $n > 1$ we have $n/\varphi(n) > \sigma(n)/n$. The two ratios in this inequality are a principal focus of this volume. For example they occur in Chapters 5, 7 and 9. It is interesting to see that for superabundant numbers the "gap" between the ratios has a nice expression in terms of the maximum prime divisor of n. The next result of Nazardonyavi and Yakubovich was in a preprint [122], but omitted in the final version [123].

Lemma 6.8 *Let n be a superabundant number with maximum prime divisor p and define*

$$\epsilon(p) := \frac{1}{\log p}\left(1 + \frac{3}{2\log p}\right).$$

Then

$$\frac{\sigma(n)}{n} > (1 - \epsilon(p))\frac{n}{\varphi(n)}.$$

Proof Consider $n = 2^{a_2} \cdots q^{a_q} \cdots p$ and note that

$$\frac{\sigma(n)}{\varphi(n)} > \prod_{q \leq p}\left(1 - \frac{1}{q^{a_q+1}}\right).$$

Take logarithms and use the inequality $\log(1 - 1/x) > -1/(x-1)$, which is valid for $x > 1$, and Lemma 6.7 to get

$$\begin{aligned}
\log\frac{\sigma(n)}{\varphi(n)} &> \sum_{q \leq p}\log\left(1 - \frac{1}{q^{a_q+1}}\right)\\
&> -\sum_{q \leq p}\frac{1}{q^{\log p/\log q} - 1} = -\sum_{q \leq p}\frac{1}{p-1}\\
&= -\frac{\pi(p)}{p-1} > -\frac{p}{(p-1)\log p}\left(1 + \frac{1.2762}{\log p}\right)\\
&> -\frac{1}{\log p}\left(1 + \frac{3}{2\log p}\right),
\end{aligned}$$

where for the final inequality we need $p \geq 23$ and we have used the estimate $\pi(x) < (x/\log x)(1 + 1.2762/\log x)$ which is valid for $x > 1$ [49]. Taking exponentials and using $e^{-x} > 1 - x$, which is true for all $x > 0$, we get

$$\frac{\sigma(n)}{\varphi(n)} > 1 - \frac{1}{\log p}\left(1 + \frac{3}{2\log p}\right) = 1 - \epsilon(p).$$

When $p < 23$ we note that the set of superabundant numbers with maximum prime in this range is finite, and we are thus able to complete the proof by computation; see Section 3.3. □

6.3 Colossally Abundant Numbers

The structure of the set of colossally abundant numbers is informed by the concept of *critical epsilon values*. This idea was introduced by Erdős and Nicolas [56] in their ground-breaking paper of 1975 on this subject. However, the essential facts were already present in the section on colossally abundant numbers in [3] published in 1944. The definition of colossally abundant number given in the earlier paper was a little different, but here we will give only the more modern version.

We give the definition which is now regarded as standard: recall that we say that N is a **colossally abundant number** if for some $\epsilon > 0$ the function $\sigma(n)/n^{1+\epsilon}$ attains its maximum at N. Therefore, for such an N and its associated ϵ we have for all $n \in \mathbb{N}$

$$\frac{\sigma(n)}{n^{1+\epsilon}} \le \frac{\sigma(N)}{N^{1+\epsilon}}.$$

For a given $\epsilon > 0$ there is at least one colossally abundant number for that given ϵ. This follows from bounds on $\sigma(n)/n$ which show $\sigma(n)/n^{1+\epsilon} \to 0$ as $n \to \infty$. We have already seen in the introduction to this chapter that every colossally abundant number is superabundant, so any result proved for that type of number can be used if needed.

Let p be a prime. Define

$$E_p := \left\{ \log_p \left(1 + \frac{1}{\sum_{1 \le i \le j} p^i} \right) : j \in \mathbb{N} \right\} \quad \text{and} \quad E := \bigcup_p E_p.$$

The set E can be written with elements in decreasing order:

$$\begin{aligned} E &= \left\{ \frac{\log(\frac{3}{2})}{\log(2)}, \frac{\log(\frac{4}{3})}{\log(3)}, \frac{\log(\frac{7}{6})}{\log(2)}, \frac{\log(\frac{6}{5})}{\log(5)}, \frac{\log(\frac{15}{14})}{\log(2)}, \frac{\log(\frac{13}{12})}{\log(3)}, \frac{\log(\frac{31}{30})}{\log(2)}, \dots \right\} \\ &= \{0.5849\dots, 0.2618\dots, 0.2223\dots, 0.1132\dots, 0.09953\dots, \\ &\qquad 0.07285\dots, 0.04730\dots, \dots\}. \end{aligned}$$

We enumerate the elements of E in this order $(\epsilon_i)_{i \in \mathbb{N}}$ so

$$\epsilon_1 = \log_2(\tfrac{3}{2}) > \epsilon_2 = \log_3(\tfrac{4}{3}) > \epsilon_3 = \log_2(\tfrac{7}{6}) > \epsilon_4 = \log_5(\tfrac{6}{5}) > \cdots$$

and set $\epsilon_0 := \infty$. The sequence E is called the **critical epsilon values**.

Let $\alpha \ge 0$ be an integer and p a prime. Define $F(p,0) := \infty$ and for $\alpha > 0$ set

$$F(p,\alpha) := \frac{\log[(p^{\alpha+1} - 1)/(p(p^\alpha - 1))]}{\log p} = \log_p \left(1 + \frac{1}{p^\alpha + \cdots + p} \right),$$

so $F(p,\alpha)$ is strictly decreasing as p and α increase (see Figure 6.2).

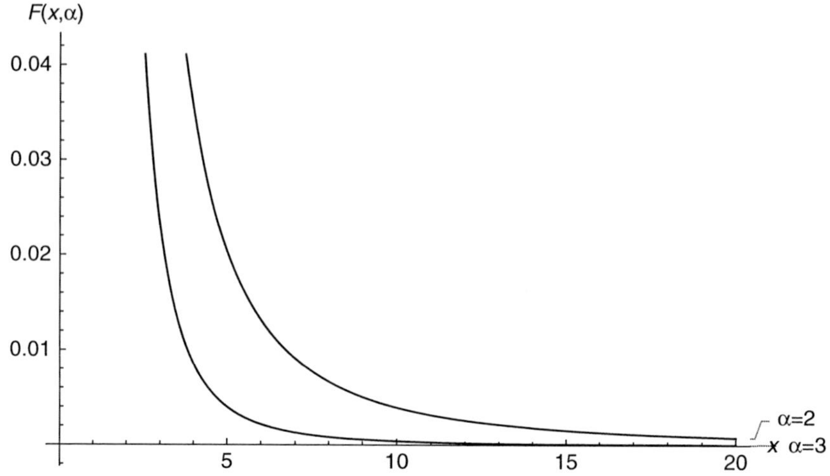

Figure 6.2 The functions $F(3,\alpha)$ and $F(2,\alpha)$ showing $F(3,\alpha) < F(2,\alpha)$.

Lemma 6.9 *[56, proposition 3] If N is colossally abundant for ϵ and p is any prime with $p^\alpha \| N$ then*

$$F(p,\alpha+1) \le \epsilon \le F(p,\alpha).$$

Conversely, if N is a positive integer such that for all primes p with $p^\alpha \| N$ we have $F(p,\alpha+1) \le \epsilon \le F(p,\alpha)$, then N is colossally abundant for ϵ.

Proof If $\alpha \ge 0$ let $n = Np$. Then

$$\frac{\sigma(Np)}{\sigma(N)} \le \left(\frac{Np}{N}\right)^{1+\epsilon} = p^{1+\epsilon}.$$

Also

$$\frac{\sigma(Np)}{\sigma(N)} = \frac{\sigma(p^{\alpha+1})}{\sigma(p^\alpha)} = \frac{p^{\alpha+2}-1}{p^{\alpha+1}-1} = p\left(1 + \frac{1}{p(p^\alpha + \cdots + 1)}\right),$$

so $1 + 1/(p(p^\alpha + \cdots + 1)) \le p^\epsilon$ which gives $F(p,\alpha+1) \le \epsilon$. Similarly if $\alpha > 0$ we can use $n = N/p$ to obtain $F(p,\alpha) \ge \epsilon$. This completes the proof. □

Note that we have $E_p = \{F(p,\alpha) : \alpha \ge 1\}$ and that for each $\eta > 0$ there are at most a finite number of elements of $E = \bigcup_p E_p$ larger than η.

Now define for given $\epsilon > 0$ a strictly decreasing sequence of positive real numbers $(x_k)_{\ge 1}$, depending on ϵ through the equations $F(x_k,k) = \epsilon$, and set $x = x_1$ and $y = x_2$.

Lemma 6.10 *[56, proposition 4(a)] If $\epsilon \notin E$, i.e. ϵ is* ***not*** *a critical epsilon value, then $\sigma(n)/n^{1+\epsilon}$ has a unique maximum at the colossally abundant* *integer denoted n_ϵ which has exponents for each prime q*

$$\alpha_q(\epsilon) = \left\lfloor \log_q\left(\frac{q^{1+\epsilon}-1}{q^\epsilon - 1}\right)\right\rfloor - 1.$$

In other words, $\alpha_q(\epsilon) = k$ when $x_{k+1} < q < x_k$ for $k \geq 1$ and $\alpha_q(\epsilon) = 0$ if $q > x_1$.

Proof Let q be a fixed prime number. Since $\epsilon \notin E_q$ and the function $F(q,k)$ is strictly decreasing in k, there exists a unique integer α such that

$$F(q,\alpha+1) < \epsilon < F(q,\alpha). \tag{6.2}$$

Solving these inequalities:

$$\begin{aligned}
1 + \frac{1}{q^{\alpha+1}+\cdots+q} &< q^\epsilon < 1 + \frac{1}{q^\alpha + \cdots + q}\\
\frac{1}{q^{\alpha+1}-1} &< \frac{q(q^\epsilon - 1)}{q-1} < \frac{1}{q^\alpha - 1}\\
q^\alpha - 1 &< \frac{q-1}{q(q^\epsilon-1)} < q^{\alpha+1} - 1\\
q^\alpha &< \frac{q^{\epsilon+1}-1}{q(q^\epsilon - 1)} < q^{\alpha+1}\\
q^{\alpha+1} &< \frac{q^{\epsilon+1}-1}{q^\epsilon - 1} < q^{\alpha+2}.
\end{aligned}$$

Therefore

$$\alpha + 1 < \log_q\left(\frac{q^{\epsilon+1}-1}{q^\epsilon-1}\right) < \alpha + 2 \quad \Longrightarrow \quad \alpha_q(\epsilon) = \left\lfloor \log_q\left(\frac{q^{1+\epsilon}-1}{q^\epsilon-1}\right)\right\rfloor - 1.$$

If $x_{k+1} < q < x_k$ we have using (6.2)

$$\begin{aligned}
F(x_k,\alpha+1) &< F(q,\alpha+1) < F(x_k,k) = \epsilon\\
&= F(x_{k+1},k+1) < F(q,\alpha) < F(x_{k+1},\alpha).
\end{aligned}$$

Thus $F(x_k,\alpha+1) < F(x_k,k)$ and $F(x_{k+1},k+1) < F(x_{k+1},\alpha)$. Therefore since $F(x,\alpha)$ is strictly decreasing in α, we get $k < \alpha+1$ and $\alpha < k+1$, so we must have $\alpha = k$, showing the alternative expression $\alpha_q(\epsilon) = k$ is true also.

To demonstrate the final part of the lemma, note that we have $\epsilon = F(x_1, 1)$ which implies $x_1^\epsilon = 1 + 1/x_1$ so

$$\log_{x_1}\left(\frac{x_1^{\epsilon+1}-1}{x_1^\epsilon - 1}\right) = \log_{x_1}\left(\frac{x_1(1+1/x_1)-1}{1/x_1}\right) = \log_{x_1} x_1^2 = 2.$$

But we have $q > x_1$ which gives $\log_q((q^{1+\epsilon}-1)/(q^\epsilon-1)) < 2$ and thus

$$\left\lfloor \log_q\left(\frac{q^{1+\epsilon}-1}{q^\epsilon-1}\right)\right\rfloor \leq 1,$$

so therefore $\alpha_q(\epsilon) = 0$.

The converse is immediate, via contradiction, because (6.2) determines the α and they determine N. This completes the proof. □

Using the expression of Lemma 6.10 it is easy to generate colossally abundant numbers. For example

$$\begin{aligned} \epsilon = 0.1 &\implies n_\epsilon = 2^2 \cdot 3 \cdots 5 \\ \epsilon = 0.01 &\implies n_\epsilon = 2^6 \cdot 3^2 \cdot 5^2 \cdot 7 \cdots 29 \\ \epsilon = 0.001 &\implies n_\epsilon = 2^9 \cdot 3^5 \cdot 5^3 \cdot 7^3 \cdot 11^2 \cdot 13^2 \cdot 17^2 \cdot 19 \cdots 181 \\ \epsilon = 0.0001 &\implies n_\epsilon = 2^{12} \cdot 3^7 \cdot 5^5 \cdot 7^4 \cdot 11^3 \cdot 17^2 \cdots 27^2 \cdot 53 \cdots 541 \end{aligned}$$

and

$$\epsilon = 0.00001 \implies$$
$$n_\epsilon = 2^{16} \cdot 3^{10} \cdot 5^6 \cdot 7^5 \cdot 11^4 \cdot 13^4 \cdot 17^3 \cdots 29^3 \cdot 31^2 \cdots 139^2 \cdot 149 \cdots 541,$$

where the symbol "$\cdots$" means a product of a contiguous set of primes in increasing order all to the same power as those on its left and right.

Lemma 6.11 *[56, proposition 4(b)] For each $i \geq 1$, n_ϵ is constant on the open interval $(\epsilon_i, \epsilon_{i-1})$. These constant values, for different values of i, are all distinct.*

Proof If ϵ_i is a critical epsilon value we have $\epsilon_i = F(q,\alpha)$ for some prime q and positive integer α. A calculation then shows that

$$1 + \alpha = \log_q \left(\frac{q^{1+\epsilon_i} - 1}{q^{\epsilon_i} - 1} \right),$$

so $g(\epsilon) := \log_q((q^{1+\epsilon_i} - 1)/(q^{\epsilon_i} - 1))$ is an integer. Similarly if for given q this value is an integer then the corresponding ϵ is a critical ϵ value. Therefore if $\epsilon_{i+1} < \epsilon_i$ are two successive critical epsilon values, for each q the value of $\alpha_q(\epsilon)$ does not change when $\epsilon \in (\epsilon_i, \epsilon_{i-1})$, so n_ϵ is constant on each of these intervals.

Because the function $g(\epsilon)$ is strictly increasing as ϵ decreases, if

$$\epsilon_{i+1} < \epsilon' < \epsilon_i < \epsilon < \epsilon_{i-1}$$

then $\alpha_q(\epsilon') > \alpha_q(\epsilon)$. This completes the proof. □

We denote by n_i the constant value of n_ϵ on the interval $(\epsilon_i, \epsilon_{i-1})$ so

$$n_1 = 1,\ n_2 = 2,\ n_3 = 6,\ n_4 = 12,\ n_5 = 60,\ \ldots$$

giving, by Lemma 6.10, the initial segment of the colossally abundant numbers.

It follows from Lemma 6.11 that $n_{i+1} > n_i$, so in particular the values of the (n_i) are all distinct. Therefore the number of the colossally abundant numbers corresponding to the sequence $(n_i)_{i\geq 1}$ is infinite.

Lemma 6.12 *[56, proposition 4(c)] If the sets $(E_q)_{q\in\mathbb{P}}$ are pairwise disjoint, the set of colossally abundant numbers is equal to the set of n_i with $i \geq 1$. The function $n \to \sigma(n)/n^{1+\epsilon_i}$ attains its maximum at the two points n_i and n_{i+1}.*

Proof Choose $\epsilon = \epsilon_i = F(q,\beta)$. For $p \neq q$, since the E_q are pairwise disjoint, $\epsilon \notin E_p$ so $\alpha_p(\epsilon)$ is given by the expression of Lemma 6.10. By Lemma 6.9 the exponent of q can be chosen to be β or $\beta - 1$. In the first case we get n_{i+1} and in the second n_i with

$$q^{1+\epsilon_i} = \frac{q^{\beta+1}-1}{q^{\beta}-1} \quad \Longleftrightarrow \quad \frac{\sigma(n_i)}{n_i^{1+\epsilon_i}} = \frac{\sigma(n_{i+1})}{n_{i+1}^{1+\epsilon_i}},$$

so the function attains its maximum at these two points. □

Figure 6.3 shows two maxima at $n_4 = 12$ and $n_5 = 60$. Figure 6.4, with $\epsilon = 0.1$ such that

$$\frac{\log(\frac{15}{14})}{\log 2} < \epsilon < \frac{\log(\frac{6}{5})}{\log 5} \quad \Longrightarrow \quad \epsilon \in (\epsilon_5, \epsilon_4),$$

shows one maximum at the colossally abundant number $n_5 = 60$.

Lemma 6.13 *[56, proposition 4(d)] If the sets $(E_q)_{q\in\mathbb{P}}$ are not pairwise disjoint, for each $\epsilon_i \in E_q \cap E_r$, the function $n \to \sigma(n)/n^{1+\epsilon_i}$ attains its maximum at the four points*

$$n_i,\ qn_i,\ rn_i,\ qrn_i = n_{i+1}.$$

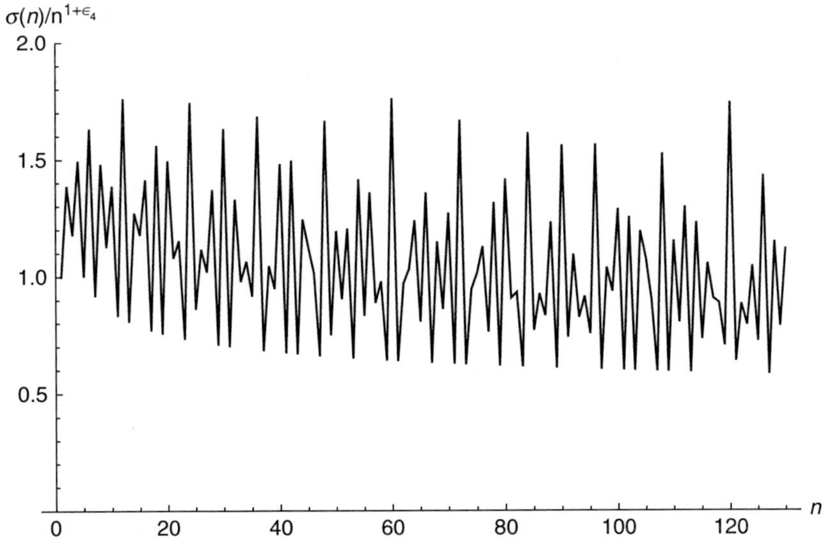

Figure 6.3 The function $n \to \sigma(n)/n^{1+\epsilon_4}$.

Table 6.1 Indexing for the first eight colossally abundant numbers.

i	n_i	ϵ_i	Value of ϵ_i	$\sigma(n_i)/n_i^{1+\epsilon_i}$
1	1	$\log_2(3/2)$	0.5849	1.00000
2	2	$\log_3(4/3)$	0.2618	1.25102
3	6	$\log_2(7/6)$	0.2223	1.34269
4	12	$\log_5(6/5)$	0.1132	1.76086
5	60	$\log_2(15/14)$	0.09953	1.86281
6	120	$\log_3(13/12)$	0.07285	2.11659
7	360	$\log_2(31/30)$	0.04730	2.46012
8	2520	$\log_2(63/62)$	0.02308	3.09998

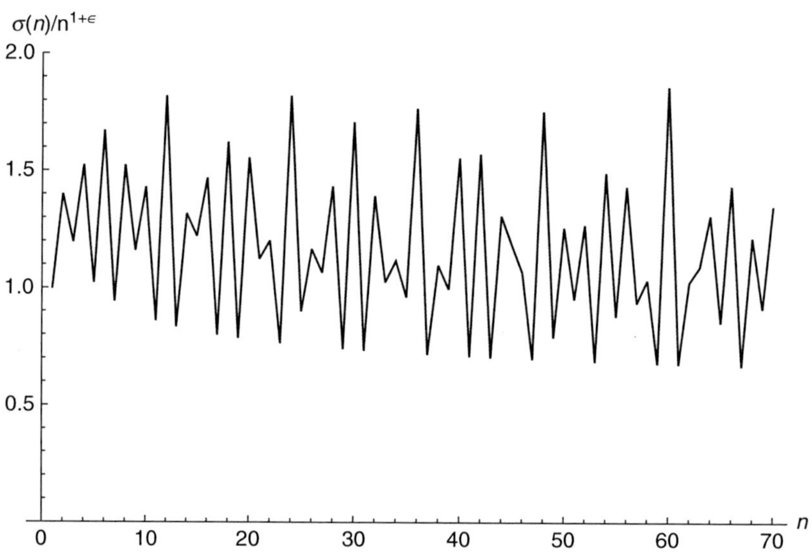

Figure 6.4 The function $n \to \sigma(n)/n^{1+0.1}$.

Proof Let $\epsilon = \epsilon_i = F(p, \alpha) \in E$. First we **claim** ϵ is irrational. If not we could write $\epsilon = a/b$ for integers a and b so

$$p^a = \left(1 + \frac{1}{p^\alpha + \cdots + p}\right)^b,$$

so the left-hand side is an integer and the right-hand side not an integer. (The theorem of Gel'fond and Schneider, that if α and β are algebraic numbers with $\alpha \neq 0$ or 1 and if β is not a rational number, then any value of α^β is a transcendental number, would show that in fact ϵ is a transcendental number.)

The theorem of Lang, given in Lemma 6.14 using

$$(u,v,w) = (\log p, \log q, \log r),$$

$a = \epsilon$ and $b = 1$, implies that if p,q,r are distinct prime numbers such that p^{ϵ}, q^{ϵ} and r^{ϵ} are all algebraic then ϵ would be rational, which is impossible. But if $\epsilon \in E_p$, p^{ϵ} is rational, so we conclude that $E_p \cap E_q \cap E_r = \emptyset$.

If there exist two sets E_q and E_r which are not disjoint and we choose

$$\epsilon_i = F(q,\beta) = F(r,\gamma) \in E_q \cap E_r$$

and a prime p satisfying $p \neq q$ and $p \neq r$, the maximum power of p dividing N is determined by Lemma 6.10. The power of q is β or $\beta - 1$ and that of r is γ or $\gamma - 1$, giving by Lemma 6.9 the four given possibilities. This completes the proof. □

Note unsolved problem (2) at the end of this chapter: no pair of distinct primes p,q are known with $E_p \cap E_q \neq \emptyset$, so we are not able to illustrate Lemma 6.13.

Lemma 6.14 *[106, chapter 2] If u,v,w are three complex numbers, linearly independent over $\mathbb{Q}$, and a,b are two complex numbers also linearly independent over $\mathbb{Q}$, then at least one of the six numbers*

$$e^{au},\ e^{av},\ e^{aw},\ e^{bu},\ e^{bv},\ e^{bw}$$

is a transcendental number.

Lemma 6.15 *The quotient of two consecutive colossally abundant numbers is either a prime or the product of two distinct primes.*

Proof **(1)** Let n and n' with $n < n'$ be two consecutive colossally abundant numbers with associated epsilon values ϵ and ϵ', so for some $i \in \mathbb{N}$ there is a critical epsilon value ϵ_i with $\epsilon' < \epsilon_i < \epsilon$. Then, with the notation of Lemma 6.11 write

$$n = \prod_p p^{\alpha_p(\epsilon)} \quad \text{and} \quad n' = \prod_p p^{\alpha_p(\epsilon')},$$

where the products are over all primes, and for all p, $0 \le \alpha_p(\epsilon) \le \alpha_p(\epsilon')$.

(2) For each prime p define a function on $(0,1)$ by

$$f_p(t) := \log_p\left(\frac{p^{1+t}-1}{p^t-1}\right),$$

and note that $f_p(t)$ increases as t decreases, and is a continuous function of t.

(3) By Lemma 6.10 we have $\alpha_p(\epsilon) = \lfloor f_p(\epsilon) \rfloor - 1$ and $\alpha_p(\epsilon') = \lfloor f_p(\epsilon') \rfloor - 1$. For fixed p, as t decreases from ϵ to ϵ', since n and n' are consecutive colossally abundant numbers and $f_p(t)$ is continuous, $\alpha_p(\epsilon')$ must differ from

$\alpha_p(\epsilon)$ by at most one, and in such a case where they differ, $f_p(\epsilon_i)$ must be a positive integer greater than 1, say $f_p(\epsilon_i) = m$. This means

$$\frac{p^{\epsilon_i+1} - 1}{p^{\epsilon_i} - 1} = p^m \quad \Longrightarrow \quad p^{\epsilon_i} = \frac{p^m - 1}{p^m - p},$$

which is a rational number.

(4) If this exponent jump occurs at three or more distinct primes p, q, r then we would have a real ϵ_i with p^{ϵ_i}, q^{ϵ_i} and r^{ϵ_i} each rational. But again by Lemma 6.14, this is impossible. This completes the proof. □

A long-standing conjecture of Erdős is that if p and q are distinct odd primes with real $\epsilon > 0$ being such that p^ϵ and q^ϵ are both rational, then ϵ is an integer. We use the result of this conjecture in RHpack to generate colossally abundant numbers using the function GenerateCA.

Now recall the definition

$$G(n) := \frac{\sigma(n)}{n \log\log n}.$$

The following lemma, originally due to Robin, is presented with a different proof. It was used in crucial ways by Briggs in [16] to extend the lower bound for any counterexample to Robin's inequality. It is also needed in the chapters which follow.

Theorem 6.16 *[142, proposition 1] If $M < N$ are consecutive colossally abundant numbers and $M \le n \le N$ then*

$$G(n) \le \max(G(M), G(N)).$$

Proof By Lemmas 6.12 and 6.13 there exists an index $i \ge 1$ such that $M = n_i$ and $N = n_{i+1}$ and such that these two numbers are critical points giving the same maximum value for the function $n \to \sigma(n)/n^{1+\epsilon_i}$.

The function $f(x) := \epsilon_i x - \log\log x$ is concave upwards for $x \ge 2$, so for $2 \le a \le \xi \le b$ we have $f(\xi) \le \max(f(a), f(b))$. Because $n_i \le n \le n_{i+1}$ this means

$$\begin{aligned}&\epsilon_i \log n - \log\log\log\log n \\ &\qquad \le \max(\epsilon_i \log n_i - \log\log\log\log n_i,\ \epsilon_{i+1} \log n_{i+1} - \log\log\log\log n_{i+1}),\end{aligned}$$

or taking exponentials

$$\frac{n^{\epsilon_i}}{\log\log n} \le \max\left(\frac{n_i^{\epsilon_i}}{\log\log n_i}, \frac{n_{i+1}^{\epsilon_i}}{\log\log n_{i+1}}\right). \tag{6.3}$$

But n_i satisfies $\sigma(n)/n^{1+\epsilon_i} \le \sigma(n_i)/n_i^{1+\epsilon_i}$ so

$$G(n) = \frac{\sigma(n)}{n \log\log n} \le \frac{\sigma(n_i)}{n_i^{1+\epsilon_i}} \frac{n^{\epsilon_i}}{\log\log n} = G(n_i) \frac{\log\log n_i}{\log\log n} \left(\frac{n}{n_i}\right)^{\epsilon_i} \tag{6.4}$$

and the same inequality holds with n_i replaced by n_{i+1}.

If we had $G(n) > \max(G(n_i), G(n_{i+1}))$, by (6.4) we would have

$$1 < \frac{\log\log n_i}{\log\log n} \left(\frac{n}{n_i}\right)^{\epsilon_i} \quad \text{and} \quad 1 < \frac{\log\log n_{i+1}}{\log\log n} \left(\frac{n}{n_{i+1}}\right)^{\epsilon_i}$$

and therefore

$$\max\left(\frac{n_i^{\epsilon_i}}{\log\log n_i}, \frac{n_{i+1}^{\epsilon_i}}{\log\log n_{i+1}}\right) < \frac{n^{\epsilon_i}}{\log\log n},$$

contradicting (6.3). Hence $G(n) \le \max(G(n_i), G(n_{i+1}))$. This completes the proof. □

6.4 Estimates for $x_2(\epsilon)$

For what follows in Chapter 7 some explicit estimates for the numbers x_i, introduced in Section 6.3, are needed. Erdős and Nicolas [56, p. 74] prove that, for $k \ge 1$ a fixed integer, we have

$$x_k^{1/k} \sim kx \quad \text{and} \quad x_2 = \sqrt{2x}\left(1 - \frac{\log 2}{2\log x} + O\left(\frac{1}{\log^2 x}\right)\right). \tag{6.5}$$

Recall the notation $x = x_1 = x_1(\epsilon)$. Here are given explicit inequalities relating to these asymptotic forms in the following lemma.

Lemma 6.17 *[142] Let $\epsilon > 0$ not be a critical epsilon value with $x_i := x_i(\epsilon)$ for $i \in \mathbb{N}$. Then*

$$x^{1/k} < x_k \qquad (x > 1,\ k \ge 2), \tag{6.6}$$

$$x_2 < \sqrt{2x} \qquad (x > 1), \tag{6.7}$$

$$\sqrt{2x}\left(1 - \frac{\log 2}{2\log x}\right) < x_2 \qquad (x \ge 1530). \tag{6.8}$$

Proof Setting $t = x^{1/k}$, we note that proving inequality (6.6) is equivalent to proving $F(x,1) = F(x_k,k) < F(t,k)$. In other words

$$\log\left(1 + \frac{1}{t^k}\right) < k \log\left(1 + \frac{1}{t + t^2 + \cdots + t^k}\right). \tag{6.9}$$

For $u > 0$ we have $1 + ku < (1+u)^k$. Hence

$$1 + \frac{1}{t^k} < \left(1 + \frac{1}{(t + t^2 + \cdots + t^k)/k}\right) < \left(1 + \frac{1}{t + t^2 + \cdots + t^k}\right)^k. \tag{6.10}$$

Thus inequality (6.10) holds, and hence inequality (6.6) is true.

To derive inequality (6.7), since it is equivalent to $F(\sqrt{2x},2) < F(x,1) = F(x_2,2)$, we must show that

$$\frac{2\log\left(1+\frac{1}{2x+\sqrt{2x}}\right)}{\log x+\log 2} < \frac{\log\left(1+\frac{1}{x}\right)}{\log x}. \tag{6.11}$$

In fact, we **claim** that

$$2\log\left(1+\frac{1}{2x+\sqrt{2x}}\right) < \log\left(1+\frac{1}{x}\right), \tag{6.12}$$

from which inequality (6.11) follows directly.

To prove inequality (6.12), substitute $y = \sqrt{2x}$ and simplify:

$$\left(1+\frac{1}{2x+\sqrt{2x}}\right)^2 \leq 1+\frac{1}{x} \quad (x>0). \tag{6.13}$$

To derive inequality (6.8) we first set $z = \sqrt{2x}(1-u/2)$ where $u = \log 2/\log x$. Because the given inequality is equivalent to $F(x_1,1) = F(x_2,2) < F(z,2)$, we need to show

$$\frac{\log\left(1+\frac{1}{x}\right)}{\log x} < \frac{\log\left(1+\frac{1}{z+z^2}\right)}{\log z}. \tag{6.14}$$

Using the inequalities $1/(a+1) < \log(1+1/a) < 1/a,\ a>1$, we find that, to prove inequality (6.14) it is sufficient to show that

$$\frac{1}{x\log x}\log\left(\sqrt{2x}\left(1-\frac{u}{2}\right)\right) < \frac{1}{1+z^2+z}, \tag{6.15}$$

which will follow if

$$(1+z+z^2)\left(1+u-\frac{u^2}{\log 2}\right) < 2x. \tag{6.16}$$

Consider

$$z^2\left(1+u-\frac{u^2}{\log 2}\right) = 2x\left(1-\frac{u}{2}\right)^2\left(1+u-\frac{u^2}{\log 2}\right). \tag{6.17}$$

For $x \geq 2^{12}$ we have $u \leq 1/12$, so the right-hand side, and hence the left-hand side, of (6.17) is bounded above by $2x(1-2.05u^2)$.

The other part of (6.16) can be bounded, since $0 < u \leq 1/12$ and $x \geq 2^{12}$, as follows:

$$(1+z)\left(1+u-\frac{u^2}{\log 2}\right) \leq \sqrt{2x}\left(1+\frac{u}{2}\right)+1+u \leq 1.06\sqrt{2x}. \tag{6.18}$$

Therefore

$$(1+z+z^2)\left(1+u-\frac{u^2}{\log 2}\right) \le 2x(1-2.05u^2)+1.054\sqrt{2x}$$
$$= 2x+1.06\sqrt{2x}-4.1x\left(\frac{\log 2}{\log x}\right)^2 < 2x,$$

where the final inequality follows since for $x \ge 2^{12}$ we have

$$1.06\sqrt{2x}-4.1x\left(\frac{\log 2}{\log x}\right)^2 < 0.$$

This proves inequality (6.8) for all $x > 2^{12}$. For $1530 \le x \le 2^{12}$, inequality (6.14) can easily be verified numerically, so the proof of the lemma is complete. □

We also need the following with $k = 3$, stated in [56] to be proved along the same lines as lemma 2(b) in [3].

Lemma 6.18 *As $\epsilon \to 0$ we have $x_k \sim (kx_1)^{1/k}$.*

Proof We have for $x = x_1$ and $y = x_k$

$$\epsilon = \frac{\log\left(1+\dfrac{1}{y+y^2+\cdots+y^k}\right)}{\log y} = \frac{\log\left(1+\dfrac{1}{x}\right)}{\log x}.$$

Note that $\epsilon \to 0+$ if and only if $x \to \infty$ if and only if $y \to \infty$. Hence, using $\log(1+1/x) \sim 1/x$ as $x \to \infty$ we get

$$y^k \log y \sim x \log x. \tag{6.19}$$

Taking logarithms gives $k\log y \sim \log x$. Inserting this in (6.19) then gives $y^k \sim kx$, completing the proof of the lemma. □

6.5 Unsolved Problems

(1) Show that the ratio of any two consecutive colossally abundant numbers is always prime – see Table A.5.
(2) Show that the sets $(E_q)_{q\in\mathbb{P}})$ are disjoint (Robin [142]).
(3) If p and q are distinct primes with x real and p^x and q^x simultaneously rational, then x is an integer (Erdős and Alaoglu [3]).
(4) Erdős and Alaoglu [3] state that $q > 11$ implies $q^{a_q} < 2^{a_2}$ when n is superabundant. Generalize this result.
(5) If n is superabundant then we have $q^{a_q} \sim p\log p/\log q$ when $q \to \infty$ and $p \sim \log n$ when $n \to \infty$ [3].

(6) For fixed k, the function $t \rightarrow F(t,k)$ is strictly decreasing for all $t > 0$, and tends to zero when $t \rightarrow \infty$. Therefore, as $k \rightarrow \infty$, if n_ϵ is colossally abundant with $x_k := x_k(\epsilon)$, $k \in \mathbb{N}$,

$$x_k \sim (kx)^{1/k}.$$

This result is in [3]. Prove it and then give an error estimate.

(7) It was proved in [3] that the number of superabundant numbers $N(x)$ up to $x > 0$ satisfies

$$N(x) \ll \frac{\log x \log\log x}{(\log\log\log x)^2}.$$

Improve this bound and find a lower bound for $N(x)$.

(8) Find "book proofs" for some of the more difficult theorems and lemmas in [3].

(9) Find an RH equivalence expressible in terms of superabundant numbers.

7
Robin's Theorem

7.1 Introduction

This chapter has as its focus Robin's theorem, an explicit inequality involving the sum-of-divisors function, valid on an explicit range, its validity being equivalent to RH. It forms part of what we call the **Ramanujan–Robin criterion**. We include Ramanujan because this great Indian mathematician first demonstrated the truth of the inequality assuming RH, albeit the asymptotic form.

The life of Srinivasa Ramanujan (pronounced Ra-manu-jan) requires no introduction, since it has been celebrated in books, plays and film. His vast mathematical contributions have been treated in depth in the monumental works of Bruce Berndt, including [10].

Guy Robin was supervised by Jean-Louis Nicolas (the Nicolas of Chapter 5) at the Université de Limoges, and received his Ph.D. in 1983 with a thesis entitled "Grandes valeurs de fonctions arithmétiques et problèmes d'optimisation en nombres entiers". One of his students, Pierre Dusart, is making important contributions to the ongoing development of this approach to RH. According to [128], Robin did not learn of Ramanujan's unpublished result, showing Robin's inequality was necessary if RH was true, until several years after he had discovered his theorem and proof.

We have already seen some introductory and historical information regarding $\sigma(n)$ in Chapter 6. It should come as no surprise that, because of the close relationships between the sum-of-divisors function and Euler's totient function $\varphi(n)$, and the equivalences to RH developed in Chapter 5, there are equivalences to RH which can be expressed in terms of $\sigma(n)$.

Paul Erdős was interested in common properties of $\sigma(n)$ and $\varphi(n)$ throughout his career, and authored or co-authored many papers exploring this connection. For example Alaoglu and Erdős [3] state that both $\sigma(n)$ and $\varphi(n)$, for every $\epsilon > 0$, apart from a set of density zero, are divisible by every prime less than $(\log\log n)^{1-\epsilon}$ and by "relatively few" primes larger than $(\log\log n)^{1+\epsilon}$.

Aside from being multinomials in the prime factors of n, and obeying the inequality $\sigma(n)/n \leq n/\varphi(n)$, at first glance these functions have very little in common. Their definitions and genesis are completely different. Nevertheless, properties of one function are frequently mirrored by the properties of the other one. As an illustration, their average orders satisfy

$$\sum_{n\leq x}\sigma(n) \asymp \sum_{n\leq x}\varphi(n),$$

where $f(x) \asymp g(x)$, for positive functions $f(x)$ and $g(x)$, means that there are numbers $\alpha > 0$, $\beta > 0$ with, as $x \to \infty$, $\alpha g(x) \leq f(x) \leq \beta g(x)$. Moreover the same is true [53, 58] for the size of the sets of values

$$\#\{\sigma(n) : n \leq x\} \asymp \#\{\varphi(n) : n \leq x\}.$$

In addition, the normal order of $\omega(\sigma(n))$ is the same as the normal order of $\omega(\varphi(n))$, namely $(\log\log n)^2/2$. Recall that we say arithmetic function $f(n)$ has **normal order** $\rho(n)$, for some non-negative non-decreasing function $\rho(n)$, if for every $\epsilon > 0$ we have

$$\lim_{x\to\infty} \frac{\#\left\{n \leq x : \left|\dfrac{f(n)}{\rho(n)} - 1\right| < \epsilon\right\}}{x} = 1.$$

See [75] or [95] for the development of this topic.

Work has also been done on special values of these functions; for example see [27].

Recently there has been much work on "joint properties" shared by these two functions. For instance there exists an infinite number of squarefree integers n such that $\sigma(n)\varphi(n)$ is a square [29]. In a similar vein Ford, Luca and Pomerance [61] proved that the two functions have an infinite number of values in common, that is $\sigma(a) = \varphi(a')$ for an infinite set of integer pairs (a,a') (see also Ford and Pollack [62, 63]). The case of when $\varphi(n)$ divides $\sigma(n)$ has also been explored [28, 30].

Now we focus on the material in this chapter. Consider the growth of $\sigma(n)/n$. On the one hand, whenever n is a prime we have $\sigma(n)/n = 1 + 1/n$, so the range is arbitrarily close to 1. On the other hand, $\sigma(n)$ cannot grow too quickly. Grönwall [69] showed that

$$\limsup_{n\to\infty} \frac{\sigma(n)}{n\log\log n} = e^{\gamma}. \tag{7.1}$$

We do not use this result in this chapter, but it is proved and used in Chapter 9 as Theorem 9.2. What is interesting is the manner in which this limit superior is approached. Ramanujan [136], assuming the Riemann hypothesis, proved that $\sigma(n)/(n\log\log n)$ is asymptotically bounded above by e^{γ}. That is he showed:

Theorem 7.2 *[136, 137] Assume the Riemann hypothesis is true. Then there exists an* $n_0 \geq 1$ *such that*

$$\frac{\sigma(n)}{n \log\log n} < e^{\gamma}, \tag{7.2}$$

for all $n \geq n_0$.

Robin [142], using explicit estimates for arithmetic functions given by Rosser and Schoenfeld [145, 153, 154], set out in Chapter 3, gave an explicit value of n_0 in Theorem 7.2. In [142, proposition 3, p. 204], he proves:

Theorem 7.11 *[142, theorem 1] Assume the Riemann hypothesis is true. Then*

$$\sigma(n) < e^{\gamma} n \log\log n \quad (n \geq 5041). \tag{7.3}$$

Moreover the values of $1 < n \leq 5040$ *for which the inequality in (7.3) fails are listed in the set A, where*

$$A := \{1,2,3,4,5,6,8,9,10,12,16,18,20,24,30,36,48,\\ 60,72,84,120,180,240,360,720,840,2520,5040\}. \tag{7.4}$$

The inequality in expression (7.3) is known as "Robin's inequality". Note that in this volume the term does not include the lower bound $n \geq 5041$ for the range of validity, which can sometimes cause confusion. We attempt to make the range explicit whenever the term is used.

Since Grönwall's result (7.1) is unconditional, one may expect an explicit inequality similar to inequality (7.2) in which limit of the right-hand side was e^{γ}. Robin proved such a result (with a different constant) in:

Theorem 7.13 *[142, theorem 2]*

$$\frac{\sigma(n)}{n} \leq e^{\gamma} \log\log n + \frac{2}{3 \log\log n} \quad (n \geq 3). \tag{7.5}$$

One can compare Ramanujan's theorem, Theorem 7.2, and Robin's theorem, Theorem 7.11, to see that, were Robin's inequality (7.2) to fail for some $n \geq 5041$, it could only fail slightly.

However, inequality in (7.3) fails infinitely often if the Riemann hypothesis is false! Robin [142, section 4] uses an oscillation result first proved by Nicolas [126] to prove the following:

Theorem 7.15 *[142, theorem 1] Assume the Riemann hypothesis is false, and let* θ *be the supremum over all real parts of the non-trivial zeros* $\rho = \beta + i\gamma$, *so* $\theta > \frac{1}{2}$. *For any number* $b \in (1-\theta, \frac{1}{2})$, *there exists a positive constant c such that*

$$\frac{\sigma(n)}{n \log\log n} > e^{\gamma}\left(1 + \frac{c}{(\log n)^b}\right), \tag{7.6}$$

for infinitely many values of n.

One may therefore combine Theorems 7.11 and 7.15 to give the following equivalence for RH, which is known also as "Robin's theorem":

Theorem 7.16 (Ramanujan–Robin criterion) *[142, theorem 1] A necessary and sufficient condition for the Riemann hypothesis is that*

$$\sigma(n) < e^{\gamma} n \log\log n \quad (n \geq 5041). \tag{7.7}$$

Because of the function $\sigma(n)$, the ratio of the left side to the right side of inequality (7.3) is far from smooth.

We see in Figure 7.1 that, even for small values of $n > 5040$, the ratio is frequently close to 1.0.

Lagarias [100] found a modification of Theorem 7.16, namely:

Theorem 7.18 (Lagarias' criterion) *[100, theorem 1.1] Let* $H_n = \sum_{j=1}^{n} j^{-1}$ *represent the nth harmonic number in* $\mathbb{Q}$*. A necessary and sufficient condition for the Riemann hypothesis is that*

$$\sigma(n) \leq H_n + \exp(H_n)\log(H_n) \tag{7.8}$$

is true for all $n \geq 1$.

To summarize this chapter: Section 7.2 gives the proof of Ramanujan's theorem, which assumes RH. In Section 7.3 a lower bound for $\prod_{p \leq x}(1 - 1/p)$ is derived. Section 7.4 gives an upper bound for $\prod_{p \leq x}(1 - 1/p^2)$. In Section 7.5 there is a lower bound for $\log\log N$ and in Section 7.8 an

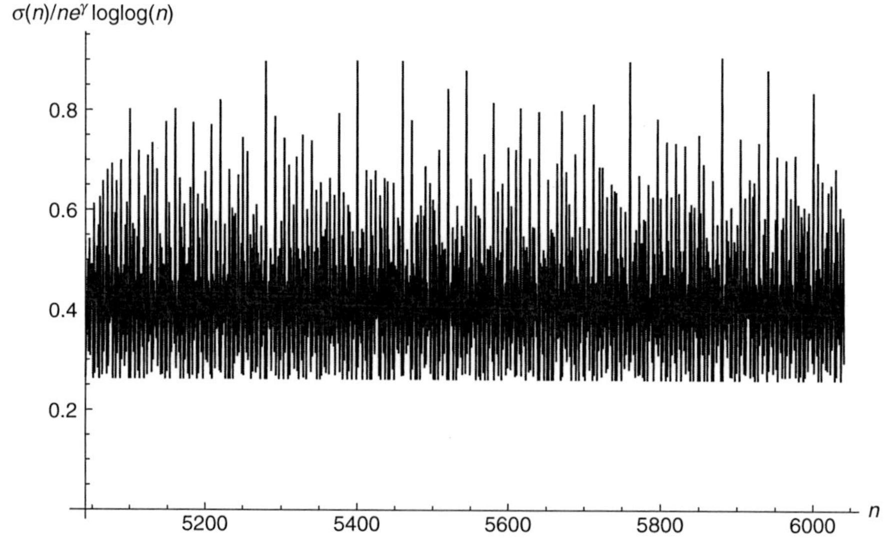

Figure 7.1 The values of $\sigma(n)/(ne^{\gamma}\log\log n)$ for $5041 \leq n \leq 6041$.

Figure 7.2 Timothy Trudgian.

upper bound. Section 7.6 includes the proof of Robin's theorem in the case RH is true. Section 7.7 derives Robin's unconditional upper bound for $\sigma(n)/n$. Section 7.9 gives the proof of Robin's theorem when RH is false. In Section 7.10, we give the proof of Lagarias' theorem, which depends on Robin's theorem. In Section 7.11 an unconditional upper bound related to Lagarias' formulation is presented, and in the final Section 7.12, some ideas relating to the sum of unitary divisors of an integer are discussed.

The substantial contributions of Tim Trudgian (Figure 7.2) to this volume, especially this essential chapter, are gratefully acknowledged.

The following RHpack (see Appendix B) functions relate to the material in this chapter: RobinsInequalityQ, PlotRobinsInequality, SigmaOverN, LagariasInequalityQ and UnitaryDivisorSigma.

To obtain an overview of the dependences between the results in this chapter, the reader is invited to consult Figures 7.3, 7.4, 7.5 and 7.6.

7.2 Ramanujan's Theorem Assuming RH

Before preceding to a proof of Theorem 7.11 we first prove Theorem 7.2. Thereafter, even though a moderate amount of work is required, all that we need do is to calculate certain error terms explicitly. Recall that we have defined

$$G(n) := \sigma(n)/(n\log\log n), \quad n > e.$$

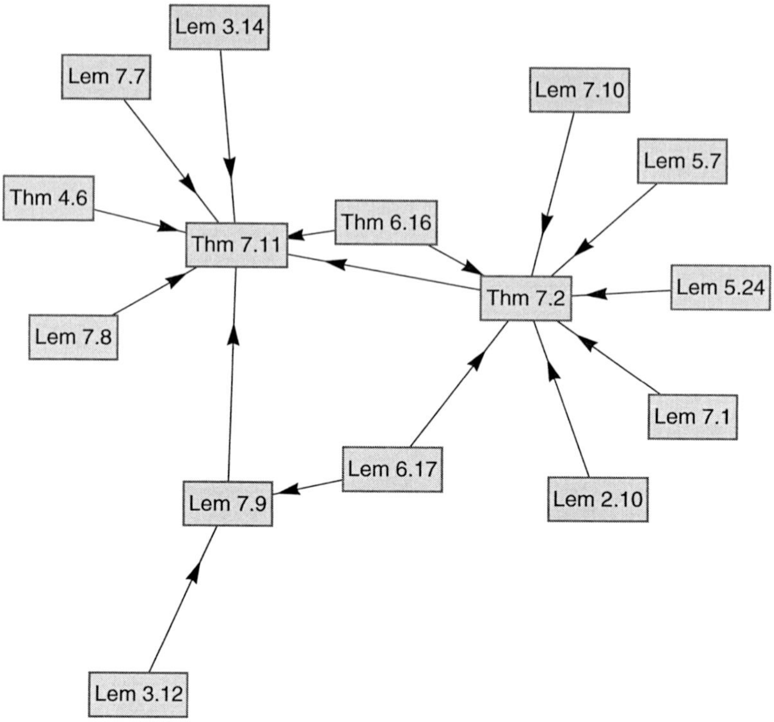

Figure 7.3 Relationships between some of the results in Chapter 7.

First a simple but essential lemma:

Lemma 7.1 *If a counterexample to Robin's inequality exists for some* $n >$ 5040*, then so does a counterexample which is colossally abundant.*

Proof Let N be a counterexample to Robin's inequality with $N > 5040$. It follows from the results of Chapter 6 that there are an infinite number of colossally abundant numbers (n_i) with $n_i \to \infty$. If each one satisfied Robin's inequality then $G(n_i) < e^\gamma$ for all i. Thus for some i we would have $n_i \leq N \leq n_{i+1}$. By Theorem 6.16, we would have

$$G(N) \leq \max(G(n_i), G(n_{i+1})) < e^\gamma,$$

so N would also satisfy Robin's inequality, which is false. Hence at least one of the n_i cannot satisfy Robin's inequality, giving a colossally abundant counterexample. This completes the proof. □

Looking ahead, Lemma 8.9 of Akbary and Friggstad will show that if a counterexample exists to Robin's inequality greater than 5040, then the first such example must be a superabundant number.

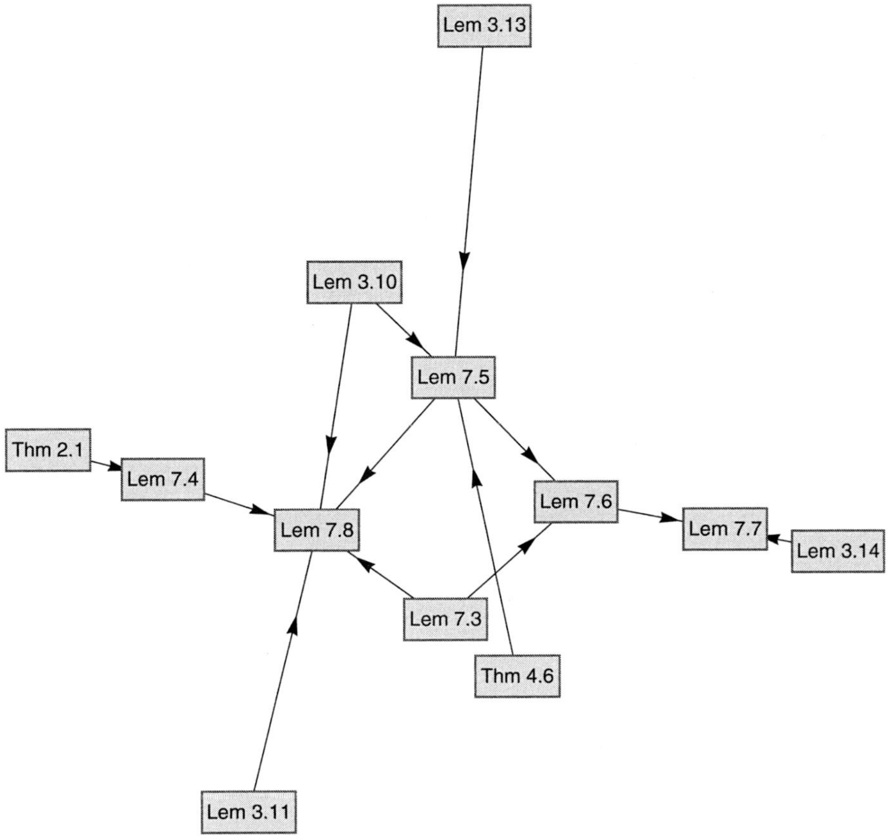

Figure 7.4 Further relationships between some of the results in Chapter 7.

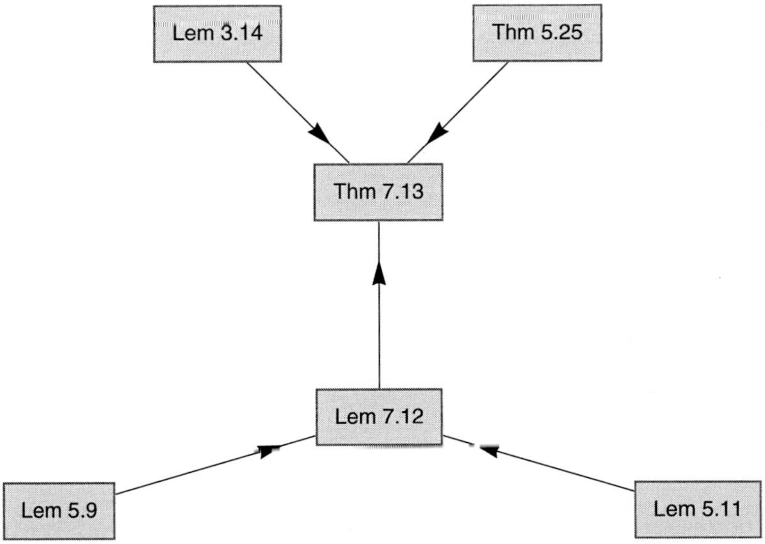

Figure 7.5 Additional relationships between some of the results in Chapter 7.

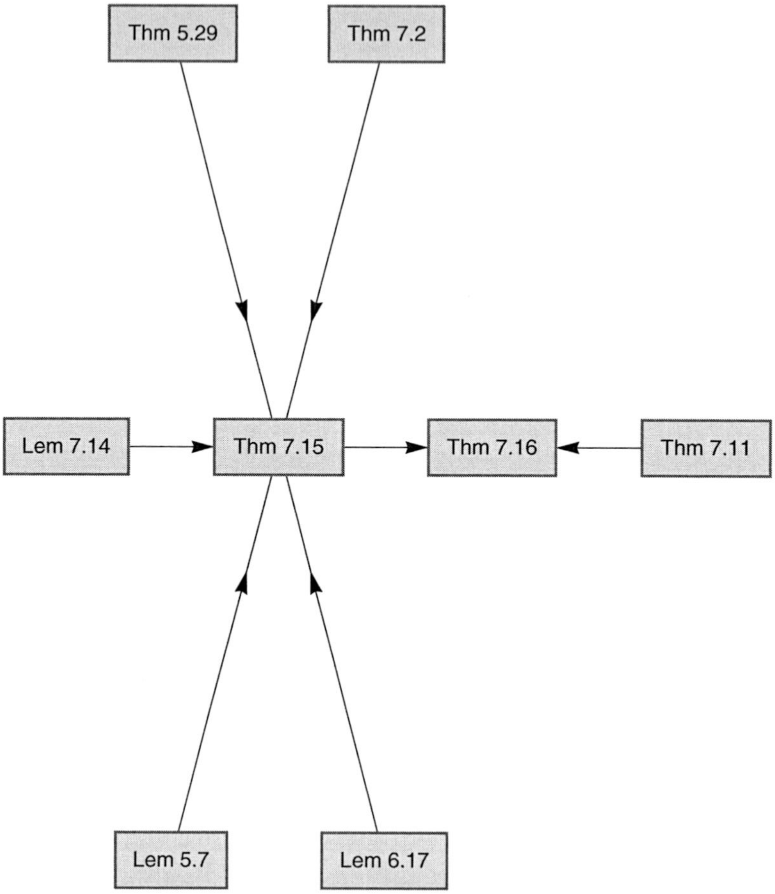

Figure 7.6 Further relationships between some of the results in Chapter 7.

Now we present the statement of Theorem 7.2. Note that the proof depends on the result of Lemma 7.10, which is proved in Section 7.5.

Theorem 7.2 (Ramanujan) *[136, 137] Assume the Riemann hypothesis is true. Then there exists an* $n_0 \geq 1$ *such that*

$$\frac{\sigma(n)}{n \log\log n} < e^{\gamma}, \tag{7.9}$$

for all $n \geq n_0$.

Proof Assume there exists an arbitrarily large counterexample N to inequality (7.9), which by Lemma 7.1 is colossally abundant. Then, by Theorem 6.16 and the results of Chapter 6 we have

$$\frac{\sigma(N)}{N} = \prod_{x_2 < p \leq x} \left(1 + \frac{1}{p}\right) \prod_{x_3 < p \leq x_2} \left(1 + \frac{1}{p} + \frac{1}{p^2}\right) \times \cdots, \tag{7.10}$$

where the x_i are defined in Chapter 6. By Lemma 6.17, we have $x_2 < \sqrt{2x}$, so that we can bound (7.10) by

$$\frac{\sigma(N)}{N} \le \frac{\prod_{x_2<p\le x}(1-1/p^2)}{\prod_{p\le x}(1-1/p)} \le \frac{\prod_{\sqrt{2x}<p\le x}(1-1/p^2)}{\prod_{p\le x}(1-1/p)}. \tag{7.11}$$

We now proceed to estimate the numerator in the right-hand side of (7.11). By Lemma 5.7 with $\alpha = 2$ we get

$$\sum_{p>x} \frac{1}{p^2} = \frac{1}{x\log x} + O\left(\frac{1}{x\log^2 x}\right).$$

Using the Maclaurin series for $-\log(1-z)$ and neglecting all terms other than the first two we see that

$$\begin{aligned}\prod_{\sqrt{2x}<p\le x}\left(1-\frac{1}{p^2}\right) &= \exp\left(-\sum_{\sqrt{2x}<p\le x}\frac{1}{p^2} + O\left(\frac{1}{x^{3/2}}\right)\right)\\ &= \exp\left(\frac{-\sqrt{2}}{\sqrt{x}\log x} + O\left(\frac{1}{\sqrt{x}\log^2 x}\right)\right),\end{aligned} \tag{7.12}$$

which gives an upper bound for the numerator in (7.11). To bound the denominator below, one may use the result of Nicolas, Lemma 5.24, namely

$$\prod_{p\le x}\left(1-\frac{1}{p}\right) \ge \frac{e^{-\gamma}}{\log\theta(x)}\exp\left(\frac{-(2+\beta)}{\sqrt{x}\log x} + O\left(\frac{1}{\sqrt{x}\log^2 x}\right)\right), \tag{7.13}$$

where, because we are assuming RH, by Lemma 2.10,

$$\beta := \sum_{\rho}\frac{1}{\rho(1-\rho)} = \sum_{\rho}\frac{1}{|\rho|^2} = \gamma + 2 - \log 4\pi = 0.0461\ldots.$$

We now proceed to estimate $\log\log N$. Since N is colossally abundant we have, by Lemma 7.10, for $x \ge 20000$:

$$\log\log N > \log\theta(x)\exp\left(\frac{0.986\sqrt{2}}{\sqrt{x}\log x} - \frac{0.484}{\sqrt{x}\log^2 x}\right). \tag{7.14}$$

Finally, using (7.11), (7.12), (7.13) and (7.14), and that N does not satisfy Robin's inequality, we see that

$$e^\gamma \le \frac{\sigma(N)}{N\log\log N} \le e^\gamma \exp\left(\frac{2+\beta-3\sqrt{2}}{\sqrt{x}\log x} + O\left(\frac{1}{\sqrt{x}\log^2 x}\right)\right). \tag{7.15}$$

Since $2+\beta-3\sqrt{2} < 0$, it follows that, for sufficiently large x, and hence for sufficiently large N, the right-hand side of estimate (7.15) is less than e^γ. This contradiction completes the proof of the theorem. □

7.3 Preliminary Lemmas With RH True

In this and the following two sections we prove preliminary lemmas which are needed for the proof of Theorem 7.11, which is Robin's inequality under RH for $n \geq 5041$. We seek explicit versions of the equations and estimates (7.12), (7.13) and (7.14). The first task is to bound $\prod_{p\leq x}(1-p^{-1})$ from below. The approach taken is along the same lines as that in Chapter 5.

Let $S(x) := \theta(x) - x$ and $R(x) := \psi(x) - x$ as in Chapter 5. Using Stieltjes integrals, and integrating by parts, we have

$$\sum_{p\leq x} \frac{1}{p} = \int_{2-}^{x} \frac{d\theta(t)}{t\log t}$$

$$= \log\log x - \log\log 2 + \frac{S(x)}{x\log x} + \frac{1}{\log 2} + \int_{2-}^{x} \frac{S(t)\{\log t + 1\}}{(t\log t)^2}\, dt. \quad (7.16)$$

Note that by the prime number theorem we have $S(x) = O(x/\log x)$, so the integral in (7.16) converges as $x \to \infty$. Note also we need a lower limit of 2– to correctly evaluate the Stieltjes integral at a point of discontinuity. Letting

$$B := \frac{1}{\log 2} - \log\log 2 + \int_{2}^{\infty} S(t)\frac{\log t + 1}{(t\log t)^2}\, dt, \quad (7.17)$$

we can rewrite (7.16) as

$$\sum_{p\leq x} \frac{1}{p} = \log\log x + B + \frac{S(x)}{x\log x} - \int_{x}^{\infty} \frac{S(t)\{\log t + 1\}}{(t\log t)^2}\, dt,$$

and so

$$\sum_{p\leq x} \frac{1}{p} = \log\log\theta(x) + B + \left(\log\log x - \log\log\theta(x) + \frac{S(x)}{x\log x}\right)$$
$$- \int_{x}^{\infty} \frac{S(t)(\log t + 1)}{(t\log t)^2}\, dt. \quad (7.18)$$

The purpose of writing the expression this way is to combine it with one which comes from the more commonly found definition for B, which can be written:

$$\sum_{p\leq x} \log\left(1 - \frac{1}{p}\right) = B - \gamma - \sum_{p\leq x} \frac{1}{p} - \sum_{p>x}\left(\log\left(1 - \frac{1}{p}\right) + \frac{1}{p}\right). \quad (7.19)$$

Then, by exponentiating (7.19), we shall obtain an explicit version of (7.13).

Before proceeding directly, we shall introduce some notation to stand for various error terms analogous to those used in Chapter 5. Let

$$L_1(x) = \int_{x}^{\infty} \frac{S(t)(\log t + 1)}{(t\log t)^2}\, dt = J_1(x) - K_1(x), \quad (7.20)$$

where

$$J_1(x) = \int_x^\infty \frac{R(t)(\log t + 1)}{(t\log t)^2}\,dt \quad \text{and} \quad K_1(x) = \int_x^\infty \frac{(\psi(t) - \theta(t))(\log t + 1)}{(t\log t)^2}\,dt. \tag{7.21}$$

The following Lemma 7.3 has the same derivation using integration by parts as that of Lemma 5.16, which has a more restricted formulation.

Lemma 7.3 *[142, lemma 1] For $n \geq 1$ and ρ a complex number with $0 < \Re\rho < 1$, set*

$$F_{\rho,n}(x) := \int_x^\infty \frac{t^\rho}{t^{n+1}} \frac{n\log t + 1}{\log^2 t}\,dt. \tag{7.22}$$

Then

$$F_{\rho,n}(x) = \frac{n}{n-\rho}\frac{x^{\rho-n}}{\log x} + r_{\rho,n}(x), \tag{7.23}$$

where

$$r_{\rho,n}(x) := \frac{\rho}{(\rho - n)^2}\left(-\frac{x^{\rho-n}}{\log^2 x} + 2\int_x^\infty \frac{t^{\rho-n-1}}{\log^3 t}\,dt\right). \tag{7.24}$$

If $\Re\rho = \frac{1}{2}$ then for $n \geq 1$ we have $|\rho| \leq |\rho - n|$ and thus

$$|r_{\rho,n}(x)| \leq \frac{1}{|\rho| x^{n-1/2}\log^2 x}\left(1 + \frac{4}{(2n-1)\log x}\right). \tag{7.25}$$

This completes the statement.

We shall apply Lemma 7.3 where ρ is a non-trivial zero of the Riemann zeta function. In particular, if we assume the Riemann hypothesis is true we have $\Re\rho = \frac{1}{2}$.

The next lemma bounds not only the first integral appearing in (7.21), but also a generalization, which we shall need in Section 7.4.

Lemma 7.4 *[142, lemma 2] Assume the Riemann hypothesis is true. For $n \geq 1$ and for $x \geq 2$, let*

$$J_n(x) = \int_x^\infty R(t)\frac{n\log t + 1}{t^{n+1}\log^2 t}\,dt \tag{7.26}$$

and recall $\beta := \sum_\rho 1/|\rho|^2$, the sum over the imaginary zeros of $\zeta(s)$. Then

$$-\frac{\beta\left(n + \frac{1}{\log x} + \frac{4}{(2n-1)\log^2 x} + \frac{2\log 2\pi}{\beta\sqrt{x}}\right)}{x^{n-1/2}\log x} \leq J_n(x) \leq \frac{\beta\left(n + \frac{1}{\log x} + \frac{4}{(2n-1)\log^2 x}\right)}{x^{n-1/2}\log x}. \tag{7.27}$$

Proof Consider first the explicit formula in Theorem 2.1 (or see for example [80, theorem 29]):

$$\psi(x) = x - \sum_\rho \frac{x^\rho}{\rho} - \log 2\pi - \frac{1}{2}\log\left(1 - \frac{1}{x^2}\right) \tag{7.28}$$

and write

$$\int_x^\infty g(t)\left\{\psi(t) - t + \log 2\pi + \frac{1}{2}\log\left(1 - \frac{1}{t^2}\right)\right\} dt = -\sum_\rho \int_x^\infty \frac{g(t)t^\rho}{\rho}\, dt \tag{7.29}$$

where

$$g(t) := \frac{n\log t + 1}{t^{n+1}\log^2 t}. \tag{7.30}$$

For $n \geq 1$ the dominated convergence theorem justifies the interchange of summation and integration on the right-hand side of (7.29). Recalling that $R(t) = \psi(t) - t$ we may write

$$J_n(x) = J_n^{(1)}(x) + J_n^{(2)}(x), \tag{7.31}$$

where

$$\begin{aligned} J_n^{(1)}(x) &= -\sum_\rho \int_x^\infty \frac{t^\rho}{\rho}\frac{n\log t + 1}{t^{n+1}\log^2 t}\, dt, \\ J_n^{(2)}(x) &= -\int_x^\infty \left(\log 2\pi + \frac{1}{2}\log\left(1 - \frac{1}{t^2}\right)\right)\frac{n\log t + 1}{t^{n+1}\log^2 t}\, dt. \end{aligned} \tag{7.32}$$

For $t \geq x \geq 2$ we have $0 < -\frac{1}{2}\log(1 - t^{-2}) < \log 2\pi$, and it then follows that

$$-\frac{2\log 2\pi}{x^n \log x} < J_n^{(2)}(x) < 0. \tag{7.33}$$

Next use Lemma 7.3 with $\rho = \frac{1}{2} + i\gamma$ to get

$$J_n^{(1)}(x) = \frac{n}{x^{n-1/2}\log x}\sum_\rho \frac{x^{i\gamma}}{\rho(\rho - n)} - \sum_\rho \frac{r_\rho(x)}{\rho}. \tag{7.34}$$

Using the upper bound (7.25), the bounds which have been derived for $J_n^{(2)}(x)$, and the estimate

$$\left|\sum_\rho \frac{x^{i\gamma}}{\rho(\rho - n)}\right| \leq \sum_\rho \frac{1}{|\rho|^2} = \beta, \tag{7.35}$$

completes the proof of the lemma. □

We now proceed to bound the integral $K_1(x)$ in (7.21). In order to do this we require the following lemma to bound the difference $\psi(x) - \theta(x)$. This gives the same result as Lemma 5.18 (*b*), but the proof is different. Note that the result improves on [153, (5.3*)], which, instead of the estimate (7.37), has

$$0.998697\sqrt{x} < \psi(x) - \theta(x) < 1.001093\sqrt{x} + 3x^{1/3}. \tag{7.36}$$

Lemma 7.5 *[142, lemma 3] Assume the Riemann hypothesis is true. Then for all $x \geq 121$*

$$0 < \psi(x) - \theta(x) - x^{1/2} \leq \tfrac{4}{3}x^{1/3}. \tag{7.37}$$

Proof Let $g(x) = (\psi(x) - \theta(x) - x^{1/2})/x^{1/3}$. We may first check that $g(x) \leq 1.332$ for all $x \leq 4.1 \times 10^{11}$. To do this we need to examine only those primes with $p \leq x^{1/2}$, that is to say, the first 52 104 primes for which the equation is able to be verified numerically.

Since $\psi(x) = \sum_{n\geq 1} \theta(x^{1/n})$ we may write

$$\psi(x) - 2\psi(x^{1/2}) = \sum_{n\geq 1}(\theta(x^{1/n}) - 2\theta(x^{1/(2n)})) = \sum_{n\geq 1}(-1)^{n+1}\theta(x^{1/n}), \tag{7.38}$$

so that

$$\psi(x) - \theta(x) \leq \psi(x^{1/2}) + (\psi(x^{1/2}) - \theta(x^{1/2})) + \theta(x^{1/3}). \tag{7.39}$$

Next, by Lemma 3.10 (*a*), for $0 < x \leq 4.1 \times 10^{11}$ we have $\theta(x) < x$. For $4.1 \times 10^{11} \leq x \leq 1.68 \times 10^{23}$ we have $x^{1/2} \leq 4.1 \times 10^{11}$, and therefore

$$\begin{aligned}\psi(x) - \theta(x) &\leq \theta(x^{1/2}) + 2(\psi(x^{1/2}) - \theta(x^{1/2})) + \theta(x^{1/3}) \\ &\leq x^{1/2} + x^{1/3}\left(1 + \frac{2}{x^{1/12}} + \frac{2.666}{x^{1/6}}\right) \leq x^{1/2} + 1.24637x^{1/3}. \end{aligned} \tag{7.40}$$

Now consider $x \geq 1.68 \times 10^{23}$ and make use of the following results due to Rosser and Schoenfeld (recall we are assuming RH):

$$\begin{aligned}\psi(x) - \theta(x) &\leq \theta(x^{1/2}) + 3x^{1/3}, & x > 0,\\ \theta(x) &< 1.000081x, & x > 0,\\ |\psi(x) - x| &\leq \frac{1}{8\pi}\sqrt{x}\log^2 x, & x \geq 74,\end{aligned} \tag{7.41}$$

which are, respectively Lemma 3.10, Lemma 3.11 and Theorem 4.6. It follows using inequality (7.39) that

$$\begin{aligned}\psi(x) - \theta(x) &\leq x^{1/2} + x^{1/3}\left(1.000081 + \frac{\log^2 x}{32\pi x^{1/12}} + \frac{1.000081}{x^{1/12}} + \frac{3}{x^{1/6}}\right)\\ &\leq x^{1/2} + 1.33128x^{1/3}, \end{aligned} \tag{7.42}$$

which proves the first part of the lemma.

For the second part of the lemma when $121 \leq x \leq 10^{16}$ we use Lemma 3.13 (*b*),

$$\psi(x) - \theta(x) > \sqrt{x}, \tag{7.43}$$

and for $x \geq 10^{16}$ we use Theorem 4.6,

$$|\theta(x) - x| \leq \frac{1}{8\pi}\sqrt{x}\log^2 x \quad (x \geq 599), \tag{7.44}$$

and the lower bound $\psi(x) - \theta(x) > \theta(x^{1/2}) + \theta(x^{1/3})$, to show that

$$\psi(x) - \theta(x) > x^{1/2} + x^{1/3}\left(1 - \frac{\log^2 x}{32\pi x^{1/12}} - \frac{\log^2 x}{72\pi x^{1/6}}\right). \tag{7.45}$$

Since the right-hand side of the inequality (7.45) is decreasing for all $x \geq e^{24} = 2.64\ldots \times 10^{10}$, we have $\psi(x) - \theta(x) > x^{1/2} + 0.33x^{1/3}$, and this completes the proof of the lemma. □

Recall some definitions:

$$\begin{aligned}
S(x) &= \theta(x) - x,\\
\beta &:= \sum_{\rho} \frac{1}{|\rho|^2},\\
K_1(x) &:= \int_x^\infty \frac{(\psi(t) - \theta(t))}{t^2}\left(\frac{1}{\log t} + \frac{1}{\log^2 t}\right) dt,\\
J_1(x) &:= \int_x^\infty \frac{(\psi(t) - t)}{t^2}\left(\frac{1}{\log t} + \frac{1}{\log^2 t}\right) dt,\\
L_1(x) &:= J_1(x) - K_1(x).
\end{aligned}$$

Using Lemma 7.5 we are now able to bound $K_1(x)$ in (7.21), which leads to:

Lemma 7.6 *[142, lemma 4] Under the assumption of the Riemann hypothesis, for $x \geq 2$, we have*

$$\begin{aligned}
-L_1(x) &= -\int_x^\infty S(t)\frac{\log t + 1}{(t\log t)^2}\, dt\\
&\leq \frac{2 + \beta + \dfrac{\beta - 2}{\log x} + \dfrac{8 + 4\beta}{\log^2 x} + \dfrac{2}{x^{1/6}} + \dfrac{2\log 2\pi}{x^{1/2}}}{\sqrt{x}\log x}.
\end{aligned} \tag{7.46}$$

Proof Using the notation of Lemma 7.3 and the result of Lemma 7.5 we have

$$K_1(x) \leq F_{\frac{1}{2},1}(x) + \tfrac{4}{3}F_{\frac{1}{3},1}(x). \tag{7.47}$$

Using the result of Lemma 7.3 and the bound in the estimate (7.25), we bound the right side of the estimate (7.47), noting from Lemma 5.16 that $r_{\frac{1}{3},1}(x) \leq 0$, to obtain

$$K_1(x) \leq \frac{2}{\sqrt{x}\log x}\left(1 + \frac{4}{\log^2 x} + \frac{1}{x^{1/6}}\right). \tag{7.48}$$

Finally, the bounds (7.48) and (7.27) can be used to complete the proof of the lemma. □

We are finally in a position to bound $\prod_{p\le x}(1-p^{-1})$ from below. In Lemma 7.7 we find an upper bound for the multiplicative inverse of this product:

Lemma 7.7 *[142, lemma 5] For $x \ge 20000$,*

$$\prod_{p\le x}\left(1-\frac{1}{p}\right)^{-1} \le e^{\gamma}\log\theta(x)\exp\left(\frac{2+\beta}{\sqrt{x}\log x}+\frac{\alpha(x)}{\sqrt{x}\log^2 x}\right), \tag{7.49}$$

where

$$\alpha(x) := \frac{S^2(x)(\log x+1.31)}{2x^{3/2}}+(\beta-2)+\frac{8+4\beta}{\log x}+\frac{2\log x}{x^{1/6}}+\frac{2\log 2\pi\log x}{x^{1/2}}. \tag{7.50}$$

Proof We use Taylor's formula to expand $\log\log\theta(x)$ about the point x, so

$$\log\log\theta(x) = \log\log x + \frac{S(x)}{x\log x} - \frac{S^2(x)}{2}g(\xi), \tag{7.51}$$

for some $\xi \in (x,\theta(x))$ or $(\theta(x),x)$, where $g(t) = (\log t+1)/(t^2\log^2 t)$. For $x \ge 20000$ we use Lemma 3.14 (*a*):

$$\theta(x) > x - \frac{x}{8\log x} \quad (x \ge 19421). \tag{7.52}$$

Note that the function $g(t)$ is decreasing, so we may use inequality (7.52) to write $g(\xi) < g(x - x/(8\log x))$. This implies, since $\log(1-\epsilon) < -\epsilon$ for $\epsilon > 0$,

$$g(\xi) < \frac{\log x + 1 - \dfrac{1}{8\log x}}{x^2\left(1-\dfrac{1}{8\log x}\right)^2\log^2 x\left(1-\dfrac{1}{7.8\log^2 x}\right)^2}.$$

Note that because $-\log(1-\epsilon) < \epsilon/(1-\epsilon)$ for $0<\epsilon<1$ we have used:

$$\log^2\left(x\left(1-\frac{1}{8\log x}\right)\right) > \left(\log x - \frac{1}{\log x - 1}\right)^2 > \log^2 x\left(1-\frac{1}{7.8\log^2 x}\right)^2.$$

If $y := \log x$, consider for $y \ge \log(20000) > 9.90349$ the expression

$$y + 1.26 + \frac{0.44}{y} - \frac{y+1-\dfrac{1}{8y}}{\left(1-\dfrac{1}{8y}\right)^2\left(1-\dfrac{1}{7.8y^2}\right)^2}.$$

This expression is strictly positive for $y \ge \log(20000)$: to see this convert it to a polynomial in y with leading term $0.01y^7$ and then divide by y^7 to get the form $k(y) + 0.01$ where the rational function $k(y)$ satisfies $|k(y)| < 0.00007$. Therefore

$$g(\xi) < \frac{\log x + 1.26 + 0.44/\log x}{x^2 \log^2 x} < \frac{\log x + 1.31}{x^2 \log^2 x}. \tag{7.53}$$

This and (7.51) show that

$$\operatorname{loglog} \theta(x) \geq \operatorname{loglog} x + \frac{S(x)}{x \log x} - \frac{S^2(x)}{2} \frac{\log x + 1.5626}{x^2 \log^2 x}. \tag{7.54}$$

Insert this into (7.18) and (7.19) gives:

$$\begin{aligned}
&\operatorname{loglog} x - \operatorname{loglog} \theta(x) + \frac{S(x)}{x \log x} \\
&\leq \frac{S(x)^2}{2} \frac{\log x + 1.31}{x^2 \log^2 x} - \sum_{p \leq x} \log\left(1 - \frac{1}{p}\right) \\
&= -B + \gamma + \sum_{p \leq x} \frac{1}{p} + \sum_{p > x} \left(\log\left(1 - \frac{1}{p}\right) + \frac{1}{p}\right) \\
&\leq -B + \gamma + \operatorname{loglog} \theta(x) + B + \frac{S(x)^2}{2} \frac{\log x + 1.31}{x^2 \log^2 x} - L_1(x),
\end{aligned}$$

so exponentiating we get

$$\prod_{p \leq x} \left(1 - \frac{1}{p}\right)^{-1} \leq e^{\gamma} \log \theta(x) \exp\left(-L_1(x) + \frac{S(x)^2}{2} \frac{\log x + 1.31}{x^2 \log^2 x}\right).$$

Finally we are able to use Lemma 7.6 to complete the derivation of the upper bound. □

7.4 Bounding $\prod_{p \leq x}(1 - p^{-2})$ From Above With RH True

In this section we make (7.12) explicit by giving a more precise form of the sum of Lemma 5.7 in the case $\alpha = 2$. The proof takes seven steps and relies on many of the results which have been proved.

Lemma 7.8 *[142, lemma 6] Assume RH is true. For $x \geq 20000$,*

$$\prod_{\sqrt{2x} < p \leq x} \left(1 - \frac{1}{p^2}\right) \leq \exp\left(\frac{-\sqrt{2}}{\sqrt{x} \log x} + \frac{4}{\sqrt{x} \log^2 x}\right). \tag{7.55}$$

Proof **(1)** We consider separately small and large values of x. For $2 \times 10^4 \leq x \leq 4 \times 10^9$ we write

$$\prod_{\sqrt{2x} \leq p \leq x} \left(1 - \frac{1}{p^2}\right) = \frac{6}{\pi^2} \prod_{p \leq \sqrt{2x}} \left(1 - \frac{1}{p^2}\right)^{-1} \prod_{p > x} \left(1 - \frac{1}{p^2}\right)^{-1}. \tag{7.56}$$

We may verify, computationally, that

$$\frac{6}{\pi^2}\prod_{p\le\sqrt{2x}}\left(1-\frac{1}{p^2}\right)^{-1}\le\exp\left(\frac{-\sqrt{2}}{\sqrt{x}\log x}+\frac{3.7}{\sqrt{x}\log^2 x}\right), \tag{7.57}$$

for $x\in[2\times10^4,\ 4\times10^9]$. As for the second product in (7.56) we write

$$\begin{aligned}\prod_{p>x}\left(1-\frac{1}{p^2}\right)^{-1}&=\exp\left(-\sum_{p>x}\log\left(1-\frac{1}{p^2}\right)\right)\\&\le\exp\left(\sum_{p>x}\frac{1}{p^2-1}\right)\le\exp\left((1+2.51\times10^{-9})\sum_{p>x}\frac{1}{p^2}\right).\end{aligned} \tag{7.58}$$

We can estimate the final sum appearing in the upper bound of (7.58) by using Abel summation and integrating by parts so that

$$\sum_{p>x}\frac{1}{p^2}=\frac{-\theta(x)}{x^2\log x}+\int_x^\infty\theta(t)\frac{2\log t+1}{t^3\log^2 t}\,dt. \tag{7.59}$$

Using the middle inequality in the estimate (7.41), and the lower bound (7.52), we can bound the right side of (7.59) thus

$$\sum_{p>x}\frac{1}{p^2}\le\frac{1.000081}{x\log x}+\frac{1}{8x\log^2 x}\le\frac{0.070912}{\sqrt{x}\log^2 x}, \tag{7.60}$$

where we have used $\theta(x)<1.000081x$ from Lemma 3.11. Substituting this bound in estimate (7.58), then using estimates (7.57) and (7.60) we establish the lemma for $x\in[2\times10^4,\ 4\times10^9]$.

(2) For larger x we make use of the results from Lemmas 7.3, 7.4 and 7.5. We write

$$\begin{aligned}-\log\left(\prod_{x_2<p\le x}\left(1-\frac{1}{p^2}\right)\right)&\ge\sum_{\sqrt{2x}<p\le x}\frac{1}{p^2}=\int_{\sqrt{2x}}^x\frac{d\theta(t)}{t^2\log t}\\&=\int_{\sqrt{2x}}^x\frac{dt}{t^2\log t}+\int_{\sqrt{2x}}^x\frac{dS(t)}{t^2\log t}=:I_1+I_2.\end{aligned} \tag{7.61}$$

Integrating by parts gives us lower bounds on I_1 and I_2.

(3) First consider I_1:

$$\begin{aligned}\int_{\sqrt{2x}}^x\frac{dt}{t^2\log t}&=\left[-\frac{1}{t\log t}\right]_{\sqrt{2x}}^x-\int_{\sqrt{2x}}^x-\frac{-dt}{t^2\log^2 t}\\&=\left[-\frac{1}{t\log t}\right]_{\sqrt{2x}}^x+\left[\frac{1}{t\log^2 t}\right]_{\sqrt{2x}}^x+2\int_{\sqrt{2x}}^x\frac{dt}{t^2\log^3 t}.\end{aligned}$$

Hence

$$\int_{\sqrt{2x}}^{x} \frac{dt}{t^2 \log t} = -\frac{1}{x \log x} + \frac{\sqrt{2}}{\sqrt{x} \log 2x} + \frac{1}{x \log^2 x} - \frac{2\sqrt{2}}{\sqrt{x} \log^2 2x}$$
$$+ 2 \int_{\sqrt{2x}}^{x} \frac{dt}{t^2 \log^3 t}$$
$$\geq \frac{\sqrt{2}}{\sqrt{x} \log x}\left(1 - \frac{\log 2}{\log x}\right) - \frac{2\sqrt{2}}{\sqrt{x} \log^2 2x} - \frac{1}{x \log x}.$$

Therefore since for $x > 2$ we have

$$\left(\frac{\log x}{\log 2x}\right)^2 + \frac{\log x}{2\sqrt{2x}} < 1,$$

we get

$$I_1 \geq \frac{\sqrt{2}}{\sqrt{x} \log x} - \frac{\sqrt{2}(2 + \log 2)}{\sqrt{x} \log^2 x}. \tag{7.62}$$

(4) For the second integral:

$$I_2 \geq \frac{S(x)}{x^2 \log x} - \frac{S(\sqrt{2x})}{x \log 2x} + L_2(\sqrt{2x}) - L_2(x), \tag{7.63}$$

where

$$L_2(y) = \int_y^\infty S(t) \frac{2 \log t + 1}{t^3 \log^2 t}\, dt = J_2(y) - K_2(y). \tag{7.64}$$

By Lemma 7.5, we have

$$F_{\frac{1}{2},2}(y) \leq K_2(y) \leq F_{\frac{1}{2},2}(y) + \tfrac{4}{3} F_{\frac{1}{3},2}(y), \tag{7.65}$$

and so we may bound $K_2(y)$ using the bounds on $F_{\rho,n}(x)$ in Lemma 7.3. Using Lemma 7.4 to bound $J_2(y)$ we find that

$$L_2(y) \leq \frac{1}{y^{3/2} \log y}\left(2\beta + \frac{\beta + \frac{2}{9}}{\log y} + \frac{4\beta}{3 \log^2 y} - \frac{4}{3}\right). \tag{7.66}$$

It is straightforward to show that the right side of estimate (7.66) is negative for $y \geq 2$. We may similarly bound $L_2(\sqrt{2x})$ from below:

$$L_2(\sqrt{2x}) \geq J_2(\sqrt{2x}) - F_{\frac{1}{2},2}(\sqrt{2x}) - \tfrac{4}{3} F_{\frac{1}{3},2}(\sqrt{2x}),$$

which gives with $c_1 = 0.0198734$

$$L_2(\sqrt{2x}) > -\frac{c_1}{(2x)^{3/4} \log \sqrt{2x}}, \qquad x > 4 \times 10^9. \tag{7.67}$$

(5) For $4\times 10^9 < x < 5\times 10^{21}$ we have $\sqrt{2x} < 10^{11}$, so therefore by Lemma 3.10 (*a*), $S(\sqrt{2x}) < 0$. Moreover, for $x \ge 10^9$, by Theorem 4.6, we have for $c_2 = 0.012$

$$|S(x)| \le \frac{c_2 x}{\log x}. \tag{7.68}$$

Therefore by (7.59)

$$\begin{aligned} I_2 &\ge -\frac{c_2}{x\log^2 x} - \frac{c_1}{(2x)^{3/4}\log(\sqrt{2x})} \\ &= -\frac{1}{\sqrt{x}\log^2 x}\left(\frac{c_2}{\sqrt{x}} + \frac{c_1\times 2^{1/4}\log^2 x}{x^{1/4}\log(2x)}\right). \end{aligned}$$

This shows that

$$I_2 \ge -\frac{c_3}{\sqrt{x}\log^2 x}, \tag{7.69}$$

where we can take $c_3 = 0.002666$.

(6) When $x \ge 5\times 10^{21}$ we use the estimate (7.44) (we are assuming RH), the bound (7.63) and the equation (7.64) to derive

$$I_2 \ge -\frac{1}{\sqrt{x}\log^2 x}\left(\frac{\log^3 x}{8\pi x} + \frac{\log(2x)\log^2 x}{162^{3/4}\pi\sqrt[4]{x}} + \frac{2^{1/4}\log^3 x}{8\pi x^{1/4}}\right) \ge -\frac{c_4}{\sqrt{x}\log^2 x}, \tag{7.70}$$

where we take $c_4 := 0.02782$.

(7) Finally combine the estimates (7.62), (7.69) and (7.70):

$$\begin{aligned} -\log\left(\prod_{\sqrt{2x}<p\le x}\left(1-\frac{1}{p^2}\right)\right) &\ge I_1 + I_2 \\ &\ge \frac{\sqrt{2}}{\sqrt{x}\log x} - \frac{\sqrt{2}(2+\log 2)}{\sqrt{x}\log^2 x} - \frac{\max\{c_3,c_4\}}{\sqrt{x}\log^2 x} \\ &\ge \frac{\sqrt{2}}{\sqrt{x}\log x} - \frac{4}{\sqrt{x}\log^2} \end{aligned}$$

and so

$$\prod_{\sqrt{2x}<p\le x}\left(1-\frac{1}{p^2}\right) \le \exp\left(-\frac{\sqrt{2}}{\sqrt{x}\log x} + \frac{4}{\sqrt{x}\log^2 x}\right),$$

which finally completes the proof of the lemma. □

7.5 Bounding loglog$N$ From Below With RH True

In this section we bound loglog$N$ from below using RH. Recall we used this lower bound in the proof of Ramanujan's theorem, Theorem 7.2. Recall from Chapter 6: if the positive integer N is colossally abundant with associated parameter $\epsilon > 0$ then for all $m > 1$ we have

$$\frac{\sigma(N)}{N^{1+\epsilon}} \geq \frac{\sigma(m)}{m^{1+\epsilon}}.$$

Define a real number $x > 1$ as the unique solution to the equation $x^{\epsilon} = 1 + 1/x$, so x depends on N through ϵ. Set $x_1 := x$ and for $k \geq 1$ define a decreasing sequence (x_k) by

$$\log_{x_k}\left(1 + \frac{1}{x_k^k + \cdots + x_k}\right) = \epsilon.$$

Lemma 7.9 *[142, lemma 7] With the notation defined in the paragraph above, if $x \geq 20000$ then $\sum_{i\geq 2} \theta(x_i) \geq 0.998x_2$.*

Proof To prove the lemma we make use of Lemma 3.12 which gives explicit bounds of the form

$$\theta(x) > \alpha x \quad (x \geq a). \tag{7.71}$$

Let $A(x_2) := \theta(x_2) + \sum_{i\geq 3} \theta(x_i)$. By Lemma 6.17 we have $(x_2^2/2)^{1/k} < x_k$, and so

$$A(x_2) \geq \theta(x_2) + \sum_{i\geq 3} \theta\left(\left(\frac{x_2^2}{2}\right)^{1/i}\right). \tag{7.72}$$

Now, let a and b with $a < b$ be two positive real numbers. For $y \in [a,b]$ we have a bound of the form (7.71), so that

$$A(y) \geq \alpha y + \sum_{3\leq i} \theta\left(\left(\frac{a^2}{2}\right)^{1/i}\right) \geq y\left(\alpha + \frac{1}{b}\sum_{3\leq i} \theta\left(\left(\frac{a^2}{2}\right)^{1/i}\right)\right) = y\eta, \tag{7.73}$$

where $\eta = \eta(\alpha, a, b)$. The sum over all $i \geq 3$ is finite, since the summands in the expression (7.73) are zero whenever $i > \log(a^2/2)/\log 2$. We sum over all $3 \leq i \leq \lfloor \log(a^2/2)/\log 2 \rfloor$ and use explicit numerical values for the $\theta(x)$ evaluations needed on the right-hand side of (7.72). The range to be covered needs to include the minimum value of x_2, which is not less than 193.

By choosing α, a and b appropriately we can bound $A(y)$ in regions $a \leq y \leq b$. We do this in Table 7.1 for which we computed the values of η, given a, b and α. The first four values of a, b and α are from [146, p. 265, corollary] and the final values are from [145, theorem 10].

The lemma then follows from expression (7.73) and Table 7.1. □

Table 7.1 Evaluation of $\eta(\alpha,a,b)$, given α, a and b.

a	b	α	η
487 381	∞	0.998	0.99800
89 387	487 381	0.995	0.99901
32 057	89 387	0.990	1.00131
11 927	32 057	0.985	1.0022
7481	11 927	0.980	1.01433
5381	7481	0.975	1.10206
3457	5381	0.970	1.01896
2657	3457	0.965	1.02935
1481	2657	0.960	1.01876
1433	1481	0.955	1.0537
1427	1433	0.950	1.05201
853	1427	0.945	1.02161
809	853	0.94	1.06316
599	809	0.93	1.03981
557	599	0.92	1.06247
349	557	0.91	1.01651
227	349	0.89	1.02732
149	227	0.86	1.00836

Lemma 7.10 *[142, lemma 8] Assume RH is true. If N is colossally abundant, then for $x \geq 20000$ we have*

$$\log\log N \geq \log\theta(x)\exp\left(\frac{0.986\sqrt{2}}{\sqrt{x}\log x} - \frac{0.484}{\sqrt{x}\log^2 x}\right).$$

Proof From Table 7.1 let $c_1 = 0.99800$. First consider the prime factorization

$$N = \prod_{i\geq 1}\prod_{x_{i+1}<p\leq x_i} p^i.$$

Take logarithms and use Lemmas 7.9 and 3.11, which gives $\theta(x) < 1.000081x$, $x > 0$, to get

$$\begin{aligned}
\log N &= \sum_{i\geq 1}\theta(x_i) = \theta(x) + \sum_{i\geq 2}\theta(x_i)\\
&\geq \theta(x) + c_1 x_2\\
&= \theta(x)\left(1 + \frac{c_1 x_2}{\theta(x)}\right)\\
&\geq \theta(x)\left(1 + \frac{c_1 x_2}{1.000081x}\right)\\
&\geq \theta(x)\left(1 + \frac{c_2 x_2}{x}\right),
\end{aligned}$$

where we can take $c_2 = 0.997797$. Therefore we get with $c_4 = 1.00796973$

$$\log\log N \geq \log\theta(x) + \frac{c_3 x_2}{x} > \log\theta(x)\left(1 + \frac{c_4 x_2}{x\log x}\right) > \log\theta(x)\left(1 + \frac{x_2}{x\log x}\right),$$

where we have used $\theta(x) < 1.000081x$ again and $\log(1+x) > x/(x+1)$, which is true for $x > 0$.

Note that for real $m, m', b > 0$ and $t \in [0, b]$, $1 + mt \geq \exp(m't)$ holds provided $m' \leq \log(1+mb)/b$. Also since using Lemma 6.17 we get

$$\sqrt{2x}\left(1 - \frac{\log 2}{2\log x}\right) < x_2 < \sqrt{2x},$$

we can take

$$\frac{x_2}{x\log x} \leq \frac{\sqrt{2}}{\sqrt{x}\log x} < 0.00101 =: b \quad \text{at } x = 20000.$$

This gives, with $m = 1$, the choice $m' = 0.999$ and so we can then write

$$\begin{aligned}\log\log N &> \log\theta(x)\exp\left(\frac{m' x_2}{x\log x}\right)\\ &\geq \log\theta(x)\exp\left(\frac{m'\sqrt{2}}{\sqrt{x}\log x}\left(1 - \frac{\log 2}{2\log x}\right)\right)\\ &> \log\theta(x)\exp\left(\frac{0.986\sqrt{2}}{\sqrt{x}\log x}\left(1 - \frac{\log 2}{2\log x}\right)\right),\end{aligned}$$

which completes the proof of the lemma. □

7.6 Proof of Robin's Theorem With RH True

Now we are ready to prove Robin's theorem, Theorem 7.11. First the statement:

Theorem 7.11 *[142, theorem 1] Assume the Riemann hypothesis to be true. Then*

$$\sigma(n) < e^{\gamma} n \log\log n \quad (n \geq 5041). \tag{7.74}$$

Moreover the values of $1 < n \leq 5040$ *for which the inequality in (7.3) fails are the elements of the set A, where*

$$\begin{aligned}A := \{&1, 2, 3, 4, 5, 6, 8, 9, 10, 12, 16, 18, 20, 24, 30, 36, 48,\\ &60, 72, 84, 120, 180, 240, 360, 720, 840, 2520, 5040\}.\end{aligned} \tag{7.75}$$

Proof Assume N is colossally abundant. Then, inequality (7.11), which is part of the proof of Theorem 7.2, is

$$\frac{\sigma(N)}{N} \leq \frac{\prod_{\sqrt{2x} < p \leq x}(1 - 1/p^2)}{\prod_{p \leq x}(1 - 1/p)}.$$

We now use the results of three of the lemmas which have been proved above, all valid for $x \geq 20000$ and assuming RH. By Lemma 7.7,

$$\prod_{p \leq x}\left(1-\frac{1}{p}\right)^{-1} \leq e^{\gamma} \log\theta(x) \exp\left(\frac{2+\beta}{\sqrt{x}\log x}+\frac{\alpha(x)}{\sqrt{x}\log^2 x}\right),$$

where

$$\alpha(x) := \frac{S^2(x)(\log x + 1.31)}{2x^{3/2}} + (\beta-2) + \frac{8+4\beta}{\log x} + \frac{2\log x}{x^{1/6}} + \frac{2\log 2\pi \log x}{x^{1/2}},$$

by Lemma 7.8,

$$\prod_{\sqrt{2x}<p\leq x}\left(1-\frac{1}{p^2}\right) \leq \exp\left(-\frac{\sqrt{2}}{\sqrt{x}\log x}+\frac{4}{\sqrt{x}\log^2 x}\right),$$

and by Lemma 7.9,

$$\log\log N > \log\theta(x)\exp\left(\frac{0.986\sqrt{2}}{\sqrt{x}\log x}-\frac{0.484}{\sqrt{x}\log^2 x}\right).$$

These inequalities enable us to write

$$\frac{\sigma(N)}{N\log\log N} \leq e^{\gamma}\exp\left(\frac{c_1+\dfrac{c_2}{\log x}}{\sqrt{x}\log x}\right), \tag{7.76}$$

where

$$c_1 \leq -0.7642, \quad c_2 = \alpha(x) + 4.484, \tag{7.77}$$

and where $\alpha(x)$ is defined in (7.50) as part of the statement of Lemma 7.7. We will show that $c_1 + c_2/\log x \leq 0$ for certain ranges of x set out below.

(1) For $x \geq 100000$ we use $|S(x)| \leq \sqrt{x}\log^2 x/(8\pi)$ (Theorem 4.6), which is valid since we are assuming RH, to bound $S(x)$ and hence $\alpha(x)$ from above. This shows that $\alpha(x) \leq 2.27146$, thus $c_1 + c_2/\log x \leq -0.175 < 0$.

(2) For $x \in [2\times 10^4, 10^5]$ we use inequality (7.52), i.e. $0 \leq x - \theta(x) \leq x/(8\log x)$ (Lemma 3.14 (*a*)) to bound $S(x)$, so $c_1 + c_2/\log x \leq -0.055 < 0$ in this case also.

(3) Finally, for $x \leq 20000$, we calculate all of the colossally abundant numbers C_k with largest prime factor at most x. We do this using the fact that the first 10^7 colossally abundant numbers satisfy that the ratio of any two consecutive numbers is a single prime which can be determined algorithmically. (A *Mathematica* implementation of this calculation is on the page for A004990 in the *Online Encyclopedia of Integer Sequences* [158].) Once the next prime greater than 20 000, namely 20 011, occurs (at the 2348th term) we have the complete set of primes which appear in order and can easily generate the corresponding 2347 colossally abundant numbers; see Table A.5.

Recall the definition $G(n) := \sigma(n)/(n \log\log n)$. By (1), (2) and (3) above we have $G(C_k) < e^\gamma$ for all $C_k \geq 5041$. By Theorem 6.16, this shows that the theorem is true for all $n \geq 5041$. Finally, we compute $G(n)$ for all $2 \leq n \leq 5040$ to derive the set S. This completes the proof. □

7.7 An Unconditional Bound for $\sigma(n)/n$

In this section we neither assume that RH is true, nor that RH is false. The main result, Theorem 7.13, is due to Robin.

Lemma 7.12 *For $x \geq 10^4$ we have*

$$\prod_{p \leq x} \left(\frac{p}{p-1} \right) \leq e^\gamma \log x \left(1 + \frac{1}{2 \log^2 x} \right). \tag{7.78}$$

Proof When $10^4 \leq x \leq 10^8$ we apply Lemma 5.11 which is:

$$\prod_{p \leq x} \left(\frac{p}{p-1} \right) \leq e^\gamma \log x \left(1 + \frac{2}{\sqrt{x} \log x} \right), \quad 0 < x \leq 10^8. \tag{7.79}$$

Because for $x \geq 10^4$ we have $\log x \leq \sqrt{x}/10$, from inequality (7.79) it follows that inequality (7.78) is true for all $x \in [10^4, 10^8]$.

When $x \geq 10^8$ using (7.19) we get

$$\begin{aligned} &\sum_{p \leq x} \frac{1}{p} - \log\log x - B \\ &\quad = -\gamma - \log\log x - \sum_{p \leq x} \log\left(1 - \frac{1}{p}\right) + \sum_{p > x} \left(-\log\left(1 - \frac{1}{p}\right) - \frac{1}{p} \right). \end{aligned}$$

Therefore by Lemmas 5.8 and 5.9 we get

$$\begin{aligned} \sum_{p \leq x} \log\left(\left(1 - \frac{1}{p}\right)^{-1} \right) &\leq \sum_{p \leq x} \log\left(\left(1 - \frac{1}{p}\right)^{-1} \right) + \sum_{p > x} \left(-\log\left(1 - \frac{1}{p}\right) - \frac{1}{p} \right) \\ &\leq \log\left(1 + \frac{1}{2 \log^2 x}\right) + \gamma + \log\log x, \end{aligned}$$

so, exponentiating,

$$\prod_{p \leq x} \frac{p}{p-1} \leq e^\gamma \log x \left(1 + \frac{1}{2 \log^2 x}\right),$$

which completes the proof. □

We now proceed to derive Robin's unconditional bound. Note that the constant c_0 is larger than that of Robin.

Theorem 7.13 *[142, theorem 2] Without assuming the Riemann hypothesis is true or false we have*

$$\frac{\sigma(n)}{n\log\log n} \le e^{\gamma} + \frac{c_0}{(\log\log n)^2}, \quad n \ge 3, \tag{7.80}$$

where $c_0 = 2/3$.

Proof **(1)** Let $c_0 > 0$ be a constant to be determined. Let $N_k = \prod_{i=1}^{k} p_i$ be the kth primorial, and suppose that $p_k \ge 20000$, so $k \ge 2263$, and so

$$y := \log\log N_k \ge 9.8942.$$

By Lemma 3.14 (a) we have

$$\log N_k = \theta(p_k) > p_k\left(1 - \frac{1}{8\log p_k}\right). \tag{7.81}$$

Since $p_k \ge 20000$ we have $x := \log p_k \ge 9.90348$, so

$$\log\log N_k \ge \log p_k + \log\left(1 - \frac{1}{8\log p_k}\right) \ge \log p_k - \frac{1}{7.899\log p_k}$$

$$\implies \quad y \ge x - \frac{1}{7.899x}.$$

By Lemma 7.12 with x replaced by p_k, in this step we want

$$\frac{N_k}{\varphi(N_k)} \le e^{\gamma}\log p_k\left(1 + \frac{1}{2\log^2 p_k}\right) \le e^{\gamma}\log\log N_k + \frac{c_0}{\log\log N_k}.$$

Thus we need the minimum positive value of $\lambda = \delta e^{\gamma}$ such that for $y \ge x - 1/(7.899)$ we have

$$e^{\gamma}\log p_k + \frac{e^{\gamma}}{2\log p_k} \le e^{\gamma}\log\log N_k + \frac{\lambda}{\log\log N_k} \quad \implies \quad x + \frac{1}{2x} \le y + \frac{\delta}{y}.$$

Since $y + \delta/y$ is increasing we can use the minimum $y = x - 1/(7.899x)$ to get

$$x + \frac{1}{2x} - (x - \frac{1}{7.899x}) \le \frac{\delta}{x - \frac{1}{7.899x}} \quad \implies \quad \lambda \ge 0.66591,$$

so $c_0 = 2/3$ will work.

(2) Note next that if $N_k \le n < N_{k+1}$ then at most k distinct primes divide n. If not and $q_1, \ldots, q_l$ are those primes with $l \ge k+1$, then since if $p_1, \ldots, p_l$ are the first l primes in increasing order we must have for each j, $p_j \le q_j$, so

$$N_{k+1} = p_1 \cdot p_{k+1} \le q_1 \cdots q_l \le n,$$

which is false. Therefore

$$\frac{\varphi(N_k)}{N_k} = \prod_{j=1}^{k}\left(1 - \frac{1}{p_j}\right) \le \prod_{j=1}^{l}\left(1 - \frac{1}{q_l}\right) = \frac{\varphi(n)}{n}.$$

It follows that for $N_k \le n < N_{k+1}$ with $p_k \ge 20000$ we have

$$\frac{\sigma(n)}{n} < \frac{n}{\varphi(n)} \le \frac{N_k}{\varphi(N_k)} < e^\gamma \log\log N_k + \frac{c_0}{\log\log N_k}$$
$$< e^\gamma \log\log n + \frac{c_0}{\log\log n}. \tag{7.82}$$

When $p_k < 20000$, that is, when $n \le N_{2263}$, we generated the first 2377 colossally abundant numbers, in the same manner as in Theorem 7.11, and in this way checked Robin's inequality, and therefore the unconditional bound, for $5041 \le n \le N_{2263}$. For $3 \le n \le 5040$ we evaluated the terms of the inequality explicitly to obtain 0.648214, so c_0 is the minimum value of the constant computed by this method. This completes the proof. □

7.8 Bounding loglog N From Above Without RH

Before proving Theorem 7.15 we must first bound $\log\log N$ from above without using RH.

Lemma 7.14 *When N is colossally abundant we have*

$$\log\log N \le \log\theta(x)\left(1 + \frac{\sqrt{2}}{\sqrt{x}\log x} + O\left(\frac{1}{\sqrt{x}\log^2 x}\right)\right).$$

Proof By the proof of Lemma 9.13 we have

$$\begin{aligned}\log N &\le \sum_{i=1}^{\infty} \theta(x_i)\\ &\le \theta(x_1) + \theta(x_2) + O(\theta(x_3)\log x)\\ &= \theta(x)\left(1 + \frac{\theta(x_2)}{\theta(x)} + O\left(\frac{x_3 \log x}{x}\right)\right).\end{aligned}$$

By Lemma 6.17 we have $x_2 = \sqrt{2x}(1 + O(1/\log x))$ and by Lemma 6.18, $x_3 \sim (3x)^{1/3}$. In addition the prime number theorem gives $\theta(x) = x + O(x/\log x)$ as $x \to \infty$. Therefore we can derive

$$\begin{aligned}\log N &\le \theta(x)\left(1 + \frac{\sqrt{2}}{\sqrt{x}} + O\left(\frac{1}{\sqrt{x}\log x}\right) + O\left(\frac{\log x}{x^{2/3}}\right)\right)\\ &= \theta(x)\left(1 + \frac{\sqrt{2}}{\sqrt{x}} + O\left(\frac{1}{\sqrt{x}\log x}\right)\right).\end{aligned}$$

Therefore, since by the mean value theorem we have $\log\theta(x) = \log x + O(1/\log x)$, we get

$$\log\log N \le \log\theta(x)\left(1 + \frac{\frac{\sqrt{2}}{\sqrt{x}} + O\left(\frac{1}{\sqrt{x}\log x}\right)}{\log x + O\left(\frac{1}{\log x}\right)}\right)$$

$$= \log\theta(x)\left(1 + \frac{\sqrt{2}}{\sqrt{x}\log x}\left(1 + O\left(\frac{1}{\log x}\right)\right)\left(1 + O\left(\frac{1}{\log^2 x}\right)\right)\right)$$

$$\leq \log\theta(x)\left(1 + \frac{\sqrt{2}}{\sqrt{x}\log x} + O\left(\frac{1}{\sqrt{x}\log^2 x}\right)\right),$$

and the proof is complete. □

7.9 A Lower Bound for $\sigma(n)/n$ With RH False

The theorem of this section is a vital ingredient for proving Robin's theorem. It relies heavily on Nicolas' theorem, Theorem 5.29, giving an omega estimate for the logarithm of $f(x)$.

Theorem 7.15 *[142, theorem 1] Assume the Riemann hypothesis is false. Let θ be the supremum over all real parts of the non-trivial zeros $\rho = \beta + i\gamma$, and thus $\theta > \frac{1}{2}$. For any number $b \in (1 - \theta, \frac{1}{2})$, there exists a positive constant c such that*

$$\frac{\sigma(n)}{n\log\log n} > e^{\gamma}\left(1 + \frac{c}{(\log n)^b}\right), \tag{7.83}$$

for infinitely many values of n.

Proof First write, for N colossally abundant, following Lemma 6.9,

$$\frac{\sigma(N)}{N} = \prod_{p\leq x}\left(1 \quad \frac{1}{p}\right)^{-1}\prod_{k\geq 1}\prod_{x_{k+1}<p\leq x_k}\left(1 - \frac{1}{p^{k+1}}\right), \tag{7.84}$$

where, by Lemma 6.17, we have $x_{k+1} > x^{1/(k+1)}$. We can rewrite the right-hand side of (7.84) as

$$\prod_{p\leq x}\left(1 - \frac{1}{p}\right)^{-1}\prod_{x_2<p\leq x}\left(1 - \frac{1}{p^2}\right)E(x), \tag{7.85}$$

where $E(x)$ satisfies

$$\prod_{x_3<p\leq x_2}\left(1 - \frac{1}{p^3}\right)\left(1 - \frac{1}{x}\right)^{\pi(x_3)} < E(x) < 1. \tag{7.86}$$

By Lemma 5.7 with $\alpha = 3$ we have

$$\prod_{p>x}\left(1 - \frac{1}{p^3}\right)^{-1} = \prod_{p>x}\exp\left(-\log\left(1 - \frac{1}{p^3}\right)\right) = \exp\left(\sum_{p>x}\frac{1}{p^3} + O\left(\frac{1}{x^5}\right)\right),$$

so for some constant $\eta > 0$ we have

$$\prod_{p>x_3}\left(1 - \frac{1}{p^3}\right) \geq \exp\left(-\sum_{p>\eta x^{1/3}}\frac{1}{p^3} + O\left(\frac{1}{x^{5/3}}\right)\right) = \exp\left(O\left(\frac{1}{x^{2/3}\log x}\right)\right).$$

In addition we have using Chebyshev's prime number theorem and also $\log(1-1/x) > -2/x$:

$$\left(1-\frac{1}{x}\right)^{\pi(x_3)} = \exp\left(\pi(x)\log\left(1-\frac{1}{x}\right)\right)$$
$$> \exp\left(-\eta\frac{x^{1/3}}{\log x}\frac{2}{x}\right) = \exp\left(O\left(\frac{1}{x^{2/3}\log x}\right)\right).$$

Therefore

$$E(x) > \exp\left(O\left(\frac{1}{\sqrt{x}\log^2 x}\right)\right) = 1 + O\left(\frac{1}{\sqrt{x}\log^2 x}\right). \tag{7.87}$$

The estimate (7.12) of Theorem 7.2 is

$$\prod_{\sqrt{2x}<p\le x}\left(1-\frac{1}{p^2}\right) = \exp\left(-\frac{\sqrt{2}}{\sqrt{x}\log x} + O\left(\frac{1}{\sqrt{x}\log^2 x}\right)\right).$$

Therefore using (7.84), the bounds (7.86) and the estimate (7.87) we get

$$\frac{\sigma(N)}{N} \ge \prod_{p\le x}\left(1-\frac{1}{p}\right)^{-1}\prod_{\sqrt{2x}<p\le x}\left(1-\frac{1}{p^2}\right)\left(1+O\left(\frac{1}{\sqrt{x}\log^2 x}\right)\right).$$

Next recall the definition of $f(x)$ originally introduced by Nicolas and employed as the key element in Chapter 5: for $x > 0$

$$f(x) := e^{\gamma}\log\theta(x)\prod_{p\le x}\left(1-\frac{1}{p}\right).$$

Nicolas, in Theorem 5.29, proved that if the Riemann hypothesis is false there is a b with $0 < b < \frac{1}{2}$ such that

$$\log f(x) = \log\left(e^{\gamma}\prod_{p\le x}\left(1-\frac{1}{p}\right)\log\theta(x)\right) = \Omega_{\pm}(x^{-b}). \tag{7.88}$$

Therefore, using Lemma 7.14, we have

$$\frac{\sigma(N)}{e^{\gamma}N\log\log N} \ge \frac{\exp\left(\frac{-\sqrt{2}}{\sqrt{x}\log x} + O\left(\frac{1}{\sqrt{x}\log^2 x}\right)\right)}{e^{\gamma}\prod_{p\le x}\left(1-\frac{1}{p}\right)\log\theta(x)\left(1+\frac{\sqrt{2}}{\sqrt{x}\log x}+O\left(\frac{1}{\sqrt{x}\log^2 x}\right)\right)}$$
$$\ge \frac{\exp\left(\frac{-\sqrt{2}}{\sqrt{x}\log x} + O\left(\frac{1}{\sqrt{x}\log^2 x}\right)\right)}{\exp\left(\Omega_{-}\left(\frac{1}{x^b}\right)\right)}$$

$$= \exp\left(\Omega_+\left(\frac{1}{x^b}\right)\right) \geq 1 + \Omega_+\left(\frac{1}{x^b}\right).$$

We also know, from Chapter 6, that if $p \mid N$ is the largest prime divisor we have

$$\log N > \sum_{q \leq p} \log q > \frac{p}{2} > \frac{x}{4},$$

since $\sqrt{2x} < p \leq x$ by the prime number theorem and Bertrand's postulate. Therefore there is a constant $c > 0$ such that

$$\frac{\sigma(N)}{e^\gamma N \log\log N} \geq 1 + \frac{c}{(\log N)^b}.$$

Hence Robin's inequality is false for an infinite number of N and the proof of the theorem is complete. □

Finally in this section, combining Theorems 7.11 and 7.15 we get:

Theorem 7.16 (Ramanujan–Robin criterion) *A necessary and sufficient condition for the Riemann hypothesis is that*

$$\sigma(n) < e^\gamma n \log\log n \quad (n \geq 5041). \tag{7.89}$$

7.10 Lagarias' Formulation of Robin's Criterion

Jeffrey Lagarias (Figure 7.7) graduated from MIT in 1974 with a thesis entitled "The 4-part of the class group of a quadratic field". His adviser was Harold Starke, whose adviser was Derrick Lehmer. Lagarias has taught since 2002 at the University of Michigan. In 1987 he received a Lester R. Ford award from the Mathematical Association of America and in 2012 he became a fellow of the American Mathematical Society.

Lagarias [100] considered another inequality involving the sum-of-divisors function. His inequality does not directly involve Euler's constant γ or a number such as 5040. Recall the definition of the nth **harmonic number**:

$$H_n := 1 + \frac{1}{2} + \frac{1}{3} + \cdots + \frac{1}{n} \quad (n \geq 1).$$

To begin, in lemmas 3.1 and 3.2 of [100] sharp bounds on the quantity on the right-hand side of inequality (7.92), namely $H_n + \exp(H_n)\log(H_n)$, are produced. We summarize this in the following:

Lemma 7.17 *For $n \geq 3$ we have*

$$e^\gamma n \log\log n + H_n \leq H_n + \exp(H_n)\log(H_n) \leq e^\gamma n \log\log n + \frac{n}{\log n} h(n), \tag{7.90}$$

where $h(n)$ is a positive function decreasing on $[4, \infty)$ to e^γ, and $3 < h(4) < 4$.

Figure 7.7 Jeffrey Lagarias.

Proof First consider the left inequality in (7.90). We use Abel's identity [6, theorem 4.2] applied to the sum $H_n = \sum_{j=1}^{n} j^{-1}$ to obtain

$$H_n = 1 + \log n - \int_1^n \frac{\{t\}}{t^2}\, dt,$$

where $\{t\}$ denotes the fractional part of t. Hence $H_n < 1 + \log n$.

Since $0 \leq \{t\} < 1$, the integral converges and so, upon writing $\gamma = 1 - \int_1^{\infty}(\{t\}/t^2)\, dt$ we have

$$H_n = \log n + \gamma + \int_n^{\infty} \frac{\{t\}}{t^2}\, dt, \tag{7.91}$$

thus $H_n > \log n + \gamma$, so $\log H_n > \log\log n$ and $\exp(H_n) > e^{\gamma} n$, and therefore the left inequality in (7.90) follows.

Now let

$$R_n := H_n - \log(n+1) = \int_1^{n+1} \left(\frac{1}{\lfloor t \rfloor} - \frac{1}{t} \right) dt \to \gamma,$$

so for all $n \in \mathbb{N}$ we have $R_{n+1} > R_n > 0$. Therefore $H_n - \log(n+1) < \gamma$ so $H_n < \gamma + \log(n+1)$. Exponentiating we get $\exp(H_n) < e^{\gamma}(1+n)$. Taking logarithms of the equation $H_n < 1 + \log n$ we have

$$\log H_n < \log(1 + \log n) = \log\left(\log n \left(1 + \frac{1}{\log n}\right)\right) < \log\log n + \frac{1}{\log n}.$$

Therefore

$$H_n + \log(H_n)\exp(H_n) < e^{\gamma}(n+1)\left(\log\log n + \frac{1}{\log n}\right) + 1 + \log n$$

$$= ne^{\gamma}\log\log n + e^{\gamma}\left(\log\log n + \frac{1}{\log n} + \frac{n}{\log n}\right) + \log en$$
$$= ne^{\gamma}\log\log n + \frac{n}{\log n}h(n),$$

where we have set

$$h(n) := e^{\gamma} + \frac{\log n}{n} + \frac{\log^2 n}{n} + e^{\gamma}\frac{\log n \log\log n}{n} + \frac{e^{\gamma}}{n}.$$

The maximum of $h(n)$ over the positive integers is at $n = 4$, with $3 < h(4) < 4$. □

Note that lemma 3.2 in [100] has a 4 instead of $h(n)$ in the right-hand side of (7.90), but otherwise the results are equivalent.

Now that bounds involving $e^{\gamma}n\log\log n$ have been converted to bounds involving H_n, we are in a position to prove Theorem 7.18. Most of the heavy lifting has been done by Theorems 7.11, 7.13 and 7.15. Lagarias proves:

Theorem 7.18 (Lagarias' criterion) *[100, theorem 1.1] Let $H_n = \sum_{j=1}^{n} j^{-1}$ for $n \geq 1$ denote the nth harmonic number. The inequality*

$$\sigma(n) \leq H_n + \exp(H_n)\log(H_n), \quad n \geq 1, \tag{7.92}$$

is equivalent to the Riemann hypothesis.

Proof First, assume the Riemann hypothesis. Then by Theorem 7.11 and Lemma 7.17,

$$\sigma(n) \leq e^{\gamma}n\log\log n \leq \exp(H_n)\log(H_n), \quad n \geq 5041, \tag{7.93}$$

so that inequality (7.92) is satisfied for $n \geq 5041$. One can check easily on a computer that it is also true for $1 \leq n \leq 5040$.

Now assume RH is false and that inequality (7.92) is true for all $n \in \mathbb{N}$ in order that we might obtain a contradiction. By Theorem 7.15 there is a b with $0 < b < \frac{1}{2}$ such that for some $c > 0$ and an infinite set of $n \in \mathbb{N}$, we have

$$1 + \frac{c}{\log^b n} < \frac{\sigma(n)}{ne^{\gamma}\log\log n}.$$

Thus by the right-hand side of the inequality of Lemma 7.17 we get

$$\begin{aligned} ne^{\gamma}\log\log n + c\frac{ne^{\gamma}\log\log n}{\log^b n} &= ne^{\gamma}\log\log n\left(1 + \frac{c}{\log^b n}\right) \\ &< \sigma(n) \leq H_n + \exp(H_n)\log(H_n) \\ &\leq ne^{\gamma}\log\log n + \frac{4n}{\log n}. \end{aligned}$$

Thus $ce^{\gamma}\log\log n \leq 4/\log^{1-b} n$ for an infinite set of n, which is false, and the proof of the theorem is complete. □

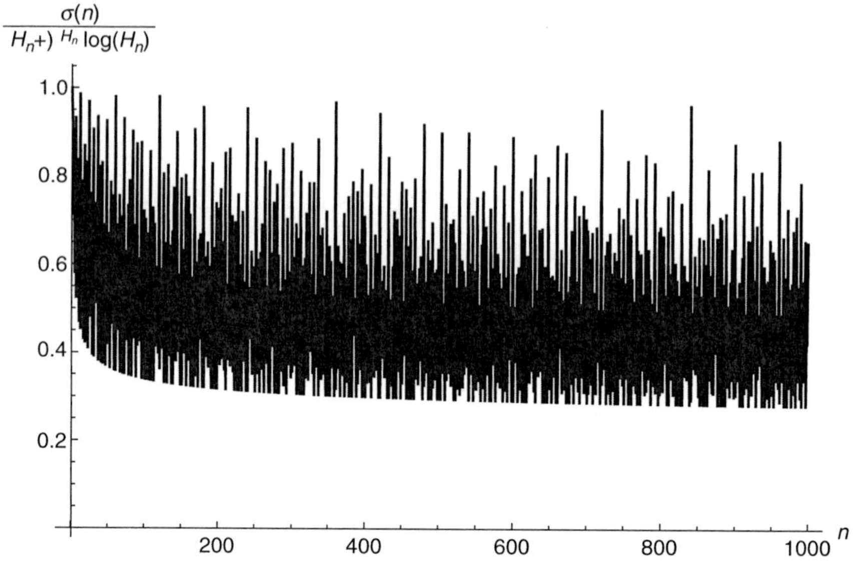

Figure 7.8 The values of $\sigma(n)/(H_n + \exp(H_n)\log(H_n))$ for $2 \le n \le 1000$.

Note that, in the expression (7.93), we did not require the addition of the term H_n. It was pointed out by Kaneko [100, p. 542] that one could instead prove that $\sigma(n) \le \exp(H_n)\log(H_n)$ for $n > 60$ is equivalent to the Riemann hypothesis.

In Figure 7.8 we see how close the left side gets to the right side of the inequality (7.92), even for small values of n.

7.11 Unconditional Results for Lagarias' Formulation

In [101] Lagarias asked whether one could show that, unconditionally,

$$\sigma(n) \le H_n + 2\exp(H_n)\log(H_n), \quad n \ge 1, \tag{7.94}$$

with equality only for $n = 1$.

In [98] this question is answered in the affirmative. A result of Ivić [81] (see also the improved result of Trudgian [170]) is that

$$\frac{\sigma(n)}{n} < 2.59 \log\log n, \quad n \ge 3. \tag{7.95}$$

Using this inequality and the expression (7.90) we see that

$$\sigma(n) < \frac{2.59}{e^\gamma}\exp(H_n)\log(H_n) < 1.455\exp(H_n)\log(H_n), \quad n \ge 3.$$

Checking the cases $1 \le n \le 2$ by hand establishes

$$\sigma(n) \le H_n + 1.455\exp(H_n)\log(H_n), \quad n \ge 1.$$

One could also use a result of Robin [142, section 6, proposition 2 (1)], which improves the right-hand side of inequality (7.95) to $2.57n\log\log n$ for $n \ge 7$.

There is an editorial comment in [102] that includes an observation by the GCHQ Problem Solving Group. They contest that one could replace the number 2 in inequality (7.94) by any constant $K > 1$ at the cost of having to check more cases by hand when K is close to one. An example is cited: when $K = 1.2$ one needs to verify inequality (7.94) for $1 \le n \le 10^6$.

Further advances have since obviated the need for such a large check. Akbary, Friggstad and Juricevic [1] showed that

$$\frac{\sigma(n)}{n} < \frac{\sigma(180)}{180\log\log 180}\log\log n \le 1.0339e^{\gamma}\log\log n \quad (n \ge 121).$$

This was improved by Akbary and Friggstad [2, p. 3] by considering superabundant numbers – see also Lemma 8.9:

$$\frac{\sigma(n)}{n} \le 1.013617e^{\gamma}\log\log n \quad (n \ge 5041).$$

This, and a computer check for $1 \le n \le 5040$, establishes the unconditional inequality given in Theorem 7.19.

Theorem 7.19 *Without any hypotheses*

$$\sigma(n) \le H_n + 1.013617\exp(H_n)\log(H_n),$$

for all $n \ge 1$.

One could reduce the size of the coefficient of $\exp(H_n)\log(H_n)$ by testing Robin's inequality against more superabundant numbers. Given the unconditional nature of Theorem 7.19, we note that if Theorem 7.18 is false, it cannot be false by much. In this sense Theorems 7.19 and 7.18 complement Theorems 7.11 and 7.13.

7.12 Unitary Divisor Sums

We say that d is a **unitary divisor** of n if $d \mid n$ and $(d,n/d) = 1$. Let

$$\sigma^*(n) = \sum_{\substack{d|n \\ (d,n/d)=1}} d$$

be the sum of all unitary divisors of n.

Robin [142, p. 210] notes that the proof of Grönwall's result, equation (7.1), can be adapted to show that

$$\limsup_{n\to\infty} \frac{\sigma^*(n)}{n\log\log n} = \frac{6e^{\gamma}}{\pi^2} = 1.08\ldots.$$

Ivić [81] showed that

$$\sigma^*(n) < \frac{28}{15} n \log\log n \quad (n \geq 31).$$

This was improved by Robin, who showed that

$$\sigma^*(n) < 1.63601 n \log\log n \quad (n \geq 31), \tag{7.96}$$

except for $n = 42$ when $\sigma^*(n) = 1.7366\ldots n \log\log n$. Trudgian [170, theorem 1.1] improved on inequality (7.96), at least for large n, showing

$$\sigma^*(n) \leq 1.3007 n \log\log n \quad (n \geq 570571). \tag{7.97}$$

The method in [170] is incapable of reducing the right-hand side of inequality (7.97) to anything less than $1.298 n \log\log n$, even if n is taken to be sufficiently large. Robin conjectured [142, proposition 1 (i), p. 210] that there are infinitely many n for which

$$\sigma^*(n) > \frac{6e^\gamma}{\pi^2} n \log\log n. \tag{7.98}$$

A related conjecture was given in proposition 1 (ii) in [142], viz. that

$$\frac{\sigma(n)}{\sigma^*(n) \log\log n} < e^\gamma,$$

for all n sufficiently large. This has been proved by Derbal [45].

7.13 Unsolved Problems

(1) Improve the power of $\log\log n$ on the right-hand side of inequality (7.80). The right-hand side of (7.80) in Theorem 7.13 cannot be improved for all n. However, for n sufficiently large, yet still explicit, one may certainly improve both on the constant $C_0 = 2/3$ and, more importantly, on the power of $\log\log n$ in the denominator. Results of the form $|\theta(x) - x| \leq c_k x / \log^k x$, for a positive constant c_k and for $k \geq 2$, could be helpful in this regard. Such results are known – see e.g. [153, theorem 8*] and [173, theorem 2].

(2) Can equation (7.97) be improved?

(3) Does equation (7.98) hold for infinitely many values of n?

(4) Can one determine an equivalence condition for the Riemann hypothesis involving $\sigma^*(n)$?

(5) Find an inequality or other equivalence to RH in terms of the number-of-divisors function $d(n)$. You may need to use the Dirichlet series

$$\sum_{n=1}^{\infty} \frac{d(n)}{n^s} = \zeta(s)^2 \quad (\Re s > 1),$$

or Mellin transform/Perron's formula

$$\sum_{n\le x} d(n) = \frac{1}{2\pi i}\int_{c-i\infty}^{c+i\infty} \zeta(z)^2 \frac{x^z}{z}\, dz \quad (c > 1).$$

(6) Attempt to show that RH is equivalent to the error term in Dirichlet's divisor problem

$$\sum_{n\le x} d(n) = x\log x + x(2\gamma - 1) + O(x^{\alpha+\epsilon}),$$

taking the value $\alpha = \frac{1}{4} + \epsilon$; but see [71] and [73].

(7) For $N \ge 2$, let $g(N)$ denote the largest value of

$$J(n) := \frac{\sigma(n) - H_n}{\exp(H_n)\log(H_n)}$$

for $2 \le n \le N$. According to Theorem 7.19 we have $g(N) \le 1.0137$ always; according to Theorem 7.18, which assumes RH, we have $g(N) \le 1$ for all N. When do we have large values of $g(N)$? We have $g(10000) = J(4) = 0.9873$. What is the next $n > 10000$ such that $J(n) > 0.9873$?

8
Numbers That Do Not Satisfy Robin's Inequality

8.1 Introduction

Recall the theorem of Robin in Chapter 7, Theorem 7.11, that the Riemann hypothesis is equivalent to **Robin's inequality** holding for all $n > 5040$:

$$\frac{\sigma(n)}{n} < e^{\gamma} \log\log n.$$

This is the **Ramanujan–Robin criterion**. Here methods are developed to derive properties of numbers which do or do not satisfy the inequality. For example, all 11-free numbers larger than 5040 satisfy it. We begin with the joint work of four authors from 2007 [38], YoungJu Choie from Korea (Figure 8.1), Nicolas Lichiardopol from France, Pieter Moree from Germany and Patrick Solé, also from France. We also include related work of Solé and Michael Planat [159] from 2011, and conclude with the work of Tim Trudgian and the present author.

The following notations for sets of integers will be used in this chapter. Let the set of natural numbers satisfying Robin's inequality be denoted R. Let

$$\begin{aligned}
A &:= \{1,2,3,4,5,6,8,9,10,12,16,18,20,24,30,36,48,\\
&\qquad 60,72,84,120,180,240,360,720,840,2520,5040\},\\
B &:= \{2,3,5,6,10,30\},\\
C &:= \{1,4,8,9,16,36\},\\
D &:= \{4,8,9,16,36,72,108,144,216,900,1800,2700,3600,\\
&\qquad 44\,100,88\,200\}.
\end{aligned}$$

Thus, by Theorem 7.11, $A \cap R = \emptyset$. In addition $B \cup C \subset A$ and $B \cap C = \emptyset$.

After some preliminaries it is shown that all squarefree integers which are not in the set B satisfy Robin's inequality. Then that all odd integers other

Figure 8.1 YoungJu Choie.

than $\{1,3,5,9,15\}$ satisfy

$$\frac{n}{\varphi(n)} < e^{\gamma} \log\log n,$$

which is called Nicolas' inequality. Unfortunately it is the negation of the Nicolas' inequality that appears in Chapter 5 in equivalences to RH involving $\varphi(n)$. Observe that Nicolas' inequality is stronger than that of Robin, since $\sigma(n)/n \le n/\varphi(n)$. The next result uses Hardy–Ramanujan numbers, which are defined below. If Robin's inequality is satisfied by all Hardy–Ramanujan integers in a range, it is satisfied by all integers in that range. These results are summarized in Theorem 8.10, namely any integer greater than 5040 and **not** satisfying Robin's inequality must be even, not squarefree and not squarefull. The smallest such integer, if it exists, must also be a Hardy–Ramanujan number and a superabundant number.

Following this work, successive improvements are given which show that counterexamples to Robin's inequality which are greater than 5040 must be divisible by a high power of at least one prime: first the theorem from [38] that all 5-free integers greater than 5040 satisfy Robin's inequality, then from [159] extending this to 7-free integers, and finally a work of Tim Trudgian and the author extending this to 11-free integers.

To summarize this chapter: Section 8.2 discusses Hardy–Ramanujan numbers and finishes with Theorem 8.10 discussed above. Then in each of the three successive sections, Sections 8.3, 8.4 and 8.5, the given cases of r-free integers are proved.

It is somewhat confusing that references to "Robin's inequality" sometimes implicitly include the lower bound $n > 5040$. In this volume by Robin's inequality we always mean

$$\frac{\sigma(n)}{n} < e^{\gamma} \log\log n,$$

and any lower bound is given explicitly.

The following RHpack (see Appendix B) variables and functions relate to the material in this chapter: SetA, SetB, SetC, SetD, HardyRamanujanQ, NthPrimeBounds, LogKthPrimorialBounds, KthApproximatePrimorial, PrimorialFormToInteger and HardyRamanujanToPrimorialForm.

8.2 Hardy–Ramanujan Numbers

We say as usual that an integer n is **squarefree** if, for all primes p, $p^2 \nmid n$. We say it is **squarefull** if $p \mid n$ implies $p^2 \mid n$. Finally we say it is a **Hardy–Ramanujan number** if it has the form for its standard factorization into primes:

$$n = p_1^{\alpha_1} \cdots p_m^{\alpha_m},$$

where $p_1 = 2$, $p_2 = 3$, etc. and, for all i with $1 \le i \le m$, p_i is the ith prime, and the exponents are non-increasing, that is to say, $\alpha_i \ge \alpha_{i+1} \ge 1$ for $1 \le i \le m-1$.

Lemma 8.1 *[38, lemma 2.1(2)] For all $x \ge 4$*

$$\sum_{p \le x} \frac{1}{p} \le \log\log x + \gamma,$$

where the sum is over primes p and γ is Euler's constant.

Proof By Lemma 5.11 we can write for $1 < x$

$$\sum_{p \le x} \frac{1}{p} \le \log\log x + B + \frac{1}{2\log^2 x},$$

where $B < 0.261498$ is Mertens' constant. Finally note that at $x = 4$ we have

$$B + \frac{1}{2\log^2 x} < \gamma$$

to complete the proof. □

Lemma 8.2 *[38, lemma 2.2 and corollary 2.1] If $a \in A$ and p is a prime number with $p \ge 7$, then other than the case $p = 7$ and $a \in \{12, 120, 360\}$, we have that ap satisfies Robin's inequality. Thus if $a \in B$ and $p \ge 7$ is prime, we have $ap \in R$.*

Proof Recall the definitions

$$\begin{aligned}A &:= \{1,2,3,4,5,6,8,9,10,12,16,18,20,24,30,36,48,\\ &\quad 60,72,84,120,180,240,360,720,840,2520,5040\},\\ B &:= \{2,3,5,6,10,30\}.\end{aligned}$$

If $p = 7$ then checking numerically, for $a \in \{12,120,360\}$, ap does not satisfy Robin's inequality. Also numerically, for each $a \in A$, $11a$ satisfies Robin's inequality. Hence, if $p \geq 11$ and $a \in A$ we get

$$\begin{aligned}\frac{\sigma(ap)}{ap} &= \left(1+\frac{1}{p}\right)\frac{\sigma(a)}{a} \leq \frac{12\sigma(a)}{11a} = \frac{\sigma(11a)}{11a}\\ &< e^{\gamma}\operatorname{loglog}(11a) \leq e^{\gamma}\operatorname{loglog}(ap),\end{aligned}$$

so ap satisfies Robin's inequality. Since $B \subset A \setminus \{12,120,360\}$, ap satisfies the inequality for all $a \in B$ and $p \geq 7$. □

We say a positive integer n satisfies **Nicolas' inequality** if

$$\frac{n}{\varphi(n)} < e^{\gamma}\operatorname{loglog} n.$$

Considering what we saw in Chapter 5, this definition is somewhat unnatural, since in Nicolas' first theorem the inequality is reversed to obtain an RH equivalence. However, we will use it since it is traditional.

Now let $S := \{3^a \cdot 5^b \cdot p^c : p \geq 7 \text{ is prime, } a,b,c \text{ are non-negative integers}\}$, so

$$S = \{1,3,5,7,9,11,13,15,17,19,21,23,\ldots\}.$$

Lemma 8.3 *[38, lemma 2.3] All elements of S other than $\{1,3,5,9\}$ satisfy Robin's inequality and all elements other than $\{1,3,5,9,15\}$ satisfy Nicolas' inequality.*

Proof For $p \geq 7$ and $n \in S \cap [31,\infty)$ we can write

$$\frac{\sigma(n)}{n} \leq \frac{n}{\varphi(n)} \leq \frac{3}{2}\cdot\frac{5}{4}\cdot\frac{p}{p-1} \leq \frac{3}{2}\cdot\frac{5}{4}\cdot\frac{7}{6} = 2.1875 < e^{\gamma}\operatorname{loglog} 31 \leq e^{\gamma}\operatorname{loglog} n.$$

The remaining elements of S can be checked numerically. □

Lemma 8.4 *[38, theorem 1.1] All* ***squarefree*** *integers other than the set B, satisfy Robin's inequality.*

Proof **(1)** Begin by noting that no element of B satisfies Robin's inequality. If $p \geq 7$ is prime then $p + 1 < 8p/7 < pe^{\gamma}\operatorname{loglog} p$, so all such primes satisfy Robin's inequality.

(2) We now proceed using induction. If $m := \omega(n)$ we have just shown the lemma is true for $m = 1$. Assume it is true for all squarefree integers n with $\omega(n) = m - 1$ for some $m \geq 2$ and let $n = q_1 \cdots q_m$ be an integer not in B and with primes q_i satisfying $q_1 < q_2 < \cdots < q_m$.

(3) Suppose first that $q_m \geq \log n$. If, on the one hand, $q_1 \cdots q_{m-1} \in B$ and $q_m \notin B$, then by Lemma 8.2, $n = (q_1 \cdots q_{m-1})q_m$ satisfies Robin's inequality. If, however, $q_m \in B$ and $n \notin B$ we must have $n = 15$, which can be shown to satisfy Robin's inequality directly.

If, on the other hand, $q_1 \cdots q_{m-1} \notin B$, we must have, by the inductive assumption,

$$(q_1 + 1)(\cdots)(q_{m-1} + 1) < e^\gamma q_1 \cdots q_{m-1} \log\log(q_1 \cdots q_{m-1}),$$

and so

$$(q_1 + 1)(\cdots)(q_m + 1) < e^\gamma q_1 \cdots q_{m-1}(q_m + 1) \log\log(q_1 \cdots q_{m-1}). \tag{8.1}$$

Note that if $0 < a < b$ we have

$$\frac{\log b - \log a}{b - a} = \frac{1}{b - a} \int_a^b \frac{dt}{t} > \frac{1}{b}.$$

Therefore setting $a = \log(q_1 \cdots q_{m-1})$ and $b = \log(q_1 \cdots q_m)$ and using the assumption $q_m \geq \log(n)$, we can write

$$\begin{aligned}\frac{\log\log n - \log\log(n/q_m)}{\log q_m} &= \frac{\log\log(q_1 \cdots q_m) - \log\log(q_1 \cdots q_{m-1})}{\log(q_1 \cdots q_m) - \log(q_1 \cdots q_{m-1})} \\ &> \frac{1}{\log(q_1 \cdots q_m)} \\ &\geq \frac{\log\log(q_1 \cdots q_{m-1})}{q_m \log q_m},\end{aligned}$$

so that

$$(q_m + 1) \log\log(q_1 \cdots q_{m-1}) \leq q_m \log\log(q_1 \cdots q_m).$$

Therefore by (8.1)

$$(q_1 + 1)(\cdots)(q_m + 1) < e^\gamma q_1 \cdots q_m \log\log(q_1 \cdots q_m),$$

so $n \in R$ and the result of the lemma follows by induction.

(4) Suppose now it is the case that $q_m < \log n$. Recall we can assume $m \geq 2$. If $q_m = 3$ then $n = 6$ but $\log 6 < 3$. If $q_m = 5$ then $n \in \{10, 15\}$, but $\log 10 < 5$ and $\log 15 < 5$. Hence we can assume $q_m \geq 7$. Note that for $a > 0$

$$\log(a + 1) - \log(a) = \int_a^{a+1} \frac{dt}{t} < \frac{1}{a},$$

so, by Lemma 8.1, we can write

$$\begin{aligned}\sum_{i=1}^{m} \log(q_i + 1) - \log(q_i) < \sum_{i=1}^{m} \frac{1}{q_i} \leq \sum_{p \leq q_m} \frac{1}{p} &\leq \gamma + \log\log q_m \\ &< \gamma + \log\log\log(q_1 \cdots q_m).\end{aligned}$$

Taking exponentials we get $\sigma(n)/n < e^\gamma \log\log n$, which completes the induction step in this case also. □

Define the **squarefree core** $\gamma(n)$ of n to be the largest squarefree integer which divides n, i.e.

$$\gamma(n) := \prod_{p|n} p.$$

Lemma 8.5 *[38, theorem 2.1] All* ***odd*** *integers other than the set* $\{1,3,5,9,15\}$, *satisfy Nicolas' inequality.*

Proof Let $m := \omega(n) \geq 1$. If $m = 1$ then, by Lemma 8.3, n satisfies Nicolas' inequality. If $m \geq 2$ then $n/\varphi(n) = \gamma(n)/\varphi(\gamma(n))$. Therefore if r is a squarefree number satisfying Nicolas' inequality so does every integer n with $\gamma(n) = r$.

Now let $n = q_1 \cdots q_m$ be squarefree with the q_i prime and $3 \leq q_1 < q_2 < \cdots < q_m$. In this case n satisfies Nicolas' inequality if and only if

$$\prod_{i=1}^{m} \frac{1}{1-1/q_i} = \prod_{i=1}^{m} \frac{q_i}{q_i - 1} < e^{\gamma} \operatorname{loglog}(q_1 \cdots q_m).$$

Because $q_1/(q_1 - 1) \leq 3/2$ and for $2 \leq i \leq m$, $q_i/(q_i - 1) < (q_{i-1} + 1)/q_{i-1}$, setting $n_1 := 2n/q_1$ gives

$$\frac{n}{\varphi(n)} < \frac{3}{2} \prod_{i=1}^{m-1} \frac{q_i + 1}{q_i} = \frac{\sigma(n_1)}{n_1}.$$

It follows that since n_1 is squarefree, by Lemma 8.4, other than elements of B, n_1 satisfies Robin's inequality, so we have

$$\frac{n}{\varphi(n)} < \frac{\sigma(n_1)}{n_1} < e^{\gamma} \operatorname{loglog} n_1 < e^{\gamma} \operatorname{loglog} n.$$

Therefore n would satisfy Nicolas' inequality. Elements of B with $m \geq 2$ can be checked numerically.

If however $m = 1$ then n would be a prime power so would be in S. The result then follows by Lemma 8.3. □

Recall the definition $\theta(x) := \sum_{p \leq x} \log p$ for $x \geq 1$. Recall that by Lemma 3.10 (d), for $x \geq 121$ we have $\theta(x) \geq 0.8x$. Recall also that by Lemma 5.12 for $x \geq 286$ we have

$$\prod_{p \leq x} \frac{p}{p-1} \leq e^{\gamma} \log x \left(1 + \frac{1}{2\log^2 x}\right).$$

These estimates enable us to derive the following upper bound for the left-hand side of that expression in terms of the product of successive primes:

Lemma 8.6 *[38, lemma 3.2] For all* $m > 4$ *we have*

$$\prod_{i=1}^{m} \frac{p_i}{p_i - 1} \leq e^{\gamma} \log(2\log(p_1 \cdots p_m)),$$

where as usual p_i is the ith prime.

Proof If $4 < m < 62$ a numerical check shows that the inequality is true, so we can assume $m \geq 62$. Note that $p_{62} = 293$. By Lemma 3.10 (*d*) we can write

$$\log(p_1 \cdots p_m) = \theta(p_m) > 0.8 p_m.$$

Therefore

$$\log(2\log(p_1 \cdots p_m)) > \log p_m + \log 1.6 \geq \log p_m + \frac{1}{\log p_m}.$$

Hence, using Lemma 5.12 with $x = p_m$, we get

$$\prod_{i=1}^{m} \frac{p_i}{p_i - 1} \leq e^\gamma \log(2\log(p_1 \cdots p_m)),$$

and the proof is complete. □

Recall the definition

$$D = \{4, 8, 9, 16, 36, 72, 108, 144, 216, 900, 1800, 2700, 3600, 44\,100, 88\,200\},$$

so all elements of D are squarefull.

Lemma 8.7 *[38, theorem 3.1] No element of D satisfies Nicolas' inequality. Every* ***squarefull*** *integer not in D satisfies Nicolas' inequality, and thus Robin's inequality.*

Proof Let $n = q_1^{\alpha_1} \cdots q_m^{\alpha_m}$ be a squarefull integer, with $q_1 < q_2 < \cdots < q_m$ and each q_j prime, and not satisfy Nicolas' inequality. Then if p_i is the ith prime

$$\prod_{i=1}^{m} \frac{p_i}{p_i - 1} \geq \prod_{i=1}^{m} \frac{q_i}{q_i - 1} = \frac{n}{\varphi(n)} \geq e^\gamma \log\log n \geq e^\gamma \log(2\log(p_1 \cdots p_m)).$$

By Lemma 8.6 we must therefore have $m \leq 4$. Hence

$$\frac{35}{8} = \frac{2}{1} \cdot \frac{3}{2} \cdot \frac{5}{4} \cdot \frac{7}{6} \geq \frac{n}{\varphi(n)} \geq e^\gamma \log\log n.$$

This implies $n \leq 116\,144$. There are 45 535 squarefull integers in this range. Each can be checked numerically to complete the proof. □

Let $n = q_1^{e_1} \cdots q_k^{e_k}$ be the standard prime factorization of the positive integer n, where in this case the primes q_i are ordered so that $e_1 \geq e_2 \geq \cdots \geq e_k$. Let $\bar{e} = (e_1, \ldots, e_k)$ and call $\bar{e}$ the **exponent pattern** of n. For example since $50 = 5^2 \times 2^1$, $\bar{e} = (2, 1)$. Define a positive integer

$$m(\bar{e}) := \prod_{i=1}^{k} p_i^{e_i},$$

where as usual p_i is the ith prime, so $m(\bar{e})$ has exponent pattern $\bar{e}$.

Lemma 8.8 *[38, proposition 5.1] If Robin's inequality holds for all Hardy–Ramanujan integers n with* $5041 \le n \le x$ *it holds for* ***all*** *integers in this same range.*

Proof **(1)** First note that for given $e > f > 0$ and $x \in (1, \infty)$ the function

$$g(x) := \frac{1 - x^{-e}}{1 - x^{-f}}$$

is strictly decreasing. To see this, observe that the numerator of the derivative is $ex^f - fx^e + f - e$ and its derivative is 0 at $x = 1$ and is negative for all $x > 1$.

(2) It follows from (1) that if p, q, f and e are positive integers with p, q prime, $p < q$ and $e < f$ then

$$\frac{\sigma(p^f q^e)}{p^f q^e} = \left(\frac{1 - p^{-f-1}}{1 - p^{-1}}\right)\left(\frac{1 - q^{-e-1}}{1 - q^{-1}}\right)$$
$$> \left(\frac{1 - p^{-e-1}}{1 - p^{-1}}\right)\left(\frac{1 - q^{-f-1}}{1 - q^{-1}}\right) = \frac{\sigma(p^e q^f)}{p^e q^f}.$$

(3) For primes p and q with $p < q$ we have $\sigma(p^e)/p^e > \sigma(q^e)/q^e$. Therefore if the exponent pattern is fixed, the maximum value of the ratio $\sigma(n)/n$ will be assumed when the primes are as small as possible, i.e. for $n = m(\bar{e})$ and if $n_1 = p_1^{\beta_1} \cdots p_m^{\beta_m}$ is any other number for which the maximum is assumed, by (2), we must have $\beta_1 \ge \cdots \ge \beta_m$. Thus

$$\frac{\sigma(m(\bar{e}))}{m(\bar{e})} = \max\left(\frac{\sigma(n)}{n} : n \text{ has factorization pattern } \bar{e}\right).$$

(4) This shows that if $\sigma(n)/n \ge e^\gamma \log\log n$, then the same inequality holds with n replaced by $m(\bar{e})$.

(5) Now we **claim** that if Robin's inequality is true for the integer $m(\bar{e})$, then it is true for every positive integer n having exponent pattern $\bar{e}$. To see this, let n be such that it has exponent pattern $\bar{e}$ and $m(\bar{e}) \le 5040$. Assume also that $n \ge 5041$ and that it does not obey Robin's inequality. Since the maximum number of prime factors of any number less than 5041 is 5, we can write using (3)

$$e^\gamma \log\log n \le \frac{\sigma(n)}{n} \le \frac{\sigma(m(\bar{e}))}{m(\bar{e})} \le \prod_{p \le 11} \frac{1 - p^{-6}}{1 - p^{-1}} < 5,$$

so $\log(5041) \le \log n < 14$. Checking n values in this range numerically shows that they satisfy Robin's inequality, a contradiction, completing the proof of the claim. Thus if $n > 5040$ and $m(\bar{e}) \le 5040$ then n obeys Robin's inequality.

If however $n > 5040$ and $m(\bar{e}) > 5040$, since $m(\bar{e})$ is a Hardy–Ramanujan number it must, by the hypothesis of the lemma, satisfy Robin's inequality. Hence, because of (3) and $n \ge m(\bar{e})$,

$$\frac{\sigma(n)}{n} \le \frac{\sigma(m(\bar{e}))}{m(\bar{e})} < e^\gamma \log\log(m(\bar{e})) \le e^\gamma \log\log n.$$

Therefore if Robin's inequality holds for all Hardy–Ramanujan integers with $5041 \leq n \leq x$ it holds for **all** integers in this same range, which completes the proof. □

Recall that a natural number n is **superabundant** if for all $1 \leq m < n$ we have $\sigma(m)/m < \sigma(n)/n$. For example, a simple computation shows that

$$\{1,2,4,6,12,24,36,48,60,120,180,240\}$$

is the set of the first 12 superabundant numbers; see Table A.3.

Lemma 8.9 *[2, theorem 3] The smallest integer* $n > 5040$ *which does* **not** *satisfy Robin's inequality must be a superabundant number.*

Proof A computation shows all integers n in the range $5040 < n \leq 10080$ satisfy Robin's inequality, and that 10080 is a superabundant number. In order that we might get a contradiction, assume there is an integer $n > 10080$ for which Robin's inequality fails, and call n the smallest such integer. Assume that n is *not* superabundant. Then there is an integer $m > 0$ such that $m < n$ and

$$\frac{\sigma(m)}{m} \geq \frac{\sigma(n)}{n} \geq e^{\gamma} \log\log n > e^{\gamma} \log\log m. \tag{8.2}$$

If $m < 10080$, since 10 080 is superabundant we would have, by (8.2),

$$\frac{\sigma(n)}{n} \leq \frac{\sigma(m)}{m} \leq \frac{\sigma(10080)}{10080},$$

and we could replace m by 10 080. Thus we may assume also $5040 < 10080 \leq m$. Again by (8.2), m does not satisfy Robin's inequality, a contradiction, since n is the smallest such integer. This completes the proof. □

Summarizing Lemmas 8.5, 8.7, 8.8 and 8.9 we have:

Theorem 8.10 *[38, theorems 1.3 and 1.4] Any integer greater than 5040* **not** *satisfying Robin's inequality is even, not squarefree and not squarefull. The smallest such integer, if it exists, must also be a Hardy–Ramanujan number and a superabundant number.*

8.3 Integers Not Divisible by the Fifth Power of Any Prime

Recall some standard definitions. If n is an integer, $P(n)$ denotes the largest prime dividing n. If $r \geq 2$ is an integer we say n is r-free if, for all primes p, $p^k \mid n$ implies $k < r$.

Lemma 8.11 *[38, theorem 1.5] All 5-free integers greater than 5040 satisfy Robin's inequality.*

Proof **(1)** Suppose there exists an integer larger than 5040 which is 5-free and does *not* satisfy Robin's inequality. We **claim** that if n is the *smallest* integer with this property then $P(n) < \log n$. To see this first let $q := P(n)$ with $n = mq$. Since n is minimal, either m satisfies Robin's inequality or it does not and $m \le 5040$ so $m \in A$. If this latter is true, because

$$\max\{n \in A : P(n) \le 5\} = 720,$$

and $5 \times 720 = 5040$, we can assume $q \ge 7$. By Lemma 8.2, since $n > 5040$ and does not satisfy Robin's inequality, we obtain a contradiction in this case. Hence m satisfies Robin's inequality so $m \ge 7$.

(2) Assume now, to get a contradiction, that we also have $q \ge \log n$. This implies

$$q \log q \ge \log n \log\log n > \log n \log\log m,$$

which, since

$$\frac{\log\log n - \log\log m}{\log q} = \frac{1}{\log n - \log m}\int_{\log m}^{\log n} \frac{dt}{t} > \frac{1}{\log n},$$

implies first

$$\frac{q}{\log n} > \frac{\log\log m}{\log q} \quad \Longrightarrow \quad \frac{q(\log\log n - \log\log m)}{\log q} > \frac{\log\log m}{\log q}$$

$$\Longleftrightarrow \quad \left(1 + \frac{1}{q}\right)\log\log m < \log\log n.$$

Then, since for all integers $a, b \ge 1$ we have $\sigma(ab) \le \sigma(a)\sigma(b)$, we get

$$\frac{\sigma(n)}{n} = \frac{\sigma(qm)}{qm} \le \left(1 + \frac{1}{q}\right)\frac{\sigma(m)}{m} < \left(1 + \frac{1}{q}\right)e^{\gamma}\log\log m < e^{\gamma}\log\log n,$$

so n satisfies Robin's inequality, a contradiction, completing the proof of the claim.

(3) Note next that the set of 5-free Hardy–Ramanujan integers $n > 5040$ with $P(n) \le 73$ is finite. An explicit computation shows each member satisfies Robin's inequality.

(4) Now let $t \ge 2$ and define for $x \ge 3$

$$R_t(x) := \prod_{p>x}\left(\frac{1}{1 - 1/p^t}\right) \quad \text{and} \quad S_t(x) := \prod_{p \le x}\left(\frac{1 - 1/p^t}{1 - 1/p}\right).$$

Then

$$\log R_t(x) = -\sum_{p>x} \log(1 - p^{-t})$$

$$= \sum_{p>x} \sum_{m\geq 1} \frac{1}{mp^{tm}}$$

$$\leq \sum_{p>x} \sum_{m\geq 1} \frac{1}{(p^m)^t}$$

$$\leq \sum_{n>x} \frac{1}{n^t} \leq \frac{t}{t-1} \frac{1}{x^{t-1}}.$$

(The reader might wish to experiment with the better upper bound given by $1/((t-1)(x-1)^{t-1})$. Therefore $\log R_t(x) \leq tx^{1-t}/(t-1)$.

(5) Because $R_t(x)/\zeta(t) = \prod_{p\leq x}(1-p^{-t})$, using the result of Step (3) we can write

$$\sum_{p\leq x} \log(1-p^{-t}) = -\log\zeta(t) + \log R_t(x) \leq -\log\zeta(t) + \frac{t}{(t-1)x^{t-1}}$$

and so, by Lemma 5.12, we get

$$S_t(x) = \sum_{p\leq x} \log\left(\frac{1-1/p^t}{1-1/p}\right)$$

$$\leq -\log\zeta(t) + \frac{tx^{1-t}}{t-1} + \gamma + \log\log x + \log\left(1 + \frac{1}{2\log^2 x}\right). \qquad (8.3)$$

(6) Next we **claim** that if m is a 5-free integer such that $P(m) < \log m$ and such that m does *not* satisfy Robin's inequality, then we must have $P(m) \leq 73$. To see this we use the estimate derived in Step (4) with $t = 5$. First we set $x = \log m$. Since m is 5-free,

$$\frac{\sigma(m)}{m} = \prod_{p|m} \frac{1-p^{-\nu_p(m)-1}}{1-p^{-1}} \leq \prod_{p|m} \frac{1-p^{-5}}{1-p^{-1}} \leq \prod_{p\leq\log m} \frac{1-p^{-5}}{1-p^{-1}} = S_5(\log m),$$

so $\sigma(m)/m \leq S_t(x)$. By the estimate (8.3),

$$\log\left(\frac{\sigma(m)}{m}\right) \leq -\log\zeta(5) + \frac{5}{4x^4} + \gamma + \log\log x + \log\left(1 + \frac{1}{\log^2 x}\right).$$

If x_0 satisfies

$$\frac{5}{4x_0^{t-1}} + \log\left(1 + \frac{1}{\log^2 x_0}\right) < \log\zeta(5),$$

then Robin's inequality would be satisfied whenever $x \geq x_0$, which is false. For example we can choose $x_0 = 196$ so here $x < 196$ and

$$e^\gamma \log\log m \leq \frac{\sigma(m)}{m} < S_5(193) < 9.2.$$

Hence we must have $\log m \le \exp(S_5(193)e^{-\gamma}) < 175$ and the largest prime less than 175 is 173.

We then repeat this process, replacing 193 with 173, and checking $\exp(S_5(173)e^{-\gamma}) \le 173$, which is true. This continues until we reach the prime 73 for which $\exp(S_5(73)) > 73$: so the process necessarily stops:

$$193 \to 173 \to 151 \to 137 \to 109 \to 103 \to 97 \to 89 \to 83 \to 79 \to 73$$

and we conclude $P(m) < \log m \le 73$.

(7) We are now able to give the final step in the proof. To obtain a contradiction, let n be the smallest 5-free integer which is greater than 5040 and which does *not* satisfy Robin's inequality. By Step (1) we have $P(n) < \log n$ and by Step (6) $P(n) \le 73$.

Now let $\bar{e}$ be the factorization pattern of n. Then necessarily $m(\bar{e})$ is 5-free and $m(\bar{e}) \le n$. Since n is minimal, by Lemma 8.8 we cannot have $5041 \le m(\bar{e}) < n$ so either $m(\bar{e}) = n$, and n is Hardy–Ramanujan, or $m(\bar{e}) \le 5040$, so by Lemma 8.8 Step (5) n satisfies Robin's inequality, which is false. Hence n is a Hardy–Ramanujan number, which by Step (3) implies n satisfies Robin's inequality. This contradiction completes the proof. □

Thus we have shown:

Theorem 8.12 *[38, theorem 1.5] Any integer greater than 5040 which does not satisfy Robin's inequality must be divisible by at least the fifth power of at least one odd prime.*

8.4 Integers Not Divisible by the Seventh Power of Any Prime

First we set out some definitions. Let $n_0 := 2263$. Recall $N_n = p_1 \cdots p_n$ for $n \ge 1$ is the nth primorial with $N_1 = 2$. Let

$$f(n) := 1 + \frac{7}{10 \log p_n \log\log N_n}$$

and set $n_1(t)$ to be the smallest integer not less than n_0 and such that for $n = n_1(t)$ we have

$$e^{2/p_n} f(n) < \zeta(t),$$

where $t \ge 2$ is an integer.

Define

$$\Psi_t(n) := n \prod_{p|n} \frac{1-p^{-t}}{1-p^{-1}},$$

and set $Q_t(n) := \Psi_t(n)/(n \log\log n)$. Recall the definitions

$$G(n) = \frac{\sigma(n)}{n \log\log n} \quad \text{and} \quad R_t(x) = \prod_{p>x} (1-p^{-t})^{-1}.$$

In addition recall the definition of a **colossally abundant** integer from Chapter 6: n is colossally abundant if for some $\epsilon > 0$ we have for all $m > 1$

$$\frac{\sigma(n)}{n^{1+\epsilon}} \geq \frac{\sigma(m)}{m^{1+\epsilon}}.$$

The theorem of Solé and Planat, described below, depends critically on computations of Briggs [16], which produced all colossally abundant numbers up to $10^{10^{10}}$, and verified that all of them larger than 5041 satisfied Robin's inequality. This computational fact is used also in Section 8.5 below. Recall the result of Lemma 7.1: if there exists an $n > 5040$ for which Robin's inequality is false, then there exists such an n which is colossally abundant.

Lemma 8.13 *[159, proposition 4] Let $n \geq 2263$ and $t \geq 2$. Then*

$$\frac{\Psi_t(N_n)}{N_n} \leq \frac{\exp(\gamma + 2/p_n)}{\zeta(t)}\left(\log\log N_n + \frac{7}{10\log p_n}\right).$$

Proof **(1)** Let $N_n \leq m < N_{n+1}$. Since $m < N_{n+1}$ the number of distinct prime divisors of m, $\omega(\gamma(m))$, must be less than or equal to n. If written in ascending order, the ith of these prime divisors must be not less than p_i so

$$\frac{\Psi_t(m)}{m} \leq \frac{\Psi_t(N_n)}{N_n}.$$

Also for $n \geq 2$, we have $0 < \log\log N_n \leq \log\log m$ which implies

$$Q_t(m) = \frac{\Psi_t(m)}{m\log\log m} \leq \frac{\Psi_t(N_n)}{N_n \log\log N_n} = Q_t(N_n).$$

(2) By Lemma 8.11 Step (4) we have

$$\log R_t(p_n) \leq \frac{t}{(t-1)p_n^{t-1}} \leq \frac{2}{p_n},$$

which implies

$$\prod_{p > p_n}(1 - p^{-t})^{-1} \leq \exp\left(\frac{2}{p_n}\right).$$

(3) By Lemma 3.14 we have for $p_n \geq 2 \times 10^4$,

$$\log N_n = \theta(p_n) = \sum_{p \leq p_n} \log p > p_n\left(1 - \frac{1}{8\log p_n}\right).$$

Thus, because for $x < \frac{1}{2}$, $\log(1-x) > -x$,

$$\log\log N_n > \log p_n + \log\left(1 - \frac{1}{8\log p_n}\right) > \log p_n - \frac{1}{8\log p_n},$$

and therefore

$$\log p_n < \log\log N_n + \frac{1}{8\log p_n}.$$

(4) Now write

$$\frac{\Psi_t(N_n)}{N_n} = \prod_{i=1}^{n} \frac{(1-p_i^{-t})}{1-p_i^{-1}} = \frac{\prod_{p>p_n}(1-p^{-t})}{\zeta(t)} \prod_{p\le p_n}\left(1-\frac{1}{p}\right)^{-1}.$$

Therefore by Steps (1), (2) and (3) and Lemma 5.12, for $n > \pi(20000) = 2262$ we get

$$\begin{aligned}\frac{\Psi_t(N_n)}{N_n} &\le \frac{\exp(2/p_n)}{\zeta(t)} e^{\gamma}\left(\log p_n + \frac{1}{2\log p_n}\right)\\ &\le \frac{\exp(\gamma + 2/p_n)}{\zeta(t)}\left(\log\log N_n + \frac{5}{8\log p_n}\right)\\ &\le \frac{\exp(\gamma + 2/p_n)}{\zeta(t)}\left(\log\log N_n + \frac{7}{10\log p_n}\right).\end{aligned}$$

This completes the proof of the lemma. □

Recall the definition of $n_1(t)$: it is the smallest integer n with $n \ge n_0$ and such that $\exp(2/p_n) f(n) < \zeta(t)$ where $f(n) := 1 + 7/(10\log p_n \log\log N_n)$.

Lemma 8.14 *[159, corollary 9] For $n \ge n_1(t)$ and $M \ge N_n$ we have $Q_t(M) < e^{\gamma}$.*

Proof Since $n \ge n_1(t)$ we have

$$e^{2/p_n}\left(1 + \frac{7}{10\log p_n}\right) < \zeta(t),$$

which implies by Lemma 8.13

$$Q_t(N_n) = \frac{\Psi_t(N_n)}{N_n \log\log N_n} \le \frac{\exp(\gamma + 2/p_n)}{\zeta(t)}\left(1 + \frac{7}{10\log p_n \log\log N_n}\right) < e^{\gamma}.$$

Because $M \ge N_n$, for some $n_2 \ge n \ge n_1(t)$ we have $N_{n_2} \le M < N_{n_2+1}$. Therefore, by Step (1) of Lemma 8.13 we have

$$Q_t(M) \le Q_t(N_{n_2}) < e^{\gamma},$$

and the proof of the lemma is complete. □

Theorem 8.15 *[159, theorem 10] If an integer N is 7-free and $N > 5040$ then it satisfies Robin's inequality.*

Proof By Step (1) of Lemma 8.13 and Lemma 8.14, if $n \ge n_1(7) = 1292$ and $N \ge N_n$ then, since n is 7-free,

$$\sigma(N) \le \Psi_7(N) < Ne^{\gamma} \log\log N,$$

so all such integers $N \geq N_{n_1(7)} < 3.51 \times 10^{4550}$ satisfy Robin's inequality. By the computations of Briggs [16, p. 253] Robin's inequality is true for

$$5040 < N \leq 10^{10^{10}},$$

and therefore true for all 7-free integers $N > 5040$. □

8.5 Integers Not Divisible by the 11th Power of Any Prime

This section differs from our usual practice of proving all, or at least most, of the preliminary lemmas. It also relies on some unpublished material of Dusart and some computational results of Briggs, which could prove difficult to reproduce. It is proved that if there is an $n \geq 5041$ for which $\sigma(n) \geq e^{\gamma} n \log\log n$, then n must be divisible by the 11th power of some prime.

Recall that Robin, in Theorem 7.16, proved that the Riemann hypothesis is equivalent to the **Ramanujan–Robin criterion**, namely the inequality

$$\sigma(n) < e^{\gamma} n \log\log n \tag{8.4}$$

being true for all $n > 5040$. Recall also that we say that a number is t-free if it is not divisible by the tth power of any prime. Choie, Lichiardopol Moree and Solé [38] showed that (8.4) is true for all 5-free integers; Solé and Planat [159] showed that (8.4) is true for all 7-free integers. We have presented these results in Theorems 8.12 and 8.15, respectively. Therefore if there is some $n \geq 5041$ for which $\sigma(n) \geq e^{\gamma} n \log\log n$, then n must be divisible by the seventh power of some prime.

The purpose of this section is to derive the following result.

Theorem 8.16 *[31] If there is some $n \geq 5041$ for which $\sigma(n) \geq e^{\gamma} n \log\log n$, then n must be divisible by the 11th power of some prime.*

It is easy to check that the only two positive integers $n \leq 5040$ that are divisible by an 11th power of a prime, namely 2^{11} and 2^{12}, both satisfy Robin's inequality. In other words:

Corollary 8.17 *The Riemann hypothesis is equivalent to Robin's inequality being satisfied by all integers greater than 2 that are divisible by the 11th power of at least one prime.*

Solé and Planat prove their results using primorials. Recall that the nth primorial is defined as $N_n = \prod_{i=1}^{n} p_i$, where p_n denotes the nth prime with $p_1 = 2$. For an integer $t \geq 2$ define as before

$$\Psi_t(n) := n \prod_{p|n} \left(1 + \frac{1}{p} + \cdots + \frac{1}{p^{t-1}}\right), \quad Q_t(n) := \frac{\Psi_t(n)}{n \log\log n}.$$

Solé and Planat note that for a t-free integer n one has $\sigma(n) \leq \Psi_t(n)$. Using their method, for a given t, first find an integer $n_1(t)$ such that $R_t(N_{n_1(t)}) < e^{\gamma}$

and $N_{n_1(t)} < 10^{10^{10}}$. Solé and Planat in Lemma 8.14 showed that for all $N > N_{n_1(t)}$ we have $Q_t(N) < e^\gamma$. Since Briggs [16] has proved that Robin's inequality is true for all $5041 \le n \le 10^{10^{10}}$, this then shows that Robin's inequality is true for all such t-free integers $n \ge 5041$.

The main idea of this section is to use explicit estimates on sums over primes to bound the primorials. This makes for an easy computation, and one which is used to verify Theorem 8.16.

First the function $Q_t(n)$ is estimated using bounds on primorials. For $t \ge 2$ we have by Step (4) of Lemma 8.13

$$Q_t(N_n) = \frac{1}{\log\log N_n} \prod_{k=1}^{n} \frac{1 - p_k^{-t}}{1 - p_k^{-1}} = \frac{\prod_{p > p_n}(1 - p^{-t})^{-1}}{\zeta(t) \log\log N_n} \prod_{p \le p_n} (1 - p^{-1})^{-1}. \quad (8.5)$$

The first product on the right-hand side of the expression (8.5) is estimated using Lemma 6 of Solé and Planat, namely,

$$\prod_{p \ge p_n} (1 - p^{-t})^{-1} \le \exp(2/p_n), \quad (8.6)$$

for all $n \ge 2$. Although this could be improved, such an improvement would have negligible influence on the final result.

To estimate the second product on the right-hand side of (8.5), the following result

$$\prod_{p \le x} (1 - p^{-1})^{-1} \le e^\gamma \log x \left(1 - \frac{1}{5 \log^2 x}\right)^{1}, \quad x \ge 2973, \quad (8.7)$$

given in the unpublished result of Dusart [50, theorem 6.12] is used. This improves on Lemma 5.12.

Now to apply the estimates (8.6) and (8.7) with $x = p_n$, an explicit bound on the nth prime is needed. It is possible to proceed without an explicit bound, though this increases greatly the computation time when n is large. In fact, if one considers the function

$$f(x) = \frac{\exp(2/x) \log x}{\left(1 - \frac{1}{5 \log^2 x}\right)}, \quad (8.8)$$

which is increasing for all $x \ge 5$, one sees that it is sufficient to consider only an explicit upper bound on p_n. Consider

$$p_n \le b_1(n) := n(\log n + \log\log n - \tfrac{1}{2}), \quad n > 20, \quad (8.9)$$

and

$$p_n \le b_2(n) := n(\log n + \log\log n - 0.9484), \quad n \ge 39017, \quad (8.10)$$

due respectively to Rosser and Schoenfeld [145, (3.11)] and Dusart [49, section 4].

Now we bound the factor $\log\log N_n$ in expression (8.5) and make use of the function $\theta(x) = \sum_{p \le x} \log p$. Since $\log N_n = \sum_{i=1}^{n} \log p_i = \theta(p_n)$, bounds on N_n can be derived using bounds on $\theta(p_n)$. These results are given in the following lemma.

Lemma 8.18 *For* $k \ge 198$,

$$k\left(\log k + \log\log k - 1 + \frac{\log\log k - 2.1454}{\log k}\right)$$
$$\le \log N_k \le k\left(\log k + \log\log k - 1 + \frac{\log\log k - 2}{\log k}\right). \tag{8.11}$$

Proof The left inequality follows from Robin [141, theorem 7] and is valid for all $k \ge 3$. The right inequality follows from Massias and Robin [114, theorem B(v)] and is valid for all $k \ge 198$. □

Note that the constant "2" in the right inequality in (8.11) cannot be improved. One could reduce the "2.1454" that appears in the left inequality at the expense of taking a much larger k. As shown below at the end of this section, this has very little influence on our problem.

Now we use the estimates (8.6)–(8.10) and Lemma 8.18 to estimate the right-hand side of expression (8.5). We have

$$Q_t(N_n) \le \begin{cases} e^{\gamma} g_1(t,n), & 430 \le n \le 39016, \\ e^{\gamma} g_2(t,n), & n \ge 39017, \end{cases} \tag{8.12}$$

where

$$g_i(t,n) = \frac{f(b_i(n))}{\zeta(t)\log\left\{n\left(\log n + \log\log n - 1 + \frac{\log\log n - 2.1454}{\log n}\right)\right\}},$$

for $i = 1,2$. We require $n \ge 430$ in the bounds (8.12) to ensure that the conditions in the estimate (8.7) and Lemma 8.18 are met since $p_{429} = 2971$ and $p_{430} = 2999$. Following Solé and Planat, define $n_1(t)$ to be the least value of $n \ge 430$ for which $g_i(t,n) < 1$. Since $g_i(t,n)$ is decreasing in n, we will have $g_i(t,n) < 1$ for all $n \ge n_1(t)$. For such an $n_1(t)$ consider the size of the associated $N_{n_1(t)}$. The results are summarized in Table 8.1. Exact values of $N_{n_1(t)}$ for $t = 11$ and 12 are not given owing to the computational complexity of calculating the nth primorial exactly.

As has been previously noted, Briggs [16] has proved that Robin's inequality is true for $5041 \le n \le 10^{10^{10}}$, by demonstrating numerically it holds for all colossally abundant numbers in this range, and then invoking Theorem 6.16. To prove a statement such as "Robin's inequality holds for all

Table 8.1 Values of $n_1(t)$ and $N_{n_1(t)}$.

t	$n_1(t)$	$N_{n_1(t)}$	Upper bound on $N_{n_1}(t)$ using Lemma 8.18
6	430	3.3×10^{1273}	1.4×10^{1276}
7	1847	3.3×10^{6836}	2.7×10^{6851}
8	39 017	4.9×10^{202520}	2.3×10^{202725}
9	39 017	4.9×10^{202520}	2.3×10^{202725}
10	234 372	1.2×10^{1416098}	1.8×10^{1416984}
11	48 304 724	—	2.8×10^{411504586}
12	162 914 433 505	—	$>10^{10^{12}}$

t-free integers", one must show that $N_{n_1(t)} \le 10^{10^{10}}$. The last entry in Table 8.1 shows that, at present, it is impossible to consider $t = 12$ without a new idea.

Conclusion. Here is a brief discussion of the possibility of proving that Robin's inequality is satisfied by all 12-free integers. Dusart [50] (unpublished) has considered some improved versions of the estimates (8.7) and (8.11), namely

$$\prod_{p\le x}(1-p^{-1})^{-1} < e^{\gamma}\log x\left(1+\frac{1}{5\log^2 x}\right), \qquad x \ge 2973,$$

and

$$k\left(\log k + \log\log k - 1 + \frac{\log\log k - 2.04}{\log k}\right) \le \theta(p_k), \quad p^k \ge 10^{15}.$$

These results appear respectively as Theorem 6.12 and Proposition 6.2 in [50]. Even with these improvements one still has $N_{n_1(12)} > 10^{10^{12}}$. Without injecting new ideas into the argument, one would have to increase the range of Briggs' computations beyond $10^{10^{10}}$.

To extend the computation one need only check those numbers that are colossally abundant and are divisible by the 11th power of some prime. Presumably, this is a very thin set of numbers. Checking only these numbers may motivate an extension of Briggs' computations and hence the possibility of extending the results in this section to $t = 12$.

8.6 Unsolved Problems

(1) Extend the method of Solé and Planat to $t = 12$ and beyond.

(2) Being divisible by the 11th power of at least one prime is perhaps only one type of constraint that can be derived for numbers greater than 5040 which violate Robin's inequality. This problem is to find new constraints for such numbers.

(3) Show that any counterexample to Robin's inequality (greater than 5040) must be colossally abundant.

9
Left, Right and Extremely Abundant Numbers

9.1 Introduction

This chapter explores further properties of numbers for which Robin's inequality fails. It is based on two papers [34, 35], by the same three authors published in 2011 and 2012: Geoffrey Caveney, Jean-Louis Nicolas and Jonathan Sondow. Two new equivalences to RH are derived. It also includes an equivalence from 2013 of Nazardonyavi and Yakubovich.

Recall the definition of Grönwall's function:

$$G(n) := \frac{\sigma(n)}{n \log\log n}, \quad n > 1,$$

and note the important property which is used in this chapter and proved in Section 9.2:

Theorem 9.2 (Grönwall's theorem) *[69; 75, theorem 323] We have that*

$$\limsup_{n\to\infty} G(n) = e^{\gamma}.$$

We say a composite positive integer n is **left abundant** if $G(n/p) \leq G(n)$ for all prime divisors $p \mid n$. For example, for all $p \in \mathbb{P}$, p is not left abundant since $G(1)$ is not defined. We say it is **right abundant** if $G(n) \geq G(an)$ for all integers $a > 1$. It is said to be **extraordinary** if it is both left abundant and right abundant. Finally, a natural number n is said to be **extremely abundant** if either $n = 10080$ or $n > 10080$ and $G(m) < G(n)$ for all m with $10080 \leq m < n$. Note that 10 080 is the smallest superabundant number which is greater than 5040.

Recall n is **superabundant** if for all $1 \leq m < n$ we have

$$\frac{\sigma(m)}{m} < \frac{\sigma(n)}{n},$$

and denote by $\mathcal{S}$ the set of all superabundant numbers. It follows from Theorem 9.2 that, since $\sigma(n)/n = G(n)\log\log n$, we have $\limsup_{n\to\infty}\sigma(n)/n = \infty$, so the number of superabundant numbers, the size of $\mathcal{S}$, is infinite.

The first main result is that RH is equivalent to 4 being the only extraordinary number, which is Theorem 9.8. We call this the **Caveney–Nicolas–Sondow criterion**. The theorem has the corollary that if a counterexample n to Robin's inequality exists with $n > 5040$, then so does one which is an extraordinary number.

The next main result, Lemma 9.16, is that Robin's inequality is satisfied for a very large range of integers, namely if $5040 < n < e^{19747}$ then $G(n) < e^{\gamma}$. This means any counterexample must be greater than 10^{10000}. Then in Theorem 9.17 the proof is given that the set of right abundant numbers not greater than 5040 is

$$A \setminus \{1,2,9,16,20,30,84,720,840\},$$

where A as defined at the start of Chapter 8 is the set of integers not greater than 5040 for which Robin's inequality is false. This theorem includes an equivalence for the Riemann hypothesis, namely if the Riemann hypothesis is true then there are **no other** right abundant numbers, and if the hypothesis is false then **infinitely many** right abundant numbers exist and in that case

$$\infty > \mu := \max\{G(n) : n > 5040\} > e^{\gamma}.$$

In addition, any integer M for which $G(M) = \mu$ is both colossally abundant and right abundant.

Theorem 9.19 shows that infinitely many colossally abundant numbers are also left abundant, and Theorem 9.21 that infinitely many colossally abundant numbers are **not** left abundant.

In a final section, Section 9.6, we include the theorem of Nazardonyavi and Yakubovich that the Riemann hypothesis is equivalent to the number of extremely abundant numbers being infinite. This is the **Nazardonyavi–Yakubovich criterion**.

This chapter has a complex network of dependences, and call-outs to results in other chapters. To assist the reader to find a way through, some dependences between results are presented as network flows. In Figures 9.1, 9.2 and 9.3, an arrow from vertex A to B, say, means derivation B requires the result of A.

The following RHpack (see Appendix B) functions relate to the material in this chapter: GronwallG, ExtraordinaryQ, RightAbundantQ, LeftAbundantQ and ExtremelyAbundantQ.

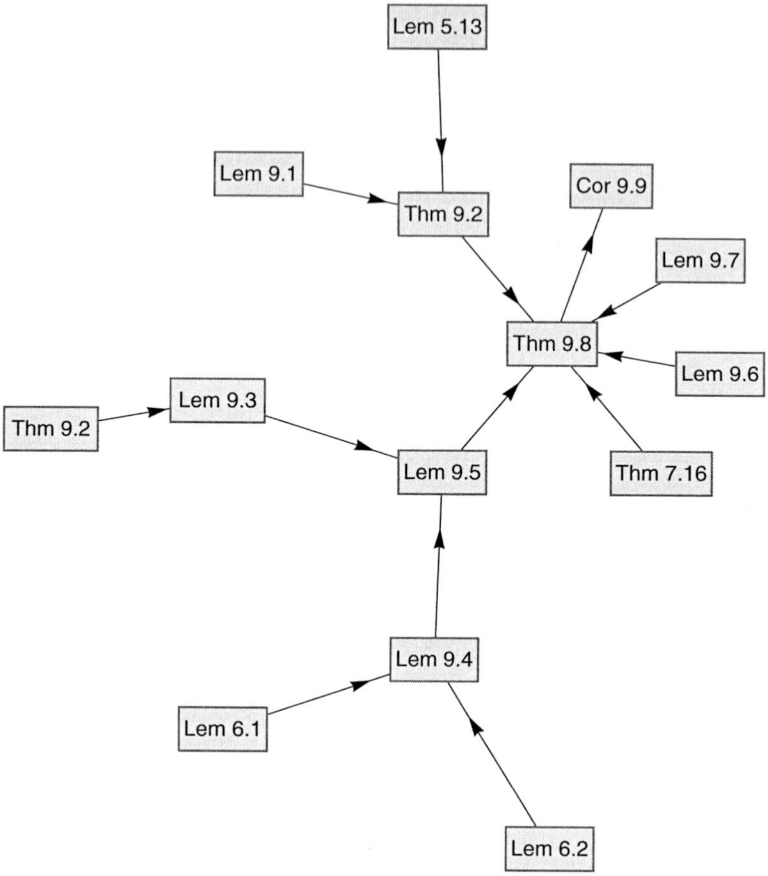

Figure 9.1 Dependences between some results in this chapter.

9.2 Grönwall's Theorem

The first preliminary result was proved by Grönwall in 1913 [69]. It underpins all of the work in this volume since Chapter 6, but is not needed explicitly until this chapter. First a standard lemma.

Lemma 9.1 *If p_n is the nth prime then as $n \to \infty$ we have $p_n \sim n \log n$.*

Proof By the prime number theorem we have $n \log p_n \sim p_n$. Taking logarithms gives

$$\log n + \log\log p_n \sim \log p_n.$$

Therefore

$$\frac{n \log n}{p_n} \sim \frac{n \log n}{n \log p_n} = \frac{\log n}{\log p_n} \sim \frac{\log p_n - \log\log p_n}{\log p_n} \sim 1,$$

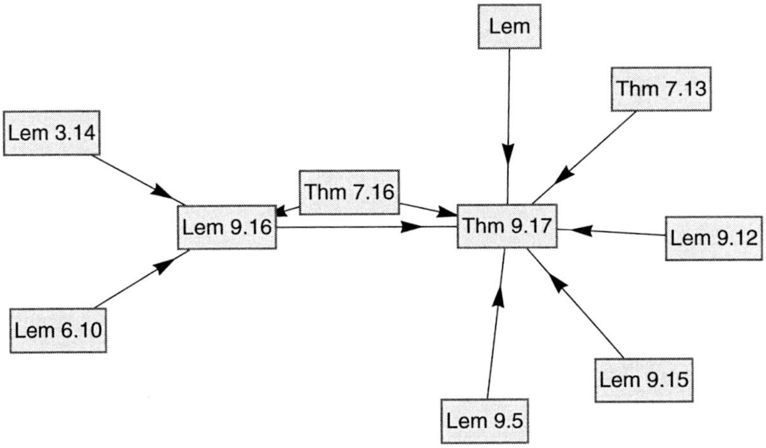

Figure 9.2 Further dependences between some results in this chapter.

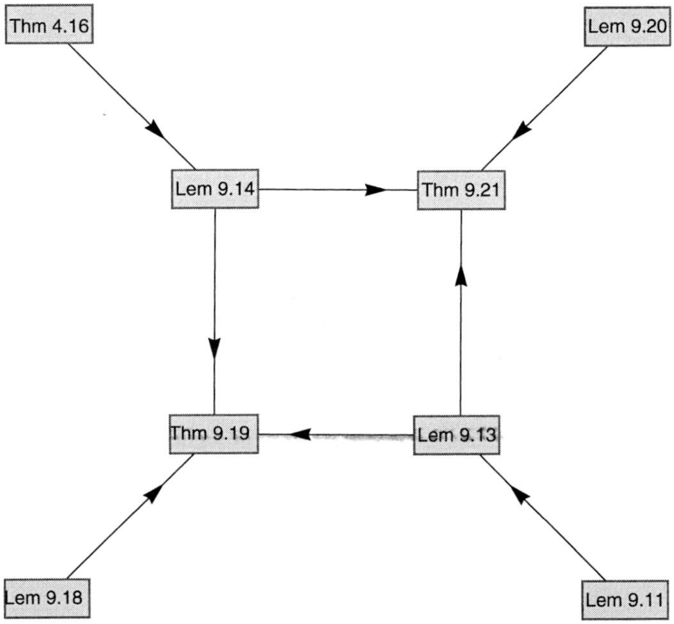

Figure 9.3 More dependences between some results in this chapter.

which completes the proof. □

Theorem 9.2 (Grönwall's theorem) *[69; 75, theorem 323] We have that*

$$\limsup_{n\to\infty} G(n) = e^{\gamma}.$$

Proof (1) Let $m \in \mathbb{N}$ and for each j let p_j be the jth prime with $p_1 = 2$. Define an integer n_m by

$$\prod_{j=1}^{n_m} p_j \leq m < \prod_{j=1}^{n_m+1} p_j \quad \Longrightarrow \quad \theta(p_{n_m}) \leq \log m < \theta(p_{n_m+1}). \tag{9.1}$$

By Lemma 9.1 we have

$$\lim_{n\to\infty} \frac{p_{n+1}}{p_n} = \lim_{n\to\infty} \frac{(n+1)\log(n+1)}{n\log n} = 1. \tag{9.2}$$

Using this limit and the prime number theorem in the form $\lim_{x\to\infty} \theta(x)/x = 1$, we can write

$$1 = \lim_{m\to\infty} \frac{\theta(p_{n_m})}{p_{n_m}} \leq \lim_{m\to\infty} \frac{\log m}{p_{n_m}} \leq \lim_{m\to\infty} \frac{\theta(p_{n_m+1})}{p_{n_m}} = \lim_{m\to\infty} \frac{p_{n_m+1}}{p_{n_m}} \frac{\theta(p_{n_m+1})}{p_{n_m+1}} = 1.$$

Thus

$$\lim_{m\to\infty} \frac{\log m}{p_{n_m}} = 1.$$

It follows that as $m \to \infty$

$$p_{n_m} = \log m(1 + o(1)). \tag{9.3}$$

(2) By Mertens' formula (Lemma 5.12) we can write, as $x \to \infty$,

$$\prod_{p\leq x} \left(1 - \frac{1}{p}\right) = \frac{1}{e^\gamma \log x}(1 + o(1)).$$

Let $x = p_m$ in this formula and use (9.3) to derive, using $\log(1 + o(1)) = o(1)$,

$$\prod_{j=1}^{n_m} \left(1 - \frac{1}{p_j}\right) = \frac{1}{e^\gamma \log p_{n_m}}(1 + o(1)) = \frac{1}{e^\gamma \log\log m}(1 + o(1)). \tag{9.4}$$

(3) Next write the prime factorization of m as

$$m = p_{j_1}^{\alpha_1} \cdots p_{j_n}^{\alpha_n},$$

where as before p_j is the jth prime, $j_1 < j_2 < \cdots < j_n$ and the exponents $\alpha_j > 0$. Then

$$\sigma(m) = m \prod_{i=1}^{n} \frac{1 - 1/p_{j_i}^{\alpha_i+1}}{1 - 1/p_{j_i}} < m \prod_{i=1}^{n} \frac{1}{1 - 1/p_{j_i}}. \tag{9.5}$$

Because $n \leq n_m$ and $p_i \leq p_{j_i}$ for all i it follows using Step (2) that

$$\prod_{i=1}^{n} \frac{1}{1 - 1/p_{j_i}} \leq \prod_{i=1}^{n} \frac{1}{1 - 1/p_i} \leq \prod_{i=1}^{n_m} \frac{1}{1 - 1/p_i} = e^\gamma \log\log m(1 + o(1)).$$

Using inequality (9.5) then gives

$$\limsup_{m\to\infty} \frac{\sigma(m)}{m \log\log m} \le e^{\gamma}. \tag{9.6}$$

(4) Next we construct a sequence of integers which shows the right-hand side of inequality (9.6) is attained. (Note that this sequence has found little use elsewhere in the work reported in this volume – which is somewhat surprising.)

Define a sequence of natural numbers $(a_n)_{n\in\mathbb{N}}$ by raising the nth primorial to a power which is determined by the logarithm of the largest prime, $\beta_n := \lfloor \log p_n \rfloor$:

$$a_n = (p_1 p_2 \cdots p_n)^{\beta_n}. \tag{9.7}$$

Then, by (9.3),

$$\begin{aligned}
\sigma(a_n) &= a_n \prod_{j=1}^{n} \frac{1 - 1/p_j^{\beta_n+1}}{1 - 1/p_j} \\
&> a_n \prod_{p} \left(1 - \frac{1}{p^{\beta_n+1}}\right) \prod_{j=1}^{n} \frac{1}{1 - 1/p_j} \\
&= \frac{a_n}{\zeta(\beta_n+1)} \prod_{j=1}^{n} \frac{1}{1 - 1/p_j} \\
&= \frac{a_n}{\zeta(\beta_n+1)} e^{\gamma} \log p_n (1 + o(1)).
\end{aligned} \tag{9.8}$$

(5) The definition of the a_n and the prime number theorem give

$$\log a_n = \beta_n \theta(p_n) = \beta_n p_n (1 + o(1)) = p_n \log p_n (1 + o(1)).$$

This implies

$$\log\log a_n = \log p_n + \log\log p_n + o(1) \quad \Longrightarrow \quad \lim_{n\to\infty} \frac{\log p_n}{\log\log a_n} = 1.$$

In addition, since $\zeta(\sigma) \to 1$ when $\sigma \to \infty$, by the expression (9.8) we get

$$\limsup_{n\to\infty} \frac{\sigma(a_n)}{a_n \log\log a_n} \ge e^{\gamma}.$$

This, together with the result of Step (3), completes the proof. □

9.3 Further Preliminary Results

Lemma 9.3 *[34, proposition 1, SA1] If $\mathcal{S}$ is the set of all superabundant numbers, then*

$$\limsup_{n\in\mathcal{S}} G(s) = e^{\gamma}.$$

Proof By Theorem 9.2 there is an increasing sequence of positive integers $(n_j)_{j\ge 1}$ such that $\lim_{j\to\infty} G(n_j) = e^\gamma$. If for any j we happen to have $n_j \in \mathcal{S}$, the set of all superabundant numbers, let $s_j := n_j$. Otherwise define

$$s_j := \max\{s \in \mathcal{S} : s < n_j\}.$$

Then if we set $T := \{s_j + 1, \ldots, n_j\}$ we would have $T \cap \mathcal{S} = \emptyset$. If we had $\sigma(s_j)/s_j < \sigma(n_j)/n_j$, then the value of $x \in T$ such that $\sigma(x)/x$ is a maximum would be in $\mathcal{S}$, which is impossible. Therefore $\sigma(s_j)/s_j \ge \sigma(n_j)/n_j$ and so $G(s_j) > G(n_j)$. It follows that

$$\limsup_{j\to\infty} G(s_j) \ge \limsup_{j\to\infty} G(n_j) = e^\gamma$$

and thus $\limsup_{s\in\mathcal{S}} G(s) = e^\gamma$. □

Lemma 9.4 *[34, proposition 1, SA2] For every fixed positive integer a, every sufficiently large $n \in \mathcal{S}$ is a multiple of a.*

Proof Let $a > 1$, let k be the largest exponent in the standard prime factorization of a, and let p be the maximum prime divisor of a. Then $a \mid (2 \cdot 3 \cdots p)^k$. Let

$$F := \{s \in \mathcal{S} : p^k \nmid s\} = \{s \in \mathcal{S} : 0 \le \nu_p(s) < k\},$$

where, as usual, $\nu_p(s)$ is the highest power of p which divides s. By Lemma 6.2 with $q = 2$ and $r = p$ we see that $\nu_2(s) < B$ for some bound B and all $s \in F$. If $q \mid s \in F$ is any prime factor then by Lemma 6.1 we have $\nu_q(s) < B$, and then, using Lemma 6.4, $q^{\nu_q(s)} < 2^{B+2}$. Using this with q the maximum prime dividing s ($q = P(s)$) gives $P(s) < 2^{B+2}$ so for all $s \in F$ we have

$$s < (2^{B+2}!)^B,$$

and therefore F is a finite set. Hence, for all s sufficiently large we have $p^k \mid s$. The result of the lemma now follows using Lemma 6.1. □

Lemma 9.5 *[34, lemma 1] If $a \in \mathbb{N}$ then $\limsup_{n\to\infty} G(an) = e^\gamma$. If n is right abundant then $G(n) \ge e^\gamma$.*

Proof The first part follows directly from Lemmas 9.3 and 9.4. For the second part, since n is right abundant we have $G(an) \le G(n)$, and so $e^\gamma = \limsup_{a\to\infty} G(an) \le G(n)$. □

Recall from the start of Chapter 8 that A denotes the set of positive integers not greater than 5040 for which Robin's inequality fails, i.e. $A := \{n \le 5040 : G(n) \ge e^\gamma\}$. Then

$$A := \{1,2,3,4,5,6,8,9,10,12,16,18,20,24,30,36,48,$$
$$60,72,84,120,180,240,360,720,840,2520,5040\},$$

Lemma 9.6 *[34, lemma 2] If $n \in A$ is greater than 5, then $G(n) < G(n/p)$ for some prime $p \mid n$.*

Proof The smallest prime factor of $n \in A \setminus \{1,2,3,4,5\}$ for which $G(n) < G(n/p)$ is given by

$$\begin{aligned}(n,p) \in \{&(6,2),(8,2),(9,3),(10,2),(12,2),(16,2),(18,3),(20,5),(24,3),\\ &(30,3),(36,2),(48,3),(60,5),(72,3),(84,7),(120,2),(180,5),\\ &(240,5),(360,5),(720,3),(840,7),(2520,7),(5040,2)\},\end{aligned}$$

so this result is a simple computation. □

Lemma 9.7 *[34, lemma 3] If $r \in A$ then there exists an $\epsilon > 0$ such that for all primes $p \geq 11$ we have $G(pr) < e^{\gamma} - \epsilon$.*

Proof Let n be a positive integer, and let p and q be odd primes not dividing n with $q < p$. Then

$$G(pn) = \frac{\sigma(pn)}{pn \log\log pn} = \frac{p+1}{p} \frac{\sigma(n)}{n \log\log pn} < \frac{q+1}{q} \frac{\sigma(n)}{n \log\log qn} = G(qn).$$

No prime $p \geq 11$ divides any element of A. Also $G(11r) < e^{\gamma}$ for all $r \in A$. Therefore $G(pr) \leq G(11r) < e^{\gamma}$. □

9.4 Riemann Hypothesis Equivalences

Recall that an extraordinary number is a positive integer N for which $G(N/p) \leq G(N)$ for all primes $p \mid N$ and $G(aN) \leq G(N)$ for all $a \geq 1$.

Theorem 9.8 (Caveney–Nicolas–Sondow first criterion) *[34, theorem 5] The Riemann hypothesis is true if and only if 4 is the only extraordinary number.*

Proof Let $N \neq 4$ be an extraordinary number. Then for all $n \in \mathbb{N}$, since N is right abundant, we have $G(N) \geq G(nN)$. Therefore, by Lemma 9.5 we have

$$G(N) \geq \limsup_{n\to\infty} G(nN) = e^{\gamma}.$$

If $N \leq 5040$ then $N \in A$. But N being extraordinary, and not equal to 4, is composite (recall the comment regarding $G(p)$ in Section 9.1) so $N > 5$, and thus by Lemma 9.6 $G(N) < G(N/p)$ for some prime factor $p \mid N$. Since N is right abundant this is impossible, so $N > 5040$ and $G(N) \geq e^{\gamma}$, and therefore by Theorem 7.16 the Riemann hypothesis is false.

Now suppose that the Riemann hypothesis is false. Then since $\limsup_{n\to\infty} G(n) = e^{\gamma}$ (Theorem 9.2) and, by Theorem 7.16, there is an N with $G(N) > e^{\gamma}$, the maximum μ and minimum N exist where

$$\mu := \max\{G(n) : n > 5040\} \quad \text{and} \quad N := \min\{n > 5040 : G(n) = \mu\}.$$

Then $\mu = G(N) \geq e^\gamma$. We **claim** that such an N is extraordinary. To see this, note first that $G(N) \geq G(n)$ for all $n \geq N$, and thus N is right abundant.

Next, if N were prime then, since $x \to (x+1)/(x \log\log x)$ is decreasing for $x \geq 5040$, we would have

$$1.7 < e^\gamma \leq G(N) = \frac{N+1}{N \log\log N} < \frac{5041}{5040 \log\log 5050} < 0.5.$$

Therefore N is composite.

Now let $p \mid N$ and set $r := N/p$. If $r > 5040$ since $r < N$ and N is the minimal integer with maximum value, we must have $G(r) < G(N)$.

If however $r \leq 5040$, when $5041 \leq n \leq 35\,280$ a computation shows that $G(n) < e^\gamma$, so $N > 35\,280 = 7 \times 5040$. Since in addition $N = rp$ and $r \leq 5040$ we must have $p \geq 11$. If $G(r) \geq e^\gamma$ we would have $r \in A$. Then Lemma 9.7 would give

$$e^\gamma > G(pr) = G(N) \geq e^\gamma$$

so $G(r) < e^\gamma \leq G(N)$. Therefore in both cases we have $G(N) > G(r) = G(N/p)$ so N is left abundant also and therefore N is an extraordinary number. □

The proof of Theorem 9.8 has the following direct corollary:

Corollary 9.9 *If there is a counterexample to Robin's inequality then the maximum* $\mu := \max\{G(n) : n > 5040\}$ *exists and the least* $N > 5040$ *with* $G(N) = \mu$ *is an extraordinary number.*

Lemma 9.10 *[34, propositions 16 and 17]* ***All*** *left abundant numbers* N *with exactly two prime factors are of the form* $N = 4$ *or* $N = 2p$ *with* $p \geq 7$ *a prime, and* ***all*** *such numbers are left abundant.*

Proof First we **claim** numbers of the form $2p$ are left abundant if and only if $p = 2$ or $p > 5$. To see this, since $G(2) < 0$, numbers of the form $2p$ are left abundant if and only if $G(p) < G(2p)$, which holds certainly for $p = 2$ but not for $p = 3$ or $p = 5$ using numerical checks. If $p > 5$ write

$$\frac{G(2p)}{G(p)} = \frac{3(p+1)}{2p \log\log 2p} \frac{p \log\log p}{p+1} = \frac{3 \log\log p}{2 \log\log 2p} > 1,$$

where the final inequality follows since $3 \log\log x > 2 \log\log 2x$ for $x \geq 7$. Hence $2p$ is left abundant for all primes $p \geq 7$.

Now we show products of two distinct odd primes, say p and q, are not left abundant. Let $p > q \geq 3$ and write

$$\frac{G(pq)}{G(q)} = \frac{(p+1) \log\log q}{p \log\log pq} < 1,$$

where we use the inequality $(x+1) \log\log y < x \log\log xy$, which is true whenever $x \geq y \geq 3$.

Finally if $p \geq 3$ then

$$\frac{G(p^2)}{G(p)} = \frac{(p^2+p+1)\log\log p}{(p^2+p)\log\log p^2} < 1.$$

This completes the proof. □

Recall that in Chapter 7, Theorem 7.13, Robin's unconditional upper bound gives for all $n > 1$ the bound $G(n) \leq e^\gamma + 1.1161/(\log\log n)^2$. We also saw in Theorem 7.15 that if RH was false there exists a real number b with $1-\theta < b < \frac{1}{2}$, where $\theta := \sup\{\beta : \zeta(\beta+i\gamma) = 0\}$, such that when $n \to \infty$ through colossally abundant numbers

$$G(n) = e^\gamma \left(1 + \Omega_\pm \left(\frac{1}{\log^b n}\right)\right).$$

Also recall from Chapter 6 we defined and used the function

$$F(z,k) := \log\left(1 + \frac{1}{z+\cdots+z^k}\right)\frac{1}{\log z}.$$

Now we need an additional construction which was made in Chapter 6. Recall we showed in Lemma 6.18 that $x_k \sim (kx)^{1/k}$ for all $k \geq 2$, where $x := x_1(\epsilon)$.

Lemma 9.11 *[35, lemma 1] In the notation of Chapter 6, let $\epsilon > 0$ and $x = x_1(\epsilon)$. For all $k \geq 2$ we have $x_k < (kx)^{1/k}$.*

Proof Let $z := (kx)^{1/k}$. If $z > 1$ and $k > 2$, using $(t+1/2)^{-1} < \log(1+1/t)$, which is true for $t > 0$, we have

$$\begin{aligned} F(z,k) &= \log\left(1 + \frac{1}{z+\cdots+z^k}\right)\frac{1}{\log z} \\ &< \frac{k}{(z+\cdots+z^k)\log z^k} \\ &< \frac{k}{(k-1+z^k)\log kx} \\ &\leq \frac{k}{(k/2+kx)\log x} \\ &< \log\left(1+\frac{1}{x}\right)\frac{1}{\log x} = F(x,1), \end{aligned}$$

so $F(z,k) < F(x,1)$. But this implies $F(x_k,k) = \epsilon = F(x,1) > F(z,k)$, and the lemma follows since $t \to F(t,k)$ is strictly decreasing on $(1,\infty)$. □

Lemma 9.12 *[35, lemma 2] Let N_0 be a colossally abundant number of parameter ϵ_0, and $N > N_0$, for all $n \geq N_0$, satisfies*

$$\frac{\sigma(n)}{n^{1+\epsilon}} \leq \frac{\sigma(N)}{N^{1+\epsilon}}$$

for some fixed $\epsilon > 0$. Then N is also colossally abundant of parameter ϵ.

Proof The given hypothesis applied to $N > N_0$ and N_0 being colossally abundant of parameter ϵ_0 give

$$\left(\frac{N}{N_0}\right)^{1+\epsilon} \le \frac{\sigma(N)}{\sigma(N_0)} \le \left(\frac{N}{N_0}\right)^{1+\epsilon_0} \quad \Longrightarrow \quad \epsilon \le \epsilon_0.$$

If $n < N_0$, again using the hypothesis and that N_0 is colossally abundant, we have

$$\begin{aligned} \frac{\sigma(n)}{n^{1+\epsilon}} &= \frac{\sigma(n)n^{\epsilon_0-\epsilon}}{n^{1+\epsilon_0}} \\ &\le \frac{\sigma(N_0)n^{\epsilon_0-\epsilon}}{N_0^{1+\epsilon_0}} \\ &\le \frac{\sigma(N_0)N_0^{\epsilon_0-\epsilon}}{N_0^{1+\epsilon_0}} \\ &\le \frac{\sigma(N)}{N^{1+\epsilon}}. \end{aligned}$$

Therefore $\sigma(n)/n^{1+\epsilon} \le \sigma(N)/N^{1+\epsilon}$, which completes the proof of the lemma. □

Lemma 9.13 *[35, lemma 3] Let N be a colossally abundant number of parameter ϵ where $\epsilon < F(2,1)$ and let $x = x_\epsilon$ be the unique solution to $F(x,1) = \epsilon$. Then*
(i) for some constant $c > 0$ we have $\log N \le \theta(x) + c\sqrt{x}$, and
(ii) if N is the largest colossally abundant number of parameter ϵ then

$$\theta(x) \le \log N \le \theta(x) + c\sqrt{x}.$$

Proof Recall that $x = x_1$ where for $k \ge 1$ we have $F(x_k,k) = \epsilon$, and that the sequence (x_k) is decreasing. Let K be the largest integer such that $x_K \ge 2$ so that for all primes q with $q \mid N$ we have $2 \le q \le x_k$ and $k = \nu_q(N) \le K$. Note also that $\nu_2(N) = K$ or $K-1$ and that if $\epsilon < F(2,1)$ then $x > 2$ and $K \ge 1$. Then

$$N = \prod_{k=1}^{K} \prod_{q \le x_k} q. \tag{9.9}$$

Therefore

$$\log N \le \theta(x_1) + \theta(x_2) + \cdots + \theta(x_K).$$

Because $t \to F(t,k)$ is decreasing we have

$$F(2,K) \ge F(x_K,K) = \epsilon = F(x_{K+1},K+1) > F(2,K+1).$$

But

$$F(2,K) = \log\left(1 + \frac{1}{2^{K+1}-2}\right)\frac{1}{\log 2} < \frac{1}{(2^{K+1}-2)\log 2} \le \frac{1}{2^K \log 2} < \frac{2}{2^K},$$

so therefore using $F(x,1) = \epsilon$ we get

$$\epsilon = \frac{\log(1+1/x)}{\log x} > \frac{1}{(x+1)\log x} \geq \frac{1}{(x+1)(x-1)} > \frac{1}{x^2}.$$

Therefore

$$\frac{2}{2^K} > F(2,K) \geq \epsilon > \frac{1}{x^2} \quad \Longrightarrow \quad K < 1 + \frac{2\log x}{\log 2}. \tag{9.10}$$

Because $k \to x_k$ is decreasing, we also get

$$\log N \leq \theta(x_1) + \theta(x_2) + K\theta(x_3).$$

By Lemma 9.11 we have $x_2 \leq (2x)^{1/2}$ and $x_3 \leq (3x)^{1/3}$ so using (9.10) and $\theta(t) \sim t$ as $t \to \infty$, we get $\log N \leq \theta(x) + c\sqrt{x}$, which proves part (i) of the lemma.

Equation (9.9) gives the largest colossally abundant number of parameter ϵ as

$$N = \prod_{k=1}^{K} \prod_{q \leq x_k} q.$$

It follows that $\theta(x) \leq \log N$, completing the proof of part (ii). □

Lemma 9.14 *[35, lemma 4] For some $c > 0$ there are infinitely many primes p such that $\theta(p) < p - c\sqrt{p}\log\log\log p$, and infinitely many other primes p such that $\theta(p) > p + c\sqrt{p}\log\log\log p$.*

Proof The proof employs Littlewood's theorem [109], Theorem 4.13, in an essential manner. In particular, that theorem gives two increasing and unbounded sequences of real values (x_n) and (y_n) and a positive constant c such that for all n

$$\theta(x_n) < x_n - 2c\sqrt{x_n}\log\log\log x_n \quad \text{and} \quad \theta(y_n) > y_n + 2c\sqrt{y_n}\log\log\log y_n. \tag{9.11}$$

In fact, the theorem refers to $\psi(x)$ rather than $\theta(x)$, but the $\theta(x)$ version is an easy corollary since, assuming RH, $\psi(x) = \theta(x) + O(\sqrt{x}\log x)$, which follows from

$$\psi(x) = \sum_{n=1}^{\infty} \theta(x^{1/n}), \quad \theta(x) = 0 \text{ for } x < 2.$$

If an infinite subsequence of the x_n is prime we can replace (x_n) by that subsequence and the first part of the lemma follows. So assume no x_n is prime and let p_n be the smallest prime larger than x_n. Then since $t \to$

$t - c\sqrt{t}\,\mathrm{logloglog}\,t$ is increasing for t sufficiently large we have

$$\begin{aligned}\theta(p_n) = \theta(x_n) + \log p_n &< x_n - 2c\sqrt{x_n}\,\mathrm{logloglog}\,x_n + \log p_n \\ &< p_n - 2c\sqrt{p_n}\,\mathrm{logloglog}\,p_n + \log p_n,\end{aligned}$$

so for n sufficiently large we get $\theta(p_n) < p_n - c\sqrt{p_n}\,\mathrm{logloglog}\,p_n$.

Similarly let p_n be the largest prime less than or equal to y_n. By (9.11) again we have

$$\theta(p_n) = \theta(y_n) > y_n + 2c\sqrt{y_n}\,\mathrm{logloglog}\,y_n > p_n + c\sqrt{p_n}\,\mathrm{logloglog}\,p_n,$$

which completes the proof. □

Lemma 9.15 *[35, lemma 6] Let $\epsilon > 0$ and for $t > e$ define*

$$g_\epsilon(t) := \epsilon \log t - \mathrm{logloglog}\,t.$$

Then there is a unique $t_0 > e$, dependent on ϵ, such that

$$\epsilon = 1/(\log t_0 \,\mathrm{loglog}\, t_0).$$

The function $g_\epsilon(t)$ is decreasing for $e < t < t_0$ and increasing for $t > t_0$.

Proof Note that

$$g'_\epsilon(t) = \left(\epsilon - \frac{1}{\log t \,\mathrm{loglog}\, t}\right)\frac{1}{t},$$

so the result follows directly provided $\epsilon = 1/(\log t_0 \,\mathrm{loglog}\, t_0)$ has a unique solution t_0. But this is true because $t \to 1/(\log t \,\mathrm{loglog}\, t)$ is strictly decreasing on (e,∞) with image $(0,\infty)$. □

Recall again from Chapter 8 that

$$\begin{aligned}A := \{&1,2,3,4,5,6,8,9,10,12,16,18,20,24,30,36,48,\\ &60,72,84,120,180,240,360,720,840,2520,5040\}\end{aligned}$$

is the set of positive integers n, not greater than 5040, for which $G(n) \geq e^\gamma$.

Lemma 9.16 *[35, p. 371] If $5040 < n < e^{19747}$ then $G(n) < e^\gamma$.*

Proof Recall from Step (3) of the proof of Theorem 7.11 in Chapter 7 that if C is the largest colossally abundant number with $P(C) < 20000$ then there is no n with $5040 < n \leq C$ such that $G(n) \geq e^\gamma$. Using Lemma 6.10 we see that each colossally abundant number includes an initial segment of primes with non-increasing exponents, and for every initial sequence of primes there is a colossally abundant number with corresponding maximum prime divisor, so $P(C) = 19993$. Therefore by Lemma 9.13 $\log C \geq \theta(20000)$. Hence can write using Lemma 3.14 (a), for $x \geq 19421$,

$$\theta(x) > x - \frac{x}{8\log x} \quad\Longrightarrow\quad \log C \geq \theta(20000) > 19747.$$

This completes the proof of the lemma. □

Recall that N is right abundant if for all $a \geq 1$ we have $G(N) \geq G(aN)$.

Theorem 9.17 (Caveney–Nicolas–Sondow second criterion) *[35, theorem 5] The set of right abundant numbers not greater than 5040 is* $A \setminus \{1,2,9,16,20,30,84,720,840\}$. *If the Riemann hypothesis is* ***true*** *then there are no other right abundant numbers. If the hypothesis is* ***false*** *then infinitely many right abundant numbers exist and in that case*

$$\infty > \mu := \max\{G(n) : n > 5040\} > e^{\gamma},$$

and, in addition, any integer M for which $G(M) = \mu$ is both colossally abundant and right abundant.

Proof **(1)** Let

$$\begin{aligned} A_2 &:= \{3,4,5,6,8,10,12,18,24,36,48,60, \\ &\qquad 72,120,180,240,360,2520,5040\} \subset A \\ &= A \setminus \{9,16,29,30,84,720,840\}, \\ A_3 &:= \{n \leq 5420 : n \text{ is right abundant}\}. \end{aligned}$$

First we show that $A_3 \subset A_2$. If N is in A_3 then, by Lemma 9.5, $G(N) \geq e^{\gamma}$ so N is in A. For $r \in A$ let $B_r := \{a \in \mathbb{N} : ar \in A,\ G(ar) > G(r)\}$ and set $a_r := \min B_r$ when B_r is non-empty. Otherwise set $a_r := 0$. A computation shows that the $r \in A$ for which $B_r \neq \emptyset$ are $\{9,16,20,30,84,720,840\}$. But N is right abundant so necessarily $B_N = \emptyset$ giving $N \in A \setminus \{9,16,20,30,84,720,840\} = A_2$.

(2) Now let $N \in A_2$. We **claim** $G(N) \geq G(aN)$ for any multiple aN of N. First consider multiples $aN \leq 5040$. If $G(aN) < e^{\gamma}$, because $N \in A$ we have $G(aN) < e^{\gamma} \leq G(N)$. Otherwise $aN \notin A$ which again gives $G(aN) < e^{\gamma} \leq G(N)$ so in both cases $G(aN) < G(N)$ when $aN \leq 5040$. (Alternatively a finite set of possibilities can be checked numerically.)

(3) Now consider multiples $aN > 5040$ with $N \in A_2$ as before. If $\log(aN) < 19747$ then Lemma 9.16 gives $G(aN) < e^{\gamma} \leq G(N)$ and we are done. If however $\log(aN) \geq 19747$ then using Theorem 7.13 we have

$$G(aN) \leq e^{\gamma} + \frac{0.667}{\log^2 19747} < 1.78849 < 1.790 < \min\{G(n) : n \in A_2\} \leq G(N).$$

Therefore $G(N) \geq G(aN)$ when $aN > 5040$, so all elements of A_3 are right abundant. This shows that $A_2 \subset A_3$, and therefore $A_3 = A_2$, completing the proof of the first part of the theorem.

(4) Now assume that the Riemann hypothesis holds. Then by Robin's theorem, Theorem 7.16, for all $n > 5040$ we have $G(n) < e^{\gamma}$, and therefore, by Lemma 9.5, n cannot be right abundant, so A_3 is precisely the set of right abundant numbers.

(5) Finally assume that the Riemann hypothesis is false. Let the real number η with $\frac{1}{2} < \eta \le 1$ be defined by

$$\eta := \sup_{\zeta(\rho)=0} \Re\rho.$$

Using Theorem 7.15 we have for some b with $1-\eta < b < \frac{1}{2}$, as $N \to \infty$ through colossally abundant numbers,

$$G(N) = e^\gamma + \Omega_+\left(\frac{1}{\log^b N}\right).$$

Therefore there are infinitely many positive integers N which are colossally abundant and satisfy $G(N) > e^\gamma$. Thus, for all real t we have $\max\{G(n) : n \ge t\} > e^\gamma$.

Let $\mu_1 := \mu = \max\{G(n) : n > 5040\}$ and let a_1 be the smallest integer and b_1 the largest integer both greater than 5040 with $G(a_1) = G(b_1) = \mu_1$. If a_n and b_n have been defined for $1 \le n < i$, we set

$$\mu_i := \max\{G(n) : n > b_{i-1}\},$$

and then let a_i be the smallest integer and b_i the largest integer, both greater than b_{i-1}, such that $G(a_i) = G(b_i) = \mu_i$. Because $\mu_i > e^\gamma = \limsup_{n\to\infty} G(n)$, there is an infinite number of distinct integers a_i, and they are such that if $n > a_i$ then $G(n) \le G(a_i)$. Therefore each a_i is right abundant, so infinitely many right abundant numbers exist.

(6) Now let an integer $M > 5040$ satisfy $G(M) = \mu$. The method of the previous paragraph can be used to show that M is necessarily right abundant. Using Lemma 9.12 with $N_0 = 5040$, $\epsilon_0 = 3/100$ (in the notation of Section 6.3, $\epsilon_7 < 0.03 < \epsilon_8$), $N = M$ and $\epsilon = 1/(\log M \log\log M)$, and employing also Lemmas 9.16 and 9.5, we get $N = M > e^{19747} > N_0$. For $n \ge N_0$, using the definition of M we get $G(n) \le G(M)$. By Lemma 9.15, because $\epsilon < N_0 < M$, on the interval $[N_0, \infty)$ the function $g_\epsilon(t)$ attains its minimum at $t = M$, so for $n \ge N_0$ we get

$$\frac{\sigma(n)}{n^{1+\epsilon}} = G(n)\frac{\log\log n}{n^\epsilon} \le \frac{G(n)}{e^{g_\epsilon(n)}} \le \frac{G(M)}{e^{g_\epsilon(n)}} \le \frac{G(M)}{e^{g_\epsilon(M)}} = \frac{\sigma(M)}{M^{1+\epsilon}}.$$

Therefore, by Lemma 9.12, M is also colossally abundant, completing the proof of the theorem. □

9.5 Comparing Colossally and Left Abundant Numbers

Lemma 9.18 *[35, lemma 7] Let N be colossally abundant of parameter ϵ and assume $p := P(N)$ with $p \ge 5$. If $\epsilon > 1/(\log(N/p)\log\log(N/p))$ then N is also left abundant.*

Proof Let $q \mid N$ be any prime divisor. Because $p \geq 5$ and N is colossally abundant we have $6p \mid N$ and so

$$\frac{N}{q} \geq \frac{N}{p} \geq 6 > e \quad \Longrightarrow \quad \operatorname{loglog}\left(\frac{N}{q}\right) > \operatorname{loglog} e = 0.$$

Again because N is colossally abundant we get

$$\frac{\sigma(N/q)}{(N/q)^{1+\epsilon}} \leq \frac{\sigma(N)}{N^{1+\epsilon}} \quad \Longrightarrow \quad \frac{\sigma(N/q)}{\sigma(N)} \leq \frac{1}{q^{1+\epsilon}}. \tag{9.12}$$

Because $\operatorname{loglog} N$ and $\operatorname{loglog}(N/q)$ are both positive we get

$$\frac{G(N/q)}{G(N)} \leq \frac{\operatorname{loglog} N}{q^{\epsilon} \operatorname{loglog}(N/q)} = \frac{(N/q)^{\epsilon} \operatorname{loglog} N}{N^{\epsilon} \operatorname{loglog}(N/q)} = \exp(g_{\epsilon}(N/q) - g_{\epsilon}(N)). \tag{9.13}$$

By the result and notation of Lemma 9.15 we see $g_{\epsilon}(t)$ is increasing for $t > t_0$. Also, from the lower bound for ϵ given in the hypothesis, we have

$$e < t_0 < N/p \leq N/q < N \quad \Longrightarrow \quad \exp(g_{\epsilon}(N/q) - g_{\epsilon}(N)) < 1,$$

giving $G(N/q) < G(N)$, so N is left abundant. □

Theorem 9.19 *[35, theorem 6] Infinitely many colossally abundant numbers are also left abundant.*

Proof Use Lemma 9.14 to find a prime p sufficiently large that it satisfies

$$\theta(p) > p + c\sqrt{p} \operatorname{logloglog} p,$$

where c is defined in Lemma 9.13, and set $\epsilon := F(p, 1)$. Let N be the largest colossally abundant number of parameter ϵ, so $F(p, 1) = \epsilon$ and $P(N) = p$ so $p \mid N$. By Lemma 9.13 we get

$$\log N \geq \theta(p) > p + c\sqrt{p} \operatorname{logloglog} p,$$

which implies

$$\log(N/p) > p + c\sqrt{p} \operatorname{logloglog} p - \log p > p + 1.$$

Because $\log(1 + t) \geq t/(1 + t)$ for $t > 0$ we have

$$\begin{aligned} \epsilon = F(p, 1) = \frac{\log(1 + 1/p)}{\log p} &\geq \frac{1}{(1 + p)\log p} > \frac{1}{(p + 1)\log(p + 1)} \\ &> \frac{1}{\log(N/p) \operatorname{loglog}(N/p)}. \end{aligned}$$

By Lemma 9.18, N is left abundant. Since there are an infinite number of primes p given by Lemma 9.14, there are an infinite number of corresponding left abundant N. This completes the proof of the theorem. □

Lemma 9.20 *[35, lemma 8] Let $p \geq 5$ be an odd prime and N the largest colossally abundant number of parameter ϵ with $\epsilon < 1/(\log N \log\log N)$. Then N is* **not** *left abundant.*

Proof Firstly we have

$$p^{\epsilon} = \frac{p+1}{p} = \frac{\sigma(p)}{p}.$$

Therefore, since $F(p,1) = \epsilon$, $P(N) = p$ and p divides N at most to the first power, inequality (9.12) with $q = p$ is an equality

$$\frac{\sigma(N/p)}{(N/p)^{1+\epsilon}} = \frac{\sigma(N)}{N^{1+\epsilon}},$$

and similarly for inequality (9.13). Defining $g_{\epsilon}(t)$ and t_0 as in Lemma 9.18 we get $g_{\epsilon}(t)$ decreasing when $t < t_0$ and

$$\frac{G(N/p)}{G(N)} = \exp(g_{\epsilon}(N/p) - g_{\epsilon}(N)).$$

Therefore the upper bound for ϵ of the hypothesis gives $N/p < N < t_0$, so $G(N) < G(N/p)$ and N is not left abundant. This completes the proof of the lemma. □

Theorem 9.21 *[35, theorem 7] There are infinitely many colossally abundant numbers which are* **not** *left abundant.*

Proof Use Lemma 9.14 to find a prime p which satisfies

$$\theta(p) < p - c\sqrt{p}\log\log\log p,$$

where c is defined in Lemma 9.13, and then let $\epsilon := F(p,1)$. Let N be the largest colossally abundant number of parameter ϵ. Using Lemma 9.13 for the upper bound for $\theta(p)$ we get, for p sufficiently large,

$$\log N \leq \theta(p) + c\sqrt{p} < p - c\sqrt{p}\log\log\log p + c\sqrt{p} < p.$$

Therefore

$$\epsilon = \frac{\log(1+1/p)}{\log p} < \frac{1}{p\log p} < \frac{1}{\log N \log\log N}.$$

By Lemma 9.20, N is not left abundant. The result of the lemma then follows since an infinite number of primes satisfy $\theta(p) < p - c\sqrt{p}\log\log\log p$. □

A final note is now given here regarding a counting result which uses many of the ideas in this chapter. A **proper left abundant number** is defined to be a left abundant number with more than two prime factors. In [35, theorem 14] the authors showed that there exists a positive constant c such that

$$Q(x) := \#\{N \leq x : N \text{ is a proper left abundant number}\} \leq \exp\left(c\frac{\log x}{\log\log x}\right).$$

9.6 Extremely Abundant Numbers

A simple observation is that the first counterexample n to Robin's inequality larger than 5040 must satisfy $G(n) \geq e^{\gamma} > G(m)$ for all m with $5040 < m < n$, and by Lemma 8.9 be superabundant. So if we note also that 10 080 is the smallest superabundant number which is larger than 5040, we see n must be extremely abundant, that is, a "champion" for the function $G(n)$ restricted to $[10080, \infty)$. Using only Grönwall's and Robin's theorems, Theorems 9.2 and 7.16 respectively, Sadegh Nazardonyavi and Semyon Yakubovich found an equivalence to the Riemann hypothesis with a very straightforward proof.

Theorem 9.22 (Nazardonyavi–Yakubovich criterion) *[123, theorem 7] The Riemann hypothesis is true if and only if the number of extremely abundant numbers is infinite.*

Proof If RH is true and the number of extremely abundant numbers is finite, let m be the largest extremely abundant number. Then for all $n > m$ we have $G(n) \leq G(m)$. But this implies $\limsup_{n\to\infty} G(n) \leq G(m) < e^{\gamma}$, where the last inequality follows from Theorem 7.16 since we are assuming RH. But this contradicts Theorem 9.2, so therefore the set of extremely abundant numbers is infinite.

If RH is false however, then by Theorem 7.16 again, there exists an $m \geq 10080$ with $G(m) \geq e^{\gamma}$. By Theorem 9.2, $\mu := \sup_{n\geq 10080} G(n)$ is finite, and there exists a maximal $m_0 \geq 10080$ such that

$$G(m_0) = \mu \geq e^{\gamma}.$$

Any integer $n > m_0$ satisfies $G(n) < \mu = G(m_0)$, so n fails to be extremely abundant. Therefore the set of extremely abundant numbers is finite, and the proof is complete. □

9.7 Unsolved Problems

(1) Show that the integer 4 is the only left abundant number which is the power of a prime [35].

(2) The smallest left abundant number that is not the power of a prime or a product of at most two primes (called proper) is [35]

$$\nu := 2^4 \cdot 3^3 \cdot 5^2 \cdot 7 \cdot 11 \cdot 13 \cdot 17 = 183\,783\,600.$$

(3) Show that the number ν defined in problem (2) is not right abundant.

(4) Show that any right abundant number $n > 5$ is even [35].

(5) Improve the upper bound for $Q(x)$ given in [35, theorem 14].

(6) Find a non-trivial lower bound for $Q(x)$.

(7) Find an upper bound for the number of extremely abundant numbers less than or equal to real positive x.

10
Other Equivalents to the Riemann Hypothesis

10.1 Introduction

In this chapter a selection of additional equivalences to RH are presented. This is to show that the techniques used to derive those involving $\psi(x)$, $\theta(x)$, $\sigma(n)$ and $\varphi(n)$ are not unique, but that other methods and equivalences have been derived. The equivalences are often surprising. Some are intricate, but others are easy consequences of other equivalences. As before, in the main, proofs are included to show depth or shallowness, and to provide the reader with techniques which could be further developed. There is one notable exception, namely the symmetric group application of Section 10.10 wherein only the introductory proofs and ideas are presented. The selection is by no means exhaustive, and tends to stray away from the arithmetic into other realms, including the analytic, which is more property the domain of Volume Two [32].

To begin, in Section 10.2 there is the interesting inequality, the **Shapiro criterion** from 1976, which is Theorem 10.1. It is based on an arithmetic function $\delta(n)$ which satisfies

$$\int_0^n \psi(t)\,dt = \log \delta(n)$$

and has the form

$$\left| \sum_{1 \le j \le \delta(n)} \frac{1}{k} - \frac{n^2}{2} \right|^2 < 36n^3, \quad n \in \mathbb{N},$$

which is equivalent to the Riemann hypothesis.

In Section 10.3 there is the famous Farey fraction equivalence. This is "doubly deep", since it depends in a non-trivial way on the equivalence of Littlewood, Theorem 4.16. Let $m := \sum_{1 \le j \le n} \varphi(j)$ and $F_n := (f_j^n : 0 \le j \le m)$ be the nth Farey series, i.e. the set of rational numbers in $[0, 1]$ in lowest terms, with denominators not greater than n, in increasing order. Then we have the theorem of Franel and Landau of 1924:

Theorem 10.6 (Franel–Landau criterion) *The Riemann hypothesis is equivalent to the statement that for all* $\epsilon > 0$, *as* $n \to \infty$,

$$\sum_{j=1}^{m} \left| f_j^n - \frac{j}{m} \right| \ll n^{1/2+\epsilon}.$$

The equivalence in Section 10.4 is also a consequence of Littlewood's theorem, Theorem 4.16. It relies on the direct relationship between the determinant of the matrix R_n describing divisibility and a sum over $\mu(j)$, values of the Möbius function. The matrix $R_n = (r_{i,j})$ is such that the element $r_{i,j}$ is zero except where $j = 1$ or $i \mid j$, in which case it is 1. The equivalence was published by Redheffer in 1976. The relationship to the sum and the following equivalence are easy to derive.

Theorem 10.9 (Redheffer criterion) *The Riemann hypothesis is equivalent to the statement that for all* $\epsilon > 0$, *as* $n \to \infty$,

$$\det R_n \ll n^{1/2+\epsilon}.$$

The next equivalence, described in Section 10.5, is an easy consequence of Theorem 10.9. It describes a property of particular finite directed graphs, the so-called "divisibility graphs". These are defined for each natural number n by labelling vertices 1 through n and placing a directed edge between vertex i and vertex j when i divides j. An additional set of edges goes from each j to vertex 1. A closed cycle in one of these graphs is called "even" if the number of edges is even, and "odd" otherwise. The number of even and odd cycles is represented by $e(G_n)$ and $o(G_n)$ respectively. We have:

Theorem 10.10 (Divisibility graph criterion) *The Riemann hypothesis is true if and only if for all* $\epsilon > 0$ *we have*

$$|o(G_n) - e(G_n)| \ll n^{1/2+\epsilon}$$

as $n \to \infty$.

The function defined by $\sum_{n=1}^{\infty}(-1)^{n+1}n^{-s}$, which converges conditionally for $\Re s > 0$, is sometimes called $\eta(s)$. It provides a very easy-to-derive equivalence which is given in Section 10.6:

Theorem 10.12 (Dirichlet eta criterion) *The Riemann hypothesis is equivalent to the zeros of* $\eta(s)$ *with* $0 < \sigma < 1$ *all having real part* $\frac{1}{2}$.

Now a different set of equivalences are described, each directly or indirectly using inequalities concerning $\zeta(s)$. In Section 10.7 a famous statement of an equivalence due originally to Speiser, who gave in 1934 a geometric/

topological but not rigorous proof [161], that RH is equivalent to all of the zeros of the derivative of $\zeta(s)$ in the critical strip being on or to the right of the critical line. Part of this statement was given rigorous justification by Levinson and Montgomery in 1974:

Theorem 10.16 (Levinson–Montgomery criterion) *If the Riemann hypothesis is true, all non-trivial zeros of the derivative of the Riemann zeta function $\zeta'(s)$ in the critical strip have real part greater than or equal to $\frac{1}{2}$. If the Riemann hypothesis is false, the number of zeros of $\zeta'(s)$ in the critical strip, with positive imaginary part less than T and real part less than $\frac{1}{2}$, is $O(\log T)$.*

A simple inequality equivalence was found by Spira in 1965:

Theorem 10.18 (Spira criterion) *The Riemann hypothesis is equivalent to the statement that if $\sigma > \frac{1}{2}$ and $t \geq 6.29073$ then $|\zeta(1-s)| > |\zeta(s)|$.*

As part of a re-examination of a lemma of Lagarias, the present author was able to improve his Riemann zeta imaginary "zero gap" explicit estimate:

Theorem 10.32 *If $t \geq 168\pi$ then $\zeta(s)$ has a zero with imaginary part satisfying $|t-\gamma| \leq 1.5$.*

Later, in Section 10.9 the details are given of work of Lagarias and Garunkstis on an inequality equivalence to RH given in terms of the logarithmic derivative of Riemann's function $\xi(s)$.

Theorem 10.35 (Garunkstis–Lagarias criterion) *Let a satisfy $\frac{1}{2} \leq a \leq 1$. If $\zeta(s)$ has no zeros for $\sigma > a$ then for $s > a$ we have*

$$\frac{\xi'(\sigma)}{\xi(\sigma)} = \inf\left\{\Re\frac{\xi'(\sigma+it)}{\xi(\sigma+it)} : t \in \mathbb{R}\right\}.$$

In Section 10.10 a summary and proofs of foundational results are presented of a completely different type of equivalence. Nonetheless it was considered that this should be included because the methods used related strongly to those used in Chapters 5 and 7. Let $g(n)$ be the maximum multiplicative order of an element of the symmetric group S_n. Then the theorem of Massias, Nicolas and Robin of 1988 is:

Theorem 10.42 (Symmetric group criterion) *The Riemann hypothesis is equivalent to the statement that for all n sufficiently large we have*

$$\log g(n) < \sqrt{\mathrm{li}^{-1}(n)},$$

where $\mathrm{li}(n)$ *is the Cauchy principal value of the logarithmic integral,*

$$\mathrm{li}(n) := \int_0^n \frac{dt}{\log t},$$

and $\mathrm{li}^{-1}(n)$ *represents its functional inverse.*

Finally in Section 10.11 we touch on the large body of work which goes under the heading the "Hilbert–Pólya conjecture". Broadly speaking, this is the idea that there is a physical system which underlies the zeta function so that the critical zeros are associated with eigenvalues of a self-adjoint transformation on a (dense subset of) a Hilbert space.

The following RHpack (see Appendix B) functions relate to the material in this chapter: FareyFractions, RedhefferMatrix, DivisibilityGraph, Plot-DivisibilityGraph, Xi, Lambda, S, IntegerHeight and ZetaZeroCount.

10.2 Shapiro's Criterion

This criterion was quite a minor part of a far-reaching summary of work on Hilbert's 10th problem by Davis, Matijasevič and Robinson [44]. Published in 1976, it contains an arithmetic inequality equivalent to RH which is superficially simpler than Robin's, and certainly much simpler to derive.

Recall the definitions. For $x \geq 1$,

$$\psi(x) := \sum_{1 \leq n \leq x} \Lambda(n) = \sum_{p^\alpha \leq x} \log p \leq x \log x,$$

$$\psi_1(x) := \int_1^x \psi(t)\, dt.$$

If $n = p^\alpha$ is a prime power define $\kappa(n) := p$. Otherwise set $\kappa(n) := 1$. For $x > 0$ define

$$\delta(x) := \prod_{n<x} \prod_{j \leq n} \kappa(j),$$

for example

$$\delta(11) = 2^{19} \cdot 3^{10} \cdot 5^6 \cdot 7^4$$

and

$$\begin{aligned}\delta(102) = {} & 2^{486} \cdot 3^{288} \cdot 5^{174} \cdot 7^{148} \cdot 11^{91} \cdot 13^{89} \cdot 17^{85} \cdot 19^{83} \\ & \cdot 23^{79} \cdot 29^{73} \cdot 31^{71} \cdot 37^{65} \cdot 41^{61} \cdot 43^{59} \cdot 47^{55} \cdot 53^{49} \\ & \cdot 59^{43} \cdot 61^{41} \cdot 67^{35} \cdot 71^{31} \cdot 73^{29} \cdot 79^{23} \cdot 83^{19} \cdot 89^{13} \cdot 97^5 \cdot 101.\end{aligned}$$

Note that with this definition we have $\delta(1) = 1$, and for $n \in \mathbb{N}$, $\psi_1(n) = \log \delta(n)$.

Theorem 10.1 (Shapiro criterion) *The inequality*

$$\left| \sum_{1 \leq j \leq \delta(n)} \frac{1}{k} - \frac{n^2}{2} \right|^2 < 36n^3, \quad n \in \mathbb{N}, \tag{10.1}$$

is equivalent to the Riemann hypothesis.

Proof (1) First assume RH is true and let S be the non-trivial zeros of $\zeta(s)$. Then for $\rho \in S$ we have $1-\rho = \overline{\rho}$ so by Lemma 2.10 we get

$$\sum_{\rho\in S} \frac{1}{|\rho|\cdot|1+\rho|} \le \sum_{\rho\in S} \frac{1}{|\rho|^2} = \sum_{\rho\in S} \frac{1}{\rho(1-\rho)} = \gamma + 2 - \log(4\pi).$$

Using the explicit formula from Theorem 2.2, by RH we have $|x^{\rho+1}| = x^{3/2}$. We also have the sum

$$\sum_{n=1}^{\infty} \frac{1}{2n(2n-1)} = \log 2,$$

so we get for $x \ge 1$

$$\left|\psi_1(x) - \frac{x^2}{2}\right| \le x^{3/2}\left(\sum_{\rho\in S} \frac{1}{|\rho|\cdot|1+\rho|} + \left|\frac{\zeta'(0)}{\zeta(0)}\right| + \left|\frac{\zeta'(-1)}{\zeta(-1)}\right| + \log 2\right).$$

Thus, by Lemma 2.5 and the evaluation $|\zeta'(-1)/\zeta(-1)| = 1.985\ldots < 2$ we can write

$$\left|\psi_1(x) - \frac{x^2}{2}\right| \le x^{3/2}(\gamma + 2 - \log(4\pi) + \log(2\pi) + 2 + \log 2) < 5x^{3/2}.$$

But for all $n \in \mathbb{N}$ we have

$$-1 + \sum_{j=1}^{n} \frac{1}{j} < \log n < \sum_{j=1}^{n} \frac{1}{j}.$$

Since for $x \ge 1$ we have $\psi_1(x) = \log \delta(x)$, we get

$$\left|\sum_{1\le j\le \delta(x)} \frac{1}{j} - \psi_1(x)\right| < 1.$$

Therefore

$$\left|\sum_{j\le\delta(x)} \frac{1}{j} - \frac{x^2}{2}\right| < 1 + 5x^{3/2} \le 6x^{3/2},$$

which gives the inequality (10.1).

(2) Now assume inequality (10.1) is true for all $n \in \mathbb{N}$. Then for x a positive integer we have $\psi_1(x) = \log \delta(x)$. Hence $|\psi_1(x) - x^2/2| < 1 + 6x^{3/2}$. Since

$$\psi(x) \le \sum_{j\le x} \log x \le x \log x \le x^{3/2},$$

this implies for real $x \ge 1$

$$\begin{aligned}\left|\psi_1(x) - \frac{x^2}{2}\right| &\le \left|\psi_1(\lfloor x\rfloor) - \frac{\lfloor x\rfloor^2}{2}\right| + \int_{\lfloor x\rfloor}^{x} \psi(t)\,dt + \frac{1}{2}|x^2 - \lfloor x\rfloor^2| \\ &< \psi(\lfloor x\rfloor) + 1 + 6x^{3/2} + x \le 9x^{3/2}.\end{aligned}$$

Therefore

$$\left|\frac{\psi_1(x) - x^2/2}{x^{s+2}}\right| \le \frac{9}{|x^{s+1/2}|}.$$

(3) Next, taking the logarithmic derivative of the Euler product for $\zeta(s)$, and using partial summation, the definition of $\psi_1(x)$ and integration by parts, we can derive for $\Re s > 1$

$$\begin{aligned}
-\frac{\zeta'(s)}{\zeta(s)} &= s\int_1^\infty \frac{\psi(x)}{x^{s+1}}\,dx\\
&= s\int_1^\infty \frac{d\psi_1(x)}{x^{s+1}}\\
&= s(s+1)\int_1^\infty \frac{\psi_1(x)}{x^{s+2}}\,dx.
\end{aligned}$$

Therefore

$$-\frac{\zeta'(s)}{\zeta(s)} - \frac{s(s+1)}{2(s-1)} = s(s+1)\int_1^\infty \frac{\psi_1(x) - x^2/2}{x^{s+2}}\,dx. \tag{10.2}$$

The inequality derived in Step (2) implies that the integral of (10.2) converges to a function of s which is holomorphic on $\Re s > \frac{1}{2}$. But this means $\zeta(s)$ has no zero in this region, so RH is true. This completes the proof. □

Note, as far as numerical checking of inequality (10.1) is concerned, that the expression

$$\left|\frac{2}{n^2}\log\delta(n) - 1\right| < \frac{12}{\sqrt{n}} \tag{10.3}$$

is better.

A subset $T \subset \mathbb{N}$ is **computable** if there is an algorithm to determine in a finite number of steps whether or not an arbitrary given natural number is a member of T [44]. From the theory of algorithms it follows that RH is decidable, i.e. its truth or negation are able to be proved. In addition, using the inequality (10.1) it would be possible to write down an explicit diophantine equation with integer coefficients which has no solutions in integers if and only if RH is true. See Davis et al. [44].

10.3 Farey Fractions

For $n \in \mathbb{N}$ let F_n be the finite sequence of rationals in the interval $[0,1]$ with denominators in $\{1,2,\ldots,n\}$, in lowest terms, sorted in increasing order. For example $F_3 = (\frac{0}{1}, \frac{1}{3}, \frac{1}{2}, \frac{2}{3}, \frac{1}{1})$. These are the so-called **Farey series**. We write f_j^n, $0 \le j \le m$, for the jth term in the finite sequence F_n, where the number of terms is given by Lemma 10.2.

This equivalence is not at all obvious, even though it relies on the classical Littlewood criterion, Theorem 4.16. Before giving the proof we need several lemmas. The proof is taken in the main from Edwards [51] which was taken from the work of Franel and Landau of 1924, based on, apparently, the work of the former [64].

Lemma 10.2 *We have for all* $n \geq 1$

$$\#F_n = 1 + \sum_{j=1}^{n} \varphi(j).$$

Proof For $1 \leq j \leq n$, the number of distinct rationals in $(0, 1]$ with lowest-terms denominator j is $\varphi(j)$. The leading 1 in the sum accounts for the rational $0 = 0/1$. □

Below in this section we set $m = \sum_{j=1}^{n} \varphi(j)$, the number of non-zero elements in F_n. Also recall the definition for $x > 0$,

$$M(x) := \sum_{n \leq x} \mu(n),$$

where $\mu(n)$ is the Möbius function.

The statement equivalent to the Riemann hypothesis says, roughly, that when n is large the Farey fractions of order n are reasonably uniformly distributed.

Lemma 10.3 *Let* $x > 1$ *be real and* $n = \lfloor x \rfloor$. *If* $m = \#F_n - 1$ *and* $M(x) = \sum_{1 \leq n \leq x} \mu(n)$, *then for any function* $g : \mathbb{R} \to \mathbb{R}$, *with* $g(x+1) = g(x)$ *for all* x, *we have*

$$\sum_{j=1}^{m} g(f_j^n) = \sum_{k=1}^{\infty} \sum_{j=1}^{k} g\left(\frac{j}{k}\right) M\left(\frac{x}{k}\right). \tag{10.4}$$

Proof Let $\chi(x)$ be the characteristic function of $[1, \infty)$, i.e. $\chi(x) = 1$ for $x \geq 1$ and $\chi(x) = 0$ for $x < 1$. Then $M(x) = \sum_{n \in \mathbb{N}} \mu(n) \chi(x/n)$, which by Möbius inversion gives $\chi(x) = \sum_{n \in \mathbb{N}} M(x/n)$. Given any rational number u/v with $0 \leq u \leq v$ and $(u, v) = 1$ then

$$g\left(\frac{u}{v}\right) = g\left(\frac{2u}{2v}\right) = g\left(\frac{3u}{3v}\right) = \cdots$$

occurs on the right-hand side of (10.4) with coefficient

$$M\left(\frac{x}{v}\right) + M\left(\frac{x}{2v}\right) + M\left(\frac{x}{3v}\right) + \cdots = \chi\left(\frac{x}{v}\right),$$

which is 1 if $v \leq x$ and 0 if $v > x$. This is exactly the coefficient of $g(u/v)$ on the left-hand side of (10.4). □

For $n \in \mathbb{N}$ let $B_n(x)$ be the nth Bernoulli polynomial. One way of defining $B_n(x)$ is that it is the unique polynomial which satisfies

$$\int_x^{x+1} B_n(t)\,dt = x^n.$$

These polynomials have many properties which are useful. The first few are

$$\begin{aligned}
B_0(x) &= 1,\\
B_1(x) &= x - \tfrac{1}{2},\\
B_2(x) &= x^2 - x + \tfrac{1}{6},\\
B_3(x) &= x^3 - \tfrac{3}{2}x^2 + \tfrac{1}{2}x.
\end{aligned}$$

Here we use also the related period-1 function $b_n(x) := B_n(x - \lfloor x \rfloor)$.

Lemma 10.4 *For all $n, k \in \mathbb{N}$ and $u \in \mathbb{R}$ we have*

$$B_n(ku) = k^{n-1}\left(B_n(u) + B_n\left(u + \frac{1}{k}\right) + \cdots + B_n\left(u + \frac{k-1}{k}\right)\right),$$

with the same identity holding with B_n replaced by b_n.

Proof First we derive the identity for $B_n(u)$. We have for any real x, changing variables and then splitting the integral domain into k subintervals of equal length,

$$\begin{aligned}
\int_x^{x+1/k} B_n(kt)\,dt &= \frac{1}{k}\int_{kx}^{kx+1} B_n(u)\,du = k^{n-1}x^n\\
&= k^{n-1}\int_x^{x+1} B_n(t)\,dt\\
&= \int_x^{x+1/k} k^{n-1}\left(B_n(t) + B_n\left(t + \frac{1}{k}\right) + \cdots + B_n\left(t + \frac{k-1}{k}\right)\right) dt.
\end{aligned}$$

Because the integrands on both sides of this equation are polynomials, they must be equal for every value of t. This gives the first identity.

Now let $0 \le u < 1/k \le 1$ so $0 \le ku < 1$ and we can write, since $u + j/k < 1$ for all $j \le k-1$,

$$\begin{aligned}
b_n(ku) = B_n(ku) &= k^{n-1}\left(B_n(u) + \cdots + B_n\left(u + \frac{k-1}{k}\right)\right)\\
&= k^{n-1}\left(b_n(u) + \cdots + b_n\left(u + \frac{k-1}{k}\right)\right),
\end{aligned}$$

so the identity in $b_n(u)$ is true for these values of u. If $0 \le u < 1$ there is an $l < k$ so $u = l/k + (u - l/k)$ with $0 \le u - l/k < 1/k$. Then, since $b_n(u)$ has period 1,

$$b_n(ku) = b_n\left(k\left(u - \frac{l}{k}\right)\right) = k^{n-1}\left(b_n\left(u - \frac{l}{k}\right) + \cdots + b_n\left(u + \frac{k-l-1}{k}\right)\right)$$

$$= k^{n-1}\left(b_n\left(u+\frac{k-l}{k}\right)+\cdots+b_n\left(u+\frac{k-l+1}{k}\right)\right)$$

$$= k^{n-1}\left(b_n(u)+b_n\left(u+\frac{1}{k}\right)+\cdots+b_n\left(u+\frac{k-1}{k}\right)\right),$$

where, for example, $b_n(u)$ occurs as the $(l+1)$th term in the sum in the second line of the derivation. This completes the proof of the lemma. □

Now define the "sawtooth function" $b(u) := b_1(u) = u - \lfloor u \rfloor - \frac{1}{2}$.

Lemma 10.5 *For all* $j,k \in \mathbb{N}$ *we have*

$$I(j,k) := \int_0^1 b(ju)b(ku)\,du = \frac{(j,k)^2}{12jk},$$

where (j,k) *on the right-hand side represents the greatest common denominator (GCD) of j and k.*

Proof First let $k = 1$. Then, since $b(u)$ has period 1, and using Lemma 10.4 with $n = 1$,

$$I(j,1) = \frac{1}{j}\int_1^j b(u)b\left(\frac{u}{j}\right)du = \frac{1}{j}\sum_{k=0}^{j-1}\int_0^1 b(k+u)b\left(\frac{k}{j}+\frac{u}{j}\right)du$$

$$= \frac{1}{j}\int_0^1 b(u)b\left(j\cdot\frac{u}{j}\right)du = \frac{1}{j}\int_0^1 (u-\tfrac{1}{2})^2\,du = \frac{1}{12j}.$$

Next in case $(j,k) = 1$ note that kv/j modulo 1 when $0 \le v \le j-1$, since k has a modulo j inverse, gives each rational v/j exactly once, so that

$$I(j,k) = \frac{1}{j}\int_0^1 b(u)b\left(j\cdot\frac{ku}{j}\right)du$$

$$= \frac{1}{j}\int_0^1 b(u)b(ku)\,du = \frac{1}{j}I(k,1) = \frac{1}{12jk}.$$

In the general case let $c = (j,k)$, the GCD, and set $j = c\alpha$ and $k = c\beta$ so $(\alpha,\beta) = 1$. Then

$$I(j,k) = \int_0^1 b(c\alpha u)b(c\beta u)\,du$$

$$= \frac{1}{c}\int_0^c b(\alpha u)b(\beta u)\,du$$

$$= \int_0^1 b(\alpha u)b(\beta u)\,du$$

$$= I(\alpha,\beta) = \frac{1}{12\alpha\beta} = \frac{c^2}{12jk}.$$

This completes the proof. □

We are now able to give the proof of the Franel–Landau criterion for RH. Note the proof's explicit dependence on Littlewood's theorem, Theorem 4.16. Recall that

$$m := \sum_{j=1}^{n} \varphi(j) = \#F_n - 1.$$

Theorem 10.6 (Franel–Landau criterion) *[51, section 12.2] The Riemann hypothesis is equivalent to the statement that for all $\epsilon > 0$, as $n \to \infty$,*

$$\sum_{j=1}^{m} \left| f_j^n - \frac{j}{m} \right| \ll n^{1/2+\epsilon}.$$

Proof (1) The proof is in five steps. First assume the statement is true. Fix $n \in \mathbb{N}$ and, for $0 \le j \le m$, write f_j instead of f_j^n. Also use the same notation as in Lemmas 10.2, 10.3 and 10.4. Let $g(u) := \exp(2\pi i u)$ in identity (10.4) to get

$$\sum_{j=1}^{m} e^{2\pi i f_j} = \sum_{k=1}^{\infty} \sum_{j=1}^{k} e^{2\pi i j/k} M\left(\frac{x}{k}\right).$$

For $1 \le j \le m$ set $\delta_j := f_j - j/m$, the so-called "discrepancy" at j. Now for $k > 1$, $\sum_{j=1}^{k} \exp(2\pi i j/k) = 0$ and for $k = 1$ the sum is $1 = k$, so the right-hand side of this expression simplifies directly to $M(x)$. Thus

$$\begin{aligned} |M(x)| &= \left| \sum_{j=1}^{m} e^{2\pi i f_j} \right| = \left| \sum_{j=1}^{m} \exp\left(2\pi i \left(\frac{j}{m} + \delta_j\right)\right) \right| \\ &= \left| \sum_{j=1}^{m} e^{2\pi i j/m}(e^{2\pi i \delta_j} - 1) + \sum_{j=1}^{m} e^{2\pi i j/m} \right| \\ &\le \sum_{j=1}^{m} |e^{2\pi i \delta_j} - 1| = 2 \sum_{j=1}^{m} |\sin(\pi \delta_j)| \le 2\pi \sum_{j=1}^{m} |\delta_j| \ll x^{1/2+\epsilon}. \end{aligned}$$

So by Theorem 4.16, the Riemann hypothesis is true.

(2) Now assume the Riemann hypothesis is true. Recall

$$b(u) := b_1(u) = u - \lfloor u \rfloor - \tfrac{1}{2},$$

and set $g(u)$ to $b(u)$ in Lemma 10.4. We get

$$b(ku) = k^0 \left(b(u) + b\left(u + \frac{1}{k}\right) + \cdots + b\left(u + \frac{k-1}{k}\right) \right).$$

Therefore, using (10.4),

$$\begin{aligned} G(u) := \sum_{j=1}^{m} b(u + f_j) &= \sum_{k=1}^{\infty} \sum_{j=1}^{k} b\left(u + \frac{j}{k}\right) M\left(\frac{x}{k}\right) \\ &= \sum_{k=1}^{\infty} b(ku) M\left(\frac{x}{k}\right). \end{aligned}$$

Next in Steps (3) and (4) we use the definition of $G(u)$ and the expression we have derived to evaluate the integral $I_2 := \int_0^1 G(u)^2\,du$.

(3) Since $F_n = \{1 - f_j : 0 \le j \le m\}$ we have

$$G(u) = \sum_{j=1}^{m} b(u + 1 - f_j) = \sum_{j=1}^{m} b(u - f_j)$$

and $b(0) = -\frac{1}{2}$, $b(0-) = \frac{1}{2}$. This shows that $G(u)$ decreases by 1 at each f_j and increases like $m \cdot u$ as u increases from f_j to f_{j+1}. Also because

$$\sum_{j=1}^{m-1}(f_j - \tfrac{1}{2}) = \sum_{j=1}^{m-1} b(f_j) = \sum_{j=1}^{m-1} b(1 - f_j) = \sum_{j=1}^{m-1}(-f_j + \tfrac{1}{2}) = -\sum_{j=1}^{m-1} b(f_j),$$

we have $\sum_{j=1}^{m-1} b(f_j) = 0$ so $\lim_{u\to 0+} G(u) = -\frac{1}{2}$. All this goes to show that we can write, for $f_j < u < f_{j+1}$,

$$G(u) = m \cdot u - j - \tfrac{1}{2}.$$

We now use this formula to evaluate the integral I_2 explicitly.

Recall $f_0 = 0$ and $f_m = 1$. Performing the integration (recall $\delta_j = f_j - j/m$), using $\delta_0 = \delta_m = 0$ and shifting the second sum in the third line,

$$\begin{aligned}
I_2 &= \sum_{j=1}^{m} \int_{f_{j-1}}^{f_j} \left(mu - j + 1 - \frac{1}{2}\right)^2 du \\
&= \sum_{j=1}^{m} \left.\frac{(mu - j + \frac{1}{2})^3}{3m}\right|_{f_{j-1}}^{f_j} \\
&= \frac{1}{3m}\sum_{j=1}^{m}\left(\left(mf_j - j + \frac{1}{2}\right)^3 - \left(mf_{j-1} - (j-1)\frac{1}{2}\right)^3\right) \\
&= \frac{1}{3m}\sum_{j=1}^{m}\left(\left(m\delta_j + \frac{1}{2}\right)^3 - \left(m\delta_j - \frac{1}{2}\right)^3\right) \\
&= m\sum_{j=1}^{m}\delta_j^2 + \frac{1}{12}.
\end{aligned}$$

(4) Now consider the representation $G(u) = \sum_{k=1}^{\infty} b(ku)M(x/k)$ from Step (2). Since this sum is finite we can interchange sums and integrals to write

$$\begin{aligned}
I_2 &= \sum_{j=1}^{\infty}\sum_{k=1}^{\infty}\left(\int_0^1 b(ju)b(ku)\,du\right) M\left(\frac{x}{j}\right) M\left(\frac{x}{k}\right) \\
&= \sum_{j=1}^{\infty}\sum_{k=1}^{\infty} \frac{(j,k)^2}{12jk} M\left(\frac{x}{j}\right) M\left(\frac{x}{k}\right),
\end{aligned}$$

where we have used Lemma 10.5 to evaluate the integrals.

(5) For given $j,k \geq 1$ let $c = (j,k)$ and set $j = \alpha c,\ k = \beta c$. By the Riemann hypothesis and Theorem 4.16, we can assume for every $\epsilon > 0$ there is an implied constant, depending only on ϵ, such that $M(x) \ll x^{1/2+\epsilon}$ for all $x > 0$. Using this to bound our expression for I_2 gives

$$\begin{aligned} I_2 &\ll \sum_{j=1}^{\infty}\sum_{k=1}^{\infty}\frac{(j,k)^2}{12jk}\left(\frac{x^2}{jk}\right)^{1/2+\epsilon} \\ &\ll x^{1+2\epsilon}\sum_{\alpha=1}^{\infty}\sum_{\beta=1}^{\infty}\sum_{c=1}^{\infty}\frac{c^2}{c^{3+2\epsilon}\alpha^{3/2+\epsilon}\beta^{3/2+\epsilon}} \\ &\ll x^{1+2\epsilon}. \end{aligned}$$

We now combine the expression for the integral I_2, we derived in Step (3) with this estimate to get

$$m\sum_{j=1}^{m}\delta_j^2 \ll x^{1+2\epsilon}.$$

Finally the Cauchy–Schwarz inequality gives

$$\sum_{j=1}^{m}|\delta_j| = \sum_{j=1}^{m}|1\cdot\delta_j| \leq \left(\sum_{j=1}^{m}1^2\right)^{1/2}\left(\sum_{j=1}^{m}\delta_j^2\right)^{1/2} = \left(m\sum_{j=1}^{m}\delta_j^2\right)^{1/2} \ll x^{1/2+\epsilon},$$

and we have verified the condition of the theorem. □

We explore the structure of the sums $\sum_{1\leq j\leq m}|f_j - j/m|$, called **Farey sums**, in Figures 10.1 and 10.2. Each plotted point in Figure 10.1 represents the sum for F_n divided by $\sqrt{n}$. Each plotted point in Figure 10.2 represents the size of an individual term in the sum.

Remark The equivalence between Farey series and RH has been one of the more popular equivalences. It has stimulated a considerable amount of additional work. Some of this involves tight estimates for functions mapped over Farey series, where the functions satisfy given conditions. Others include relations between the prime number theorem and Farey series. See for example [8, 85, 86, 87, 116, 117, 181].

There is another mysterious property connecting Farey series with zeta functions. It was observed in [18] that the sums of the values of the Hurwitz zeta function over the Farey fractions of a given order, other than zero, at a zero of the zeta function, are all zero.

10.4 Redheffer Matrix

Fix $n \in \mathbb{N}$ and define an $n \times n$ matrix with $0,1$ entries by $r_{i,j} = 1$ if $j = 1$ or $i \mid j$. Otherwise $r_{i,j}$ is set to 0. We denote by R_n the matrix with entries $r_{i,j}$,

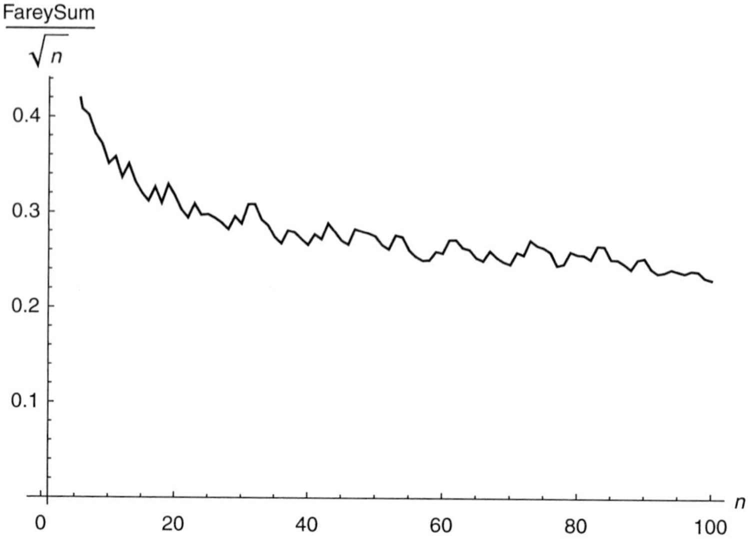

Figure 10.1 The Farey sums over $\sqrt{n}$ with $1 \leq n \leq 100$.

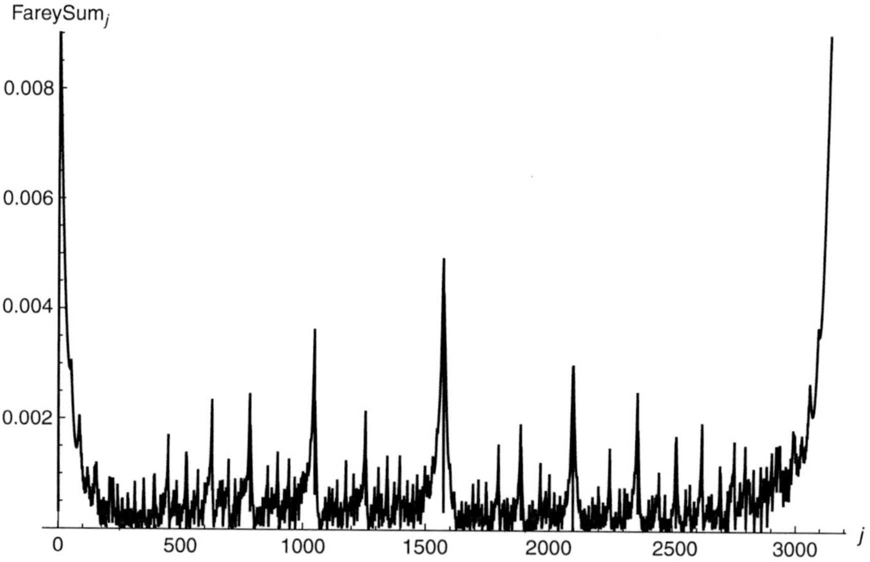

Figure 10.2 Individual terms in the Farey sum for $n = 101$.

and call it the Redheffer matrix of order n. The determinant of this matrix was first evaluated by Ray Redheffer in 1976 [138]. It was also evaluated, using a combinatorial argument, by Herbert Wilf in 2006 [180]. Here we present the method of Bryan Gillespie from 2011 [67], which better reveals the underlying structure and gives a vast generalization.

Let $f : \mathbb{N} \to \mathbb{C}$ be an arithmetic function with $f(1) \neq 0$. Then [6, theorem 2.8] there is a unique arithmetic function f^{-1}, the so-called Dirichlet inverse of f, such that

$$f * f^{-1} = f^{-1} * f = I,$$

where $I(1) = 1$ and $I(n) = 0$ for $n > 1$ is the identity for Dirichlet multiplication, which is as usual denoted "$*$". Indeed, f^{-1} can be defined recursively by $f^{-1}(1) := 1/f(1)$ and for $n > 1$

$$f^{-1}(n) := -\frac{1}{f(1)} \sum_{d|n,\ d<n} f\left(\frac{n}{d}\right) f^{-1}(d).$$

If f is completely multiplicative this looks much simpler. In this case, for all $n \geq 1$ we have [6, theorem 2.17] $f^{-1}(n) = \mu(n) f(n)$.

Given any arithmetic function $f : \mathbb{N} \to \mathbb{C}$ with $f(1) \neq 0$, and a natural number n, define an $n \times n$ matrix by $R_n(f) := (r_{i,j})$ where, for $1 \leq i, j \leq n$, $r_{i,j} = f(j/i)$ if $i \mid j$, $r_{i,j} = 1$ if $i > 1$ and $j = 1$ and $r_{i,j} = 0$ otherwise.

Lemma 10.7 *[67, theorem 2.7] Let f be an arithmetic function with $f(1) \neq 0$. Then*

$$\det R_n(f) = f(1)^n \left(1 + \sum_{j=2}^{n} f^{-1}(j)\right).$$

Proof Define an upper triangular $n \times n$ matrix with complex coefficients $S_n(f) = (s_{i,j})$ by $s_{i,j} := f(j/i)$ if $i \mid j$ and $s_{i,j} = 0$ otherwise. Then $S_n(f)$ has determinant $f(1)^n$ so is invertible in $\mathbb{C}$. Note also that if f and g are two such arithmetic functions then $S_n(f) \times S_n(g) = (t_{i,j})$ where "$\times$" is normal matrix multiplication and $t_{i,j} = 0$ if $i \nmid j$ and $t_{i,j} = (f * g)(j/i)$ if $i \mid j$. In other words $S_n(f) \times S_n(g) = S_n(f * g)$.

Therefore, if A is the $n \times n$ matrix with 0 in all positions except the leading column, which has 1 in every position except the first, then

$$S_n(f^{-1}) \times R_n(f) = S_n(f^{-1})(S_n(f) + A) = I_{n\times n} + S_n(f^{-1})A, \tag{10.5}$$

where $I_{n\times n}$ is the $n \times n$ identity matrix. Then because of their shape $S_n(f^{-1})A$ is the zero matrix except for the $(1,1)$ element, which is $\sum_{j=2}^{n} f^{-1}(j)$, and the $(i,1)$ elements with $i > 1$. Taking determinants of each side of (10.5) gives

$$f(1)^{-n} \det R_n(f) = 1 + \sum_{j=2}^{n} f^{-1}(j)$$

and the proof is complete. □

Now let $f(n) = 1$ for all $n \in \mathbb{N}$ so $f^{-1}(n) = \mu(n)$ in Lemma 10.7. We get:

Corollary 10.8 *For all $n \in \mathbb{N}$,*

$$\det R_n = \sum_{j=1}^{n} \mu(j).$$

Theorem 10.9 then follows directly from Theorem 4.16.

Theorem 10.9 (Redheffer criterion) *[138] The Riemann hypothesis is equivalent to the statement that for all $\epsilon > 0$, as $n \to \infty$,*

$$\det R_n \ll n^{1/2+\epsilon}.$$

10.5 Divisibility Graph

The divisibility matrix R_n gives rise also to a graph-theory-based equivalence to RH. These graphs are rather special, and designed to make the equivalence an easy consequence of Section 10.4.

Form a directed graph G_n on the vertices labelled $\{1,\ldots,n\}$ by inserting a directed edge from i to j if $i \mid j$ and $i \neq j$, and a directed edge from j to 1 for $2 \leq j \leq n$. We call G_n the **divisibility graph** of order n (see for example Figure 10.3, which is G_6).

A **cycle** of a directed graph is a sequence of edges such that the final point of each edge is followed by an edge with the same initial point, other than the last edge, whose final point corresponds to the initial point of the first edge.

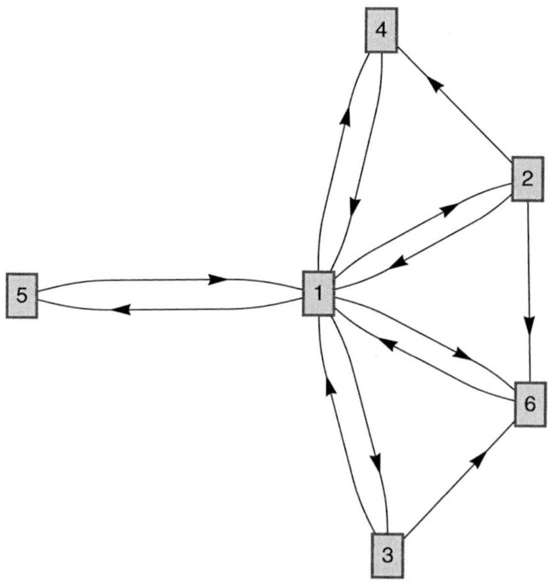

Figure 10.3 The divisibility graph for $n = 6$.

In this context, other than the first and last, the vertices in a cycle are **distinct**. A cycle is called *odd* if it consists of an odd number of edges. Similarly *even* cycles are defined. Let $o(G_n)$ be the number of odd cycles of G_n and $e(G_n)$ the number of even cycles. Then we have the following parity equivalence to RH:

Theorem 10.10 (Divisibility graph criterion) *[9] The Riemann hypothesis is true if and only if for all* $\epsilon > 0$, *as* $n \to \infty$, *we have*

$$|o(G_n) - e(G_n)| \ll n^{1/2+\epsilon}.$$

Proof Let I_n be the $n \times n$ identity matrix and set $B_n := R_n - I_n$. Let G_n be the directed graph with adjacency matrix $B_n = (b_{i,j})$ so the vertices of G_n are $\{1,\ldots,n\}$ and an edge in G_n connects i to j if and only if $b_{i,j} = 1$. Then G_n is a divisibility graph. Define a polynomial in $\mathbb{Z}[x]$ by

$$q_n(x) := \det(B_n + xI_n).$$

Each cycle

$$1 \to i_1 \to i_2 \to \cdots \to i_{k-1} \to 1$$

in G_n, starting and ending at the vertex 1, having length k and distinct $i_j \neq 1$, contributes to the determinant of $B_n + xI_n$ a term with value $(-1)^{k-1}$ times the determinant of the $(n-k)\times(n-k)$ submatrix which is obtained by deleting all rows and columns indexed $1, i_1, i_2, \ldots, i_{k-1}$ from that matrix. The submatrix is upper triangular, so its determinant is simply x^{n-k}, the product of terms on the diagonal. Every cycle includes 1. Therefore each k-cycle in G_n contributes $(-1)^{k-1}x^{n-k}$ to $\det(B_n + xI_n)$. In addition to cycles, the determinant includes the diagonal x^n and no other non-zero terms.

Now let $c(n,k)$ be the number of distinct cycles of length $2 \le k \le n$ in G_n, so

$$q_n(x) = x^n + \sum_{k=2}^{n} (-1)^{k-1} c(n,k) x^{n-k}.$$

Then

$$\det R_n = q_n(1) = 1 + \sum_{k=2}^{n} (-1)^{k-1} c(n,k) = 1 + o(G_n) - e(G_n).$$

Therefore

$$|o(G_n) - e(G_n)| = |\det(R_n) - 1|.$$

By Theorem 10.9, it follows that RH is true if and only if for all $\epsilon > 0$, $|o(G_n) - e(G_n)| \ll n^{1/2+\epsilon}$, which completes the proof. □

10.6 Dirichlet Eta Function

Perhaps the easiest equivalence to RH to derive is in terms of the so-called Dirichlet eta function. An advantage of using this function is that its Dirichlet series representation converges conditionally in the half plane $\sigma > 0$, in particular, in all of the open critical strip.

Let $s \in \mathbb{C}$ and define for $\sigma = \Re s > 0$

$$\eta(s) := (1 - 2^{1-s})\zeta(s).$$

Then we have:

Lemma 10.11 *If $\sigma > 0$ the series $\sum_{n=1}^{\infty}(-1)^{n+1}/n^s$ converges and has sum $\eta(s)$.*

Proof If $s = \sigma + it$ with σ, t real, then the series converges absolutely for $\sigma > 1$ and therefore represents a holomorphic function in that half plane. Indeed,

$$\begin{aligned}\sum_{n=1}^{\infty}\frac{(-1)^{n+1}}{n^s} &= -\sum_{n=1}^{\infty}\frac{1}{(2n)^s} + \sum_{n=1}^{\infty}\frac{1}{(2n-1)^s}\\ &= -2^{-s}\sum_{n=1}^{\infty}\frac{1}{n^s} + \sum_{n=1}^{\infty}\frac{1}{n^s} - \sum_{n=1}^{\infty}\frac{1}{(2n)^s}\\ &= -2^{1-s}\sum_{n=1}^{\infty}\frac{1}{n^s} + \sum_{n=1}^{\infty}\frac{1}{n^s}\\ &= (1 - 2^{1-s})\zeta(s) = \eta(s).\end{aligned}$$

If $\sigma > 0$ then, by the alternating real series test, the series

$$\sum_{n=1}^{\infty}\frac{(-1)^{n+1}}{n^{\sigma}}$$

converges. Therefore, by [6, theorem 11.8], it converges for all $s \in \mathbb{C}$ with $\sigma = \Re s > 0$. It diverges at $\sigma = 0$ so the abscissa of convergence of the series is $\sigma_c = 0$. Therefore we can continue $\eta(s)$ with this series representation into the half plane $\Re s > 0$. □

Theorem 10.12 (Dirichlet eta criterion) *The Riemann hypothesis is equivalent to the zeros of $\eta(s)$ with $0 < \sigma < 1$ all having real part $\frac{1}{2}$.*

Proof A simple calculation shows that none of the zeros of the factor $1 - 2^{1-s}$ have real part in $(0, 1)$. □

10.7 The Derivative of $\zeta(s)$

In this section an "almost equivalence", due originally to Speiser, is described. In 1934 [161] he gave a geometric/topological but not rigorous proof that RH is equivalent to all of the zeros of the derivative of $\zeta(s)$ in the critical strip being on or to the right of the critical line. Part of this statement was given rigorous justification by Spira in 1973, who proved that RH implies the derivative $\zeta'(s)$ has no zero in the open left half of the critical strip [163]; see also Trudgian [171]. Levinson and Montgomery went part way to proving the converse in their 1974 paper [107], by showing that there were at most $O(\log T)$ zeros of the derivative up to height T in the left half.

Let us start with some definitions:

$$\begin{aligned} N(T) &:= \#\{\rho \in \mathbb{C} : \zeta(\rho) = 0,\ 0 < \Im\rho \le T\}, \\ N^-(T) &:= \#\{\rho \in \mathbb{C} : \zeta(\rho) = 0,\ 0 < \Re\rho < \tfrac{1}{2},\ 0 < \Im\rho \le T\}, \\ N_1^-(T) &:= \#\{\rho \in \mathbb{C} : \zeta'(\rho) = 0,\ 0 < \Re\rho < \tfrac{1}{2},\ 0 < \Im\rho \le T\}. \end{aligned}$$

In what follows we also need a well-known approximation to the logarithmic derivative of the gamma function. This is defined for all $s \in \mathbb{C} \setminus \{0, -1, -2, \ldots\}$ by

$$\Gamma(s) = \int_0^\infty t^{s-1} e^{-t}\, dt.$$

Lemma 10.13 *[51, section 6.3] Let $w = u + iv \in \mathbb{C}$. Then*

$$\frac{\Gamma'(w)}{\Gamma(w)} = \log w - \frac{1}{2w} + R,$$

where $|R| \le 1/(10|w|^2)$, $|w| \ge 2$ and $u \ge 0$.

This follows for example from the representation for $\log\Gamma(s)$ given in the proof of Lemma 10.17.

The proof of the following very useful lemma, Jensen's formula, is given in Theorem B.5 in Volume Two [32] or [51, section 2.2].

Lemma 10.14 (Jensen's formula) *Let $R > 0$ and let $f(z)$ be a function which is holomorphic inside and on the boundary of the disc $\Omega = \{z \in \mathbb{C} : |z| \le R\}$ with no zeros on the boundary or at the centre. Let the zeros inside the disc be labelled $z_1, z_2, \ldots, z_n$, where a zero of multiplicity m is included m times. Then, if we let $n(r)$ represent the number of zeros inside a circle with centre 0 and radius r with no zero on the boundary:*

$$\int_0^R \frac{n(r)}{r}\, dr = n \log R - \sum_{i=1}^n \log|z_i| = \frac{1}{2\pi} \int_0^{2\pi} \log|f(Re^{i\theta})|\, d\theta - \log|f(0)|.$$

Recall Table 2.1 in Section 2.2. Let $T \ge e$ and denote by $N(T)$ the number of zeros, including multiplicity, of $\zeta(s)$ in the rectangle $[0,1] \times [0,T] \subset \mathbb{C}$. Then

$$\left| N(T) - \left(\frac{T}{2\pi} \log \frac{T}{2\pi} - \frac{T}{2\pi} + \frac{7}{8} \right) \right| < 0.122 \log T + 0.278 \log\log T + 2.510. \tag{10.6}$$

Lemma 10.15 *For all $T > 0$ we have $N_1^-(T) = N^-(T) + O(\log T)$. If it is false to say $N^-(T) \ge T + O(1)$ as $T \to \infty$ then there exists a sequence $T_j \to \infty$ such that for all j we have $N_1^-(T_j) = N^-(T_j)$.*

Proof Denote the zeros of $\zeta(s)$ in the critical strip by ρ. Then because

$$(s-1)\pi^{-s/2}\Gamma\left(\frac{s}{2}+1\right)\zeta(s) = \xi(s) = \xi(0)\prod_{\rho}\left(1 - \frac{s}{\rho}\right),$$

if $\Re s \ge 0$ and s is not a zero of zeta, taking logs, differentiating and then taking the real part, we get

$$\Re \frac{\zeta'(s)}{\zeta(s)} = -\Re \frac{1}{s-1} + \frac{1}{2}\log \pi - \frac{1}{2}\Re \frac{\Gamma'(s/2+1)}{\Gamma(s/2+1)} + \Re \sum_{\rho} \frac{1}{s-\rho}. \tag{10.7}$$

Let $s = \sigma + it$. If $\rho = \beta + i\gamma$ is a zero of $\zeta(s)$ and $\beta < \frac{1}{2}$ so is $1 - \bar{\rho} = 1 - \beta + i\gamma$ and

$$\Re\left(\frac{1}{s-\rho} + \frac{1}{s-1+\bar{\rho}}\right) = -2\left(\frac{1}{2} - \sigma\right)\frac{(t-\gamma)^2 + (\sigma - \frac{1}{2})^2 - (\frac{1}{2} - \beta)^2}{|s-\rho|^2|s-1+\bar{\rho}|^2}.$$

If we set

$$I_1 := 2\sum_{\beta<1/2} \frac{(t-\gamma)^2 + (\sigma - \frac{1}{2})^2 - (\frac{1}{2} - \beta)^2}{|s-\rho|^2|s-1+\bar{\rho}|^2} + \sum_{\beta=1/2} \frac{1}{|s-\rho|^2}, \tag{10.8}$$

then we have

$$I := \Re \sum_{\rho} \frac{1}{s-\rho} = -\left(\frac{1}{2} - \sigma\right) I_1. \tag{10.9}$$

By Lemma 10.13 we have an estimate for the logarithmic derivative of $\Gamma(w)$ where $w = u + iv$:

$$\frac{\Gamma'(w)}{\Gamma(w)} = \log w - \frac{1}{2w} + R, \quad \text{where } |R| \le \frac{1}{10|w|^2}, \ |w| \ge 2, \ u \ge 0.$$

Therefore for $|s| \ge 3$ and $\sigma \ge 0$ we can write

$$\Re \frac{\Gamma'(s/2+1)}{\Gamma(s/2+1)} = \log|1 + s/2| - \frac{\sigma+2}{|s+2|^2} + R_1, \quad \text{where } |R_1| \le \frac{2}{5|s+2|^2}. \tag{10.10}$$

By (10.6), (10.7)–(10.10) and using $\beta = \frac{1}{2}$ for $|\gamma| < 10^3$, we see that when $t = 10$ and $\sigma \in [0, \frac{1}{2}]$ we have $\Re(\zeta'(s)/\zeta(s)) < 0$. This was the method of Levinson and Montgomery. Alternatively it is much easier to check this condition numerically, where we find that in the given range $\Re(\zeta'(s)/\zeta(s)) < -0.2$.

Consider now the line $\sigma = 0$. By (10.8), using $0 < \beta \le \frac{1}{2}$ all terms in I_1 are positive, so $I < 0$. Then using (10.7) and (10.10), when $t \ge 10$, evaluating the bounds for each term in (10.7), we have $\Re(\zeta'(s)/\zeta(s)) < 0$ also.

On the line $\sigma = \frac{1}{2}$, unless we are at a zero of $\zeta(s)$, we have $I = 0$. If $\rho_0 = \frac{1}{2} + i\gamma_0$ is a zero on this line then the first sum for I_1 is positive and, in the second sum, the term $1/|s-\rho_0|^2$ can be made arbitrarily large for s sufficiently close to ρ_0. Thus on a sufficiently small semicircle with centre ρ_0 and with $\sigma < \frac{1}{2}$ we get $I_1 > 0$, and thus $I < 0$. Hence on these semicircles we also have $\Re(\zeta(s)/\zeta(s)) < 0$. Replacing the segment $\sigma = \frac{1}{2}$ by a suitably indented contour then gives $\Re(\zeta'(s)/\zeta(s)) < 0$ for $t \ge 10$ on that contour.

Now assume there exists a real sequence $(T_j)_{j\ge 0}$ with $T_j \to \infty$ such that $\Re(\zeta'(s)/\zeta(s)) < 0$ on the segment $(0, \frac{1}{2}) \times \{T_j\}$. On the indented contour around

$$[10i,\ \tfrac{1}{2} + 10i] \times [iT_j,\ \tfrac{1}{2} + iT_j]$$

we must have $\Re(\zeta'(s)/\zeta(s)) < 0$, so the change in argument of $\zeta'(s)/\zeta(s)$ is zero on the contour. By the principle of the argument, this implies the number of zeros of $\zeta'(s)$ and $\zeta(s)$ are the same inside each of these contours, and the proof is complete in this case.

Now suppose no such sequence (T_j) exists. This requires that there exists a $t_0 > 0$ such that, for all $t \ge t_0$, $\Re(\zeta'(s)/\zeta(s)) \ge 0$ for $s = \sigma + it$ and some $\sigma \in (0, \frac{1}{2})$. Because in (10.7) for t sufficiently large the first term on the right can be made arbitrarily small and the third term arbitrarily large and negative, this implies $I > 0$, so at least one term in the sum expression for I_1 must be negative. Therefore there exists a $\beta < \frac{1}{2}$ such that

$$(\tfrac{1}{2} - \beta)^2 > (t - \gamma)^2 + (\sigma - \tfrac{1}{2})^2$$

so $|t - \gamma| < \frac{1}{2}$. If $t = n$ is a sufficiently large integer, there must be at least one zero $\rho = \beta + i\gamma$ of zeta with $\beta < \frac{1}{2}$ and $|\gamma - n| < \frac{1}{2}$. This shows $N^-(T) \ge T + O(1)$.

Now one can use Lemma 10.14 to show that the change in argument of $\zeta(\sigma + it)$ and of $\zeta'(\sigma + it)$ for fixed $t \ge t_1$ and $\sigma \in [0, \frac{1}{2}]$ are both $O(\log t)$. Alternatively, this is Step (4) of Theorem I.1 of Volume Two [32]. These derivations Also use the estimates $|\zeta(s)| \ll t^A$ and $|\zeta'(s)| \ll t^A$ for $-3 \le \sigma \le 1$ and some $A > 0$ which is property (9) in Section 2.2.

Using this and the negativity $\Re(\zeta'(s)/\zeta(s))$ on the other three components of the indented rectangular contour around $[0, \frac{1}{2}] \times [10, t]$ completes the proof. □

Theorem 10.16 *[107, theorem 1; 162] If the Riemann hypothesis is true, all non-trivial zeros of the derivative of the Riemann zeta function $\zeta'(s)$ in the* *critical strip have real part greater than or equal to $\frac{1}{2}$. Even if the Riemann hypothesis is false, the number of zeros of $\zeta'(s)$ in this strip with imaginary part less than $T > 0$ and real part less than 1/2 is $O(\log T)$.*

Proof If the Riemann hypothesis holds, then for all $T > 0$, $N^-(T) = 0$ so $N^-(T) > T + O(1)$ is false for all $T \geq T_0$. Hence, by Lemma 10.15, $N_1^-(T) = N^-(T) = 0$ for all $T > 0$, so $\zeta'(s)$ has no zero with $0 < \sigma < \frac{1}{2}$.

Conversely if $N_1^-(T) = 0$ for all $T > 0$ then again, by Lemma 10.15, $N^-(T) = O(\log T)$. (Note that this is not sufficient to give the Riemann hypothesis.) □

Recently Suriajaya derived an asymptotic formula for the number of zeros of the kth derivative of $\zeta(s)$, assuming RH [164].

10.8 A Zeta-Related Inequality

The next equivalence is included because of the simplicity of its statement. The proof was published by Dixon and Spira in 1965. It comes almost directly from the functional equation for $\zeta(s)$, which makes it special when compared with the other equivalences to RH described here. Also, in this section, we include the derivation of explicit bounds for the function $|\Lambda(s)|$, which is defined in Lemma 10.17. These could be useful, especially as a basis for further work relating to the Dixon–Spira criterion.

Lemma 10.17 *[47] If $|t| \geq 2\pi + 0.1$ and $1 > \sigma > \frac{1}{2}$ then $|\zeta(1-s)| > |\zeta(s)|$ except when $\zeta(s) = 0$.*

Proof Write the functional equation for $\zeta(s)$ as $\zeta(1-s) = \Lambda(s)\zeta(s)$ where as usual

$$\Lambda(s) := 2(2\pi)^{-s}\Gamma(s)\cos\left(\frac{\pi s}{2}\right).$$

Because $|\Lambda(s)\Lambda(1-s)| = 1$ we have $|\Lambda(s)| > 0$ for all $s \in \mathbb{C}$ and $|\Lambda(\frac{1}{2} + it)| = 1$. Setting $s_0 = \frac{1}{2} + it$ define

$$h(s) := \log\left(\frac{|\Lambda(s)|}{|\Lambda(s_0)|}\right) = \log|\Lambda(s)|.$$

Then we have $h(s) > 0$ if and only if $|\Lambda(s)| > 1$, so to prove the lemma, because $|\zeta(1-s)/\zeta(s)| = |\Lambda(s)|$, we need only show $h(s) > 0$ when $1 > \sigma > \frac{1}{2}$.

Let $1 > \sigma > \frac{1}{2}$ and $|t| \geq 2\pi + 0.1$. Then

$$\begin{aligned} h(s) &= \log|\Gamma(s)| - \log|\Gamma(s_0)| + \frac{1}{2}\log\left(\frac{\cosh \pi t + \cos \pi\sigma}{\cosh \pi t}\right) - \left(\sigma - \frac{1}{2}\right)\log 2\pi \\ &\geq \left(\sigma - \frac{1}{2}\right)\left(\frac{\partial}{\partial\sigma}\log|\Gamma(\sigma + it)|\right)_{\sigma=\eta} + \frac{1}{2}\log\left(1 - \frac{|\sin \pi(\sigma - \frac{1}{2})|}{\cosh \pi t}\right) \\ &\quad - \left(\sigma - \frac{1}{2}\right)\log 2\pi, \end{aligned}$$

where we have used the mean value theorem to obtain η with $\frac{1}{2} < \eta < \sigma$.

If $0 \le x \le \frac{1}{2}$ we have $\frac{1}{2}\log(1-x) \ge -x$. Therefore for $|t| \ge \frac{1}{2}$ we can derive the lower bound

$$\frac{1}{2}\log\left(1 - \frac{|\sin\pi(\sigma-\frac{1}{2})|}{\cosh\pi t}\right) \ge -\frac{|\sin\pi(\sigma-\frac{1}{2})|}{\cosh\pi t} > -2\pi\left(\sigma-\frac{1}{2}\right)e^{-\pi|t|}.$$

Therefore, dividing by $\sigma - \frac{1}{2}$ we get

$$\frac{h(s)}{\sigma-\frac{1}{2}} > \left(\frac{\partial}{\partial\sigma}\log|\Gamma(\sigma+it)|\right)_{\sigma=\eta} - 2\pi e^{-\pi|t|} - \log 2\pi.$$

Now if we recall $B_3(x) := x(2x^2-3x+1)/2$, the Bernoulli polynomial of order 3, we can write Stirling's formula (see Volume Two [32, Appendix H]) in the form

$$\log\Gamma(s) = \left(s-\frac{1}{2}\right)\log s - s + \frac{1}{2}\log 2\pi + \frac{1}{12s} - \frac{1}{3}\int_0^\infty \frac{B_3(\{x\})}{(s+x)^3}\,dx,$$

where $\{x\}$ is the fractional part. This formula is true when s is real and positive. Define the complex logarithm such that $\log 1 = 0$, with the plane cut along the negative x-axis. With this choice the formula remains true for complex s in the open subset of $\mathbb{C}$ which is the complement of the cut.

Note that if $x \in [0,1]$ we have $|B_3(x)| \le \sqrt{3}/36$. Therefore

$$\begin{aligned}\frac{\partial}{\partial\sigma}\log|\Gamma(\sigma+it)| &= \Re\frac{d}{ds}\log\Gamma(s)\\ &= \Re\left(\log s - \frac{1}{2s} - \frac{1}{12s^2} + \int_0^\infty \frac{B_3(\{x\})}{(s+x)^4}\,dx\right)\\ &> \frac{1}{2}\log(\sigma^2+t^2) - \frac{\sigma}{2(\sigma^2+t^2)} - \frac{\sigma^2-t^2}{12(\sigma^2+t^2)^2}\\ &\quad - \frac{\sqrt{3}}{36}\int_0^\infty \frac{1}{((\sigma+x)^2+t^2)^2}\,dx.\end{aligned}$$

To bound the integral let $y := \sigma + x$ and then $\tan\theta := y/|t|$ to get

$$\begin{aligned}\int_0^\infty \frac{1}{((\sigma+x)^2+t^2)^2}\,dx &\le \int_0^\infty \frac{1}{(y^2+t^2)^2}\,dy\\ &= \frac{1}{|t|^3}\int_0^{\pi/2}\cos^2(\theta)\,d\theta = \frac{\pi}{4|t|^3}.\end{aligned}$$

Using this bound and $\frac{1}{2} < \sigma < 1$ we get

$$\frac{h(s)}{\sigma-\frac{1}{2}} > \log|t| - \frac{1}{2t^2} - \frac{\sqrt{3}\pi}{72|t|^3} - 2\pi e^{-\pi|t|} - \log 2\pi.$$

When $|t| \ge 2\pi + 0.1$ the right-hand side is strictly positive, and therefore so is $h(s)$. This completes the proof of the lemma. □

Note that, for $s = \frac{1}{2} + it$, $h(s) = 1$, and for $\sigma > \frac{1}{2}$, $\partial h(\sigma + it)/\partial\sigma$ can easily be shown, using the method of Lemma 10.17, to be quite positive.

Theorem 10.18 (Dixon–Spira criterion) *[47, 162] The Riemann hypothesis is equivalent to the statement that, if $\frac{1}{2} < \sigma < 1$ and $t \geq 6.29073$, then $|\zeta(1-s| > |\zeta(s)|$.*

Proof If the Riemann hypothesis is false and s is a zero not on the critical line, then both sides of the inequality are zero, so it is false. If the Riemann hypothesis is true, then Lemma 10.17 shows the inequality is true. □

Corollary 10.19 *For $|t| > 10$ and $1 \geq \sigma > \frac{1}{2}$ we have $t^{\sigma-1/2} > |\Lambda(\sigma+it)| > t^{(\sigma-1/2)/5}$.*

Proof First we derive an upper bound for $h(s)/(\sigma - \frac{1}{2})$, where $h(s)$ is defined in Lemma 10.17. We have

$$\log\left(1 + \frac{\cos(\pi\sigma)}{\cosh(\pi t)}\right) \leq \frac{|\sin(\pi(\sigma - \frac{1}{2}))|}{\cosh(\pi t)} < \pi\left(\sigma - \frac{1}{2}\right)e^{-\pi|t|}$$

and

$$\begin{aligned}\frac{\partial}{\partial\sigma}\log|\Gamma(\sigma+it)| &\leq \frac{1}{2}\log(\sigma^2+t^2) - \frac{\sigma}{2(\sigma^2+t^2)} - \frac{\sigma^2-t^2}{12(\sigma^2+t^2)^2} \\ &\quad + \frac{\sqrt{3}}{36}\int_0^\infty \frac{1}{((\sigma+x)^2+t^2)^2}\,dx \\ &\leq \frac{1}{2}\log(\sigma^2+t^2) + \frac{1}{12t^2} + \frac{\sqrt{3}\pi}{72|t|^3}.\end{aligned}$$

Therefore since $\sigma \leq 1$

$$\begin{aligned}\frac{h(s)}{\sigma - \frac{1}{2}} &< \frac{1}{2}\log(1+t^2) + \frac{1}{12t^2} + \frac{\sqrt{3}\pi}{72|t|^3} - \log(2\pi) \\ &\leq \log|t| + \frac{1}{12t^2} + \frac{\sqrt{3}\pi}{72|t|^3} - \log(\sqrt{2}\pi) < \log|t|\end{aligned}$$

for $|t| > 1$, giving the upper bound.

Next we derive a lower bound. Using the notation and argument of Lemma 10.17 we first evaluate the integral:

$$I(\sigma,t) := \int_0^\infty \frac{dx}{((\sigma+x)^2+t^2)^2} = \frac{\tan^{-1}(t/\sigma)}{2|t|^3} - \frac{\sigma}{2t^2(\sigma^2+t^2)}.$$

Therefore, if we define as before

$$F(\sigma,t) := -\frac{\sigma}{2(t^2+\sigma^2)} - \frac{\sigma^2-t^2}{12(t^2+\sigma^2)^2} + \frac{1}{2}\log(t^2+\sigma^2),$$

we can write

$$\frac{h(s)}{\sigma-\frac{1}{2}} > F(\sigma,t) - \frac{\sqrt{3}}{36} I(\sigma,t) - 2\pi e^{-\pi|t|} - \log 2\pi.$$

Since the integral is decreasing in σ, and $F(\sigma,t)$ is increasing, we can substitute $\sigma = \frac{1}{2}$ and obtain a lower bound

$$\frac{h(s)}{\sigma-\frac{1}{2}} > F(\tfrac{1}{2},t) - \frac{\sqrt{3}}{6} I(\tfrac{1}{2},t) - 2\pi e^{-\pi|t|} - \log 2\pi.$$

Then for all $\sigma > \frac{1}{2}$ we get $h(s)/(\sigma - \frac{1}{2}) > G(t)$ where

$$G(t) := \frac{1}{2}\log(4t^2+1) + \frac{4t^2-1}{3(4t^2+1)^2} + \frac{1}{12\sqrt{3}t^2(4t^2+1)}$$
$$- \frac{1}{4t^2+1} - \frac{\tan^{-1}(2t)}{24\sqrt{3}|t|^3} - 2\pi e^{-\pi t} - \log(4\pi).$$

Finally this expression is greater than $0.201675\log|t| > \log|t|/5$ for all $|t| \geq 10$. This completes the proof. □

10.9 The Real Part of the Logarithmic Derivative of $\xi(s)$

Lagarias [98] pointed out that RH was equivalent to a positivity criterion expressed in terms of the real part of the logarithmic derivative of the Riemann xi function $\xi(s)$, namely

$$\Re s > \frac{1}{2} \quad \Longrightarrow \quad \Re\frac{\xi'(s)}{\xi(s)} > 0.$$

This is **Lagarias' positivity criterion.**

Lagarias showed that the logarithmic derivative of $\xi(s)$ was positive for $\sigma \geq 9$ and suggested the region could be extended to $\sigma \geq 1$. This was shown by the present author [24] for all $|t|$ sufficiently large or small. This restriction did not apply to the theorem of Garunkstis [66] who showed that

$$\Re s > 1 \quad \Longrightarrow \quad \Re\frac{\xi'(s)}{\xi(s)} > 0,$$

and it is this work, following an account of the work of Lagarias, which we have included here.

Lagarias' work was set in the context of number fields and particular sets of zeros of meromorphic functions. Let $\Omega \subset \mathbb{C}$ be a discrete subset of points with multiplicity. We say Ω is an **admissible zero set** if $\Omega = \bar{\Omega}$, if corresponding points ρ and $\bar{\rho}$ have the same multiplicity denoted $m(\rho)$, and if

$$\sum_{\rho\in\Omega} \frac{1+|\Re\rho|}{1+|\rho|^2} < \infty. \tag{10.11}$$

This concept, considered along with Theorem 10.23, sheds new light on the Riemann hypothesis and its generalizations.

Lemma 10.20 *The zeros of the Riemann function $\xi(s)$ form an admissible zero set.*

Proof By Lemma 2.8, Step (2), $\sum_\rho 1/|\rho|^2$ converges. □

In what follows we need the Weierstrass product, which is developed and proved in Volume Two [32, Theorem B.13]. See also Alfors [4, chapter 5].

Lemma 10.21 (Weierstrass product) *The Weierstrass product*

$$f_\Omega(s) := s^{m(0)} \prod_{\rho\in\Omega\cap\mathbb{R}} \left(1-\frac{s}{\rho}\right)^{m(\rho)} \prod_{\rho\in\Omega,\,\Im\rho>0} \left(\left(1-\frac{s}{\rho}\right)\left(1-\frac{s}{\bar\rho}\right)\right)^{m(\rho)}$$

converges uniformly on compact subsets of $\mathbb{C}$ to an entire function which is real on the real axis, where Ω is admissible.

Finally, in this summary of preliminary results, we need an expression for the logarithmic derivative of $f_\Omega(s)$. Note that the proof uses Cauchy's well-known bound for the first derivative of a function $f(s)$ which is holomorphic in an open neighbourhood of a closed disc $B(s_0,r]$ with $|f(s)| \le M$ on this disc, namely

$$|f'(s_0)| \le \frac{M}{r}.$$

This follows directly from Cauchy's integral formula for the first derivative.

Lemma 10.22 *The Mittag–Leffler expansion*

$$\frac{f_\Omega'(s)}{f_\Omega(s)} = \sum_{\rho\in\Omega\cap\mathbb{R}} \frac{m(\rho)}{s-\rho} + \sum_{\rho\in\Omega,\,\Im\rho>0} m(\rho)\left(\frac{1}{s-\rho}+\frac{1}{s-\bar\rho}\right)$$

converges uniformly on compact subsets $K \subset V := \mathbb{C}\setminus\Omega$.

Proof (1) First we show there exists $r > 0$ such that $|s-w| > r$ for all s in K and w in $\mathbb{C}\setminus V$. If not there is a convergent sequence (s_n) in K and corresponding points w_n in $\mathbb{C}\setminus V = \Omega$ with $|s_n - w_n| < 1/n$. But this means (s_n) and (w_n) have the same limit point, s_0 say, in K. Since $\mathbb{C}\setminus V$ is closed, this gives a contradiction since $K\cap\Omega = \emptyset$, showing ρ must exist.

(2) Next we show that for any $r > 0$ if K is compact so is the union of closed discs

$$L := \bigcup_{s\in K} B(s,r].$$

To see this let $s_n = x_n + y_n$ be a sequence in L with $y_n \in B(x_n, r]$. Then since K is compact there is a convergent subsequence $x_{n_k} \to x_0$ in K. For each

$j \in \mathbb{N}$ we have $y_{n_{k_j}} \in B(x_0, r + 1/j]$ such that this subsequence converges, to y_0 say, which necessarily is in $B(x_0, r] \subset L$. Hence (s_n) has the convergent subsequence

$$s_{n_{k_j}} = y_{n_{k_j}} + y_{n_{k_j}},$$

so L is compact also.

(3) Now for $n \in \mathbb{N}$ let

$$f_{\Omega,n}(s) := s^{m(0)} \prod_{\substack{\rho \in \Omega \cap \mathbb{R} \\ |\rho| < n}} \left(1 - \frac{s}{\rho}\right)^{m(\rho)} \prod_{\substack{\rho \in \Omega,\, \Im\rho > 0 \\ |\rho| < n}} \left(\left(1 - \frac{s}{\rho}\right)\left(1 - \frac{s}{\bar{\rho}}\right)\right)^{m(\rho)}.$$

We **claim** $f'_{\Omega,n} \to f'_\Omega$ uniformly on K. Let $\epsilon > 0$ be given. By Steps (1) and (2) there is a $r > 0$ such that the corresponding L is a compact subset of V. Therefore $f_{\Omega,n} \to f_\Omega$ uniformly on L and thus there is an N_ϵ such that for all $s \in L$ and $n \geq N_\epsilon$ we have

$$|f_{\Omega,n}(s) - f_\Omega(s)| < \epsilon r.$$

In particular this inequality holds on each $B(s, r]$ for every $s \in K$, so by Cauchy's bound applied to $f_{\Omega,n}(s) - f_\Omega(s)$ we have for all $s \in K$ and $n \geq N_\epsilon$

$$|f'_{\Omega,n}(s) - f'_\Omega(s)| < \epsilon.$$

(4) Since the convergence is uniform on compact subsets K for both $f_{\Omega,n}$ and $f'_{\Omega,n}$, and $f_{\Omega,n}(s) \neq 0$ on K, convergence is uniform for the logarithmic derivative also. This completes the proof. □

Theorem 10.23 *[98, theorem 1.1] If $\Omega \subset \mathbb{C}$ is an admissible zero set with corresponding entire function $f_\Omega(s)$ then the following two conditions are equivalent for a given real θ:*

(a) All elements $\rho \in \Omega$ have $\Re\rho \leq \theta$.

(b) For $\Re s > \theta$ we have

$$\Re\left(\frac{f'_\Omega(s)}{f_\Omega(s)}\right) > 0.$$

Proof If $\Re s > \theta$ and (b) is true, then $f'_\Omega(s)/f_\Omega(s)$ is well defined at $s \in \mathbb{C}$, hence does not have a pole, so $s \notin \Omega$. Thus if $\rho \in \Omega$ we have $\Re\rho \leq \theta$, which is condition (a).

Since Ω is admissible, by Lemma 10.22 there is a Mittag–Leffler expansion for $f'_\Omega(s)/f_\Omega(s)$. If (a) is true and $\Re s > \theta$ then $\Re s > \Re\rho$ for all $\rho \in \Omega$. We **claim** that if $c > 0$ is real and $\rho = \sigma + i\gamma$, $z = x + iy$, then if $x > \sigma$ we have $\Re(c/(z - \rho)) > 0$. This follows directly from the identity

$$\Re\left(\frac{c}{z - \rho}\right) = \frac{c(x - \sigma)}{(x - \sigma)^2 + (y - \gamma)^2}.$$

It follows that every term in the expansion of $\Re(f'_\Omega(s)/f_\Omega(s))$ is positive and therefore so is the sum. Therefore condition (b) is true also. □

In Garunkstis [66] from 2002, there is a better result than that given in [24], and we give the details of this below. Let $\frac{1}{2} \le a \le 1$ and suppose that $\zeta(s)$ has no zero when $\sigma > a$. Then Garunkstis [66] showed that for all such σ we have for all $t \in \mathbb{R}$

$$\Re \frac{\xi'(\sigma+it)}{\xi(\sigma+it)} \ge \frac{\xi'(\sigma)}{\xi(\sigma)}.$$

There are a number of lemmas leading to this result; some are due to Lagarias [98].

Note that (see Volume Two [32, Theorem 4.6] or Edwards [51, chapter 2]) we can write the Weierstrass product form of $\xi(s)$ valid for all $s \in \mathbb{C}$

$$\xi(s) = \xi(0) \prod_{\rho} \left(1 - \frac{s}{\rho}\right), \tag{10.12}$$

where the product is over all of the complex zeros ρ of zeta taken by combining complex conjugate pairs.

Lemma 10.24 *[98; 99, lemma 3.1] For non-zero t, σ and γ we have*

$$\frac{1}{\sigma^2+(t+\gamma)^2} + \frac{1}{\sigma^2+(t-\gamma)^2} \ge \frac{2}{\sigma^2+\gamma^2} \quad \Longleftrightarrow \quad 3\gamma^2 \ge \sigma^2 + t^2. \tag{10.13}$$

Proof No denominator is zero, so clearing denominators of the first inequality, dividing by 2 and simplifying gives the second inequality. These steps can be reversed so the second implies the first. □

Note that if $t = 0$ the left-hand inequality is an equality.

Lemma 10.25 *[66, lemma 2] Assume all zeros $\rho = \beta + i\gamma$ of $\zeta(s)$ satisfy $\beta < a$. For $a < \sigma \le 10$ and $0 < |t| \le 21$ we have*

$$\Re\left(\frac{\xi'(\sigma+it)}{\xi(\sigma+it)}\right) > \frac{\xi'(\sigma)}{\xi(\sigma)}.$$

Proof First note that $1 - a < \beta < a$ and $|\gamma| \ge 14.134\ldots$. If we set $\sigma_0 := \sigma - \beta$ and let t satisfy $|t| \le 21$ we get $3\gamma^2 > 3(14)^2 > 10^2 + 21^2 \ge \sigma_0^2 + t^2$ so the second inequality of Lemma 10.24 holds strictly, hence so does the first. Since $\sigma > a$ we can write using (10.12)

$$\begin{aligned}\Re\left(\frac{\xi'(\sigma+it)}{\xi(\sigma+it)}\right) &= \sum_{\gamma>0}\left(\frac{\sigma-\beta}{(\sigma-\beta)^2+(t-\gamma)^2} + \frac{\sigma-\beta}{(\sigma-\beta)^2+(t+\gamma)^2}\right)\\ &> \sum_{\gamma>0} \frac{2(\sigma-\beta)}{(\sigma-\beta)^2+\gamma^2} = \frac{\xi'(\sigma)}{\xi(\sigma)}\end{aligned}$$

and the proof of the lemma is complete. □

Lemma 10.26 *[66, lemma 3] Let $\frac{1}{2} \le \beta \le a \le 1$, $\sigma > a$ and $\gamma \ge (\sigma + \frac{1}{2})/\sqrt{3}$. Then we have two inequalities:*

$$\frac{\sigma - a}{(\sigma - a)^2 + \gamma^2} + \frac{\sigma + a - 1}{(\sigma + a - 1)^2 + \gamma^2} \le \frac{\sigma - \beta}{(\sigma - \beta)^2 + \gamma^2} + \frac{\sigma + \beta - 1}{(\sigma + \beta - 1)^2 + \gamma^2},$$

$$\frac{\sigma - \beta}{(\sigma - \beta)^2 + \gamma^2} + \frac{\sigma + \beta - 1}{(\sigma + \beta - 1)^2 + \gamma^2} \le \frac{2(\sigma - \frac{1}{2})}{(\sigma - \frac{1}{2})^2 + \gamma^2}.$$

Proof Define

$$f(x) := \frac{\sigma - x}{(\sigma - x)^2 + \gamma^2} \quad \text{and} \quad g(x) := f(x) + f(1 - x)$$

with $x \in [a - 1, a]$. Then

$$f''(x) = \frac{2(\sigma - x)^3 - 6(\sigma - x)\gamma^2}{((\sigma - x)^2 + \gamma^2)^3}.$$

Because $\gamma \ge (\sigma + \frac{1}{2})/\sqrt{3}$ we have $f''(x) \le 0$ so $f(x)$ is concave down, giving $2f((x+y)/2) \ge f(x) + f(y)$. Setting $x = \beta$, $y = 1 - \beta$ gives the second inequality. Since $\frac{1}{2} < \beta \le a$, $g'(\frac{1}{2}) = 0$ and $g'(x) < 0$ for $\frac{1}{2} < x \le a$ the first inequality follows also. □

Lemma 10.27 *[66, lemma 4] For all a and σ with $a < \sigma \le 9/2$ we have*

$$0 < \frac{\xi'(\sigma)}{\xi(\sigma)} \le 0.047(\sigma - \tfrac{1}{2}).$$

Proof Taking the logarithmic derivative of the definition of $\xi(s)$ we get for $\zeta(s) \ne 0$

$$\frac{\xi'(s)}{\xi(s)} = \frac{1}{s} + \frac{1}{s - 1} - \frac{1}{2}\log\pi + \frac{1}{2}\frac{\Gamma'(s/2)}{\Gamma(s/2)} + \frac{\zeta'(s)}{\zeta(s)}.$$

From the Laurent expansion of $\zeta(s)$ about $s = 1$ (Volume Two [32, Theorem I.2] or [82]) we get, with C being Euler's constant,

$$\frac{\zeta'(s)}{\zeta(s)} = -\frac{1}{s - 1} + C + O(|s - 1|).$$

We also know $[\Gamma'(\frac{1}{2})/\Gamma(\frac{1}{2})]/2 = -C/2 - \log 2$ (using for example the representation for $\Gamma'(z)/\Gamma(z)$ given in Theorem I.1 in Volume Two [32] – and see Lemma 2.5 in this volume) so we can compute $\xi'(1)/\xi(1) = 0.0230957\ldots$.

Now let $A := \{\rho : \zeta(\rho) = 0,\ \beta = \frac{1}{2},\ \gamma > 0\}$ and $B := \{\rho : \zeta(\rho) = 0,\ \beta > \frac{1}{2},\ \gamma > 0\}$. Then by Lemma 10.26 we have for $\sigma \ge a$

$$\frac{\xi'(\sigma)}{\xi(\sigma)} = \sum_{\gamma > 0}\left(\frac{1}{\sigma - \rho} + \frac{1}{\sigma - \bar{\rho}}\right)$$

$$
\begin{aligned}
&= \sum_{\rho\in A} \frac{2(\sigma-\frac{1}{2})}{(\sigma-\frac{1}{2})^2+\gamma^2} + 2\sum_{\rho\in B}\left(\frac{\sigma-\beta}{(\sigma-\beta)^2+\gamma^2} + \frac{\sigma-1+\beta}{(\sigma-1+\beta)^2+\gamma^2}\right)\\
&\le \sum_{\rho\in A} \frac{2(\sigma-\frac{1}{2})}{(\sigma-\frac{1}{2})^2+\gamma^2} + \sum_{\rho\in B} \frac{4(\sigma-\frac{1}{2})}{(\sigma-\frac{1}{2})^2+\gamma^2}\\
&= \sum_{\gamma>0} \frac{2(\sigma-\frac{1}{2})}{(\sigma-\frac{1}{2})^2+\gamma^2}\\
&\le 2\sum_{\gamma>0} \frac{1}{\gamma^2}(\sigma-\tfrac{1}{2}).
\end{aligned}
$$

Now we find an upper bound for the coefficient of $\sigma - \frac{1}{2}$. First note by Lemma 10.26 and (10.12) that, if $\gamma_0 = 14.1\ldots$ is the imaginary part of the lowest zero of $\zeta(s)$ in the critical line,

$$
\begin{aligned}
0.04619 > 2\frac{\xi'(1)}{\xi(1)} &= \sum_{\rho\in A} \frac{2}{(1/2)^2+\gamma^2} + \sum_{\rho\in B}\left(\frac{4(1-\beta)}{(1-\beta)^2+\gamma^2} + \frac{4\beta}{\beta^2+\gamma^2}\right)\\
&\ge \sum_{\rho\in A} \frac{2}{1/4+\gamma^2} + \sum_{\rho\in B} \frac{4}{1/4+\gamma^2}\\
&\ge \sum_{\gamma>0} \frac{2}{1+\gamma^2} > 2\frac{\gamma_0^2}{1+\gamma_0^2}\sum_{\gamma>0}\frac{1}{\gamma^2},
\end{aligned}
$$

so $\sum_{\gamma>0} 2/\gamma^2 < 0.0465\ldots$ and the proof of the lemma is complete. □

Lemma 10.28 *[66, lemma 5] Suppose there are no zeta zeros in the half plane $\sigma > a$. Let $\frac{1}{2} \le a < \sigma \le \frac{9}{2}$ and $t \in \mathbb{R}$ be given. Suppose further that at least one of the following conditions is satisfied:*

(*i*) *there are two non-trivial zeros or a double zero of $\zeta(s)$ with $|t-\gamma| \le 5$, or*
(*ii*) *there is a complex zero $\rho = \beta + i\gamma$ of $\zeta(s)$ with $|t-\gamma| \le 2$. Then*

$$\Re\left(\frac{\xi'(\sigma+it))}{\xi(\sigma+it)}\right) \ge \frac{\xi'(\sigma)}{\xi(\sigma)}.$$

Proof Case (i): In case both zeros, not necessarily distinct, lie on the critical line, $\rho_1 = \frac{1}{2} + i\gamma_1$ and $\rho_2 = \frac{1}{2} + i\gamma_2$ say, then by Lemma 10.27

$$
\begin{aligned}
\Re\left(\frac{\xi'(\sigma+it)}{\xi(\sigma+it)}\right) &\ge \frac{\sigma-\frac{1}{2}}{(\sigma-\frac{1}{2})^2+(t-\gamma_1)^2} + \frac{\sigma-\frac{1}{2}}{(\sigma-\frac{1}{2})^2+(t-\gamma_2)^2}\\
&\ge \frac{2(\sigma-\frac{1}{2})}{4^2+5^2} > \frac{\xi'(\sigma)}{\xi(\sigma)}.
\end{aligned}
$$

In case one of the zeros is off the critical line, consider the two zeros $\rho_1 = \beta + i\gamma$ and $\rho = 1 - \beta + i\gamma$ with $\frac{1}{2} < \beta$ and $|t-\gamma| \le 5$. Then since $\beta < a$ using

Lemma 10.26

$$\Re\left(\frac{\xi'(\sigma+it)}{\xi(\sigma+it)}\right) \geq \frac{\sigma-\beta}{(\sigma-\beta)^2+(t-\gamma)^2} + \frac{\sigma-1+\beta}{(\sigma-1+\beta)^2+(t-\gamma)^2}$$
$$\geq \frac{\sigma-\beta}{(\sigma-\beta)^2+25} + \frac{\sigma-1+\beta}{(\sigma-1+\beta)^2+25}$$
$$\geq \frac{\sigma-a}{(\sigma-a)^2+25} + \frac{\sigma-1+a}{(\sigma-1+a)^2+25}$$
$$= h(\sigma,a)(\sigma-\tfrac{1}{2}),$$

where we define

$$h(\sigma,a) := \left(\frac{\sigma-a}{(\sigma-a)^2+25} + \frac{\sigma-1+a}{(\sigma-1+a)^2+25}\right)\left(\frac{1}{\sigma-\frac{1}{2}}\right).$$

Using *Mathematica* to evaluate the derivatives of $h(\sigma,a)$ and simplify the resulting expressions we find

$$\frac{\partial h(\sigma,a)}{\partial\sigma} = -\frac{(2\sigma-1)(a^2-11a-20+\sigma-\sigma^2)(a^2+9a-30+\sigma-\sigma^2)}{((\sigma-a)^2+25)^2((\sigma-1+a)^2+25)^2}$$

and

$$\frac{\partial h(9/2,a)}{\partial a} = \frac{32(2a-1)(16a^4-32a^3-1288a^2+1340a-29031)}{(4a^2-36a+181)^2(4a^2+28a+149)^2}.$$

Now we have assumed $a \leq \sigma \leq 9/2$ and $\frac{1}{2} < a \leq 1$ giving $\partial h(9/2,a)/\partial\sigma \leq 0$ and thus $h(\sigma,a) \geq h(9/2,a)$. In the range $a \in [\frac{1}{2},1]$, $\partial h(9/2,a)/\partial a \leq 0$ giving

$$h(9/2,a) > h(9/2,1) = 0.0483518\ldots.$$

Feeding this inequality back into the lower bound for the logarithmic derivative of $\xi(s)$ and then using Lemma 10.27 completes the proof in this case.

Case (ii): If the zero does not lie on the critical line then we have case (i) which is already covered. Hence we can write $\rho = \frac{1}{2}+i\gamma$ and assume $|t-\gamma| \leq 2$. Then by (10.12) we have

$$\Re\left(\frac{\xi'(\sigma+it)}{\xi(\sigma+it)}\right) \geq \frac{\sigma-\frac{1}{2}}{(\sigma-\frac{1}{2})^2+(t-\gamma)^2}$$
$$\geq \frac{1}{4^2+2^2}\left(\sigma-\frac{1}{2}\right) = 0.05\left(\sigma-\frac{1}{2}\right)$$
$$\geq \frac{\xi'(\sigma)}{\xi(\sigma)}$$

by Lemma 10.27 and the proof is complete. □

Table 10.1 The first 20 Riemann zeros and gaps.

Rank n	Zero $\rho = \frac{1}{2} + i\gamma_n$	Gap $\gamma_{n+1} - \gamma_n$
1	14.134725141734693790	6.887314497036861 2022
2	21.022039638771554993	3.9888179413741337706
3	25.010857580145688763	5.4140185457138244471
4	30.424876125859513210	2.5101854618796764804
5	32.935061587739189691	4.6511165710864815666
6	37.586178158825671257	3.3325408533218239302
7	40.918719012147495187	2.4083542687675043321
8	43.327073280914999519	4.6780776002521602084
9	48.005150881167159728	1.7686815965051424540
10	49.773832477672302182	3.1964890000421584622
11	52.970321477714460644	3.4759262193489341602
12	56.446247697063394804	2.9007963055389582753
13	59.347044002602353080	1.4847345220074567646
14	60.831778524609809844	4.2807655234717968166
15	65.112544048081606661	1.9672664814125670536
16	67.079810529494173714	2.4665911816798055384
17	69.546401711173979253	2.5207559633079283296
18	72.067157674481907583	3.6375330246020255858
19	75.704690699083933168	1.4401493697908722044
20	77.144840068874805373	2.1925349513745625501

Note that $\rho_{300} = 0.5 + i541.847\ldots$. Table 10.1 gives the first 20 imaginary parts of the imaginary zeros of zeta. The third column is the size of the gap to the next zero. Figure 10.4 is a plot of the first 300 coordinate gaps.

We see from the data in Table 10.1 that when $21 \le t \le 79$ there is either a zero with $|t-\gamma| \le 2$ or two zeros with $|t-\gamma| \le 5$.

There was some discussion of the error in finding the number of zeros $N(T)$ of $\zeta(s)$ in the critical strip up to a given height in Chapter 2, Section 2.3. It is a well-studied problem and well known that we can write

$$N(T) = \frac{T}{2\pi}\log\left(\frac{T}{2\pi e}\right) + \frac{7}{8} + S(T) + O\left(\frac{1}{T}\right),$$

where $S(T)$ is the argument of $\zeta(s)$ on the critical line at the point $\frac{1}{2} + iT$ which is not the ordinate of a zero. The argument is fixed by taking the value found by continuous variation from an argument value of 0 along the horizontal line from (∞, iT) to $(\frac{1}{2}, iT)$. At a zero the value is defined to be $\lim_{t\to 0+} S(t)$. This function and its properties are discussed fully by Edwards [51, chapters 8 and 9] for example. In this section we use the explicit estimate of Trudgian:

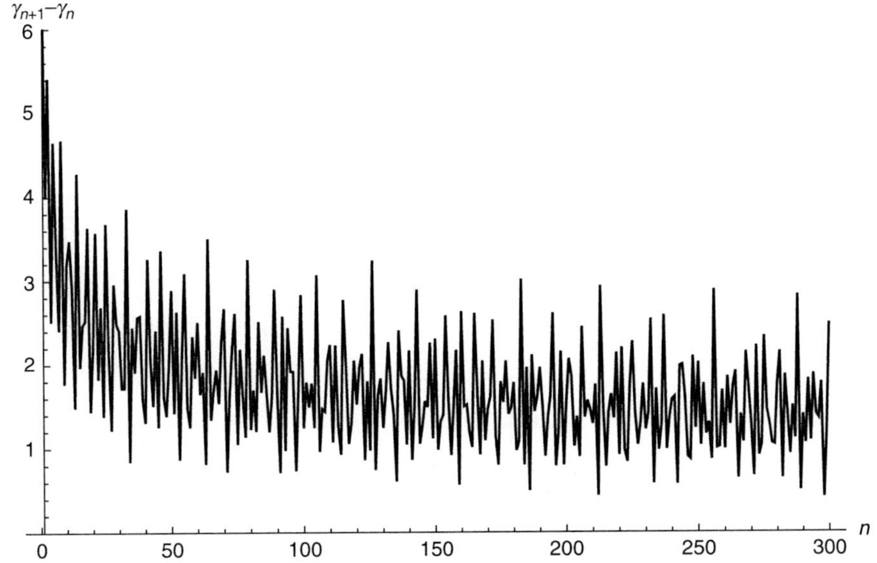

Figure 10.4 The first 300 imaginary coordinate gaps between zeros of $\zeta(s)$.

Lemma 10.29 *[168, theorem 2.2] If $t+h>t>168\pi$ then*

$$\left|\int_t^{t+h} S(u)\,du\right| \leq 2.067 + 0.059\log(t+h).$$

Another way of writing the formula for the number of zeros of $\zeta(s)$ up to height T is

$$N(T) = \frac{\vartheta(T)}{\pi} + S(T) + 1, \tag{10.14}$$

where (see [51, section 6.5])

$$\begin{aligned}\vartheta(t) &:= \frac{1}{2i}\left(\log\Gamma\left(\frac{1+2it}{4}\right) - \log\Gamma\left(\frac{1-2it}{4}\right)\right) - \frac{\log\pi}{2}t \\ &= \frac{t}{2}\log\frac{t}{2\pi e} - \frac{1}{8} + \frac{1}{48t} + O\left(\frac{1}{t^3}\right).\end{aligned}$$

Note that if γ_n and γ_{n+1} are the positive imaginary parts of the ordinates of sorted successive zeta zeros and $\gamma_n < t < \gamma_{n+1}$, then $N(t) = n$ for all such t so $S(t)$ decreases on (γ_n, γ_{n+1}). At $t = \gamma_n$, $S(t)$ has a jump discontinuity of size a positive integer. See Figure 10.5, where the discontinuities are represented by vertical line segments.

Lemma 10.30 *For $t \geq e$ we have*

$$\vartheta'(t) = \frac{1}{2}\log\left(\frac{2t}{\pi}\right) + R, \quad \textit{where } |R| \leq \frac{21}{160t^2}.$$

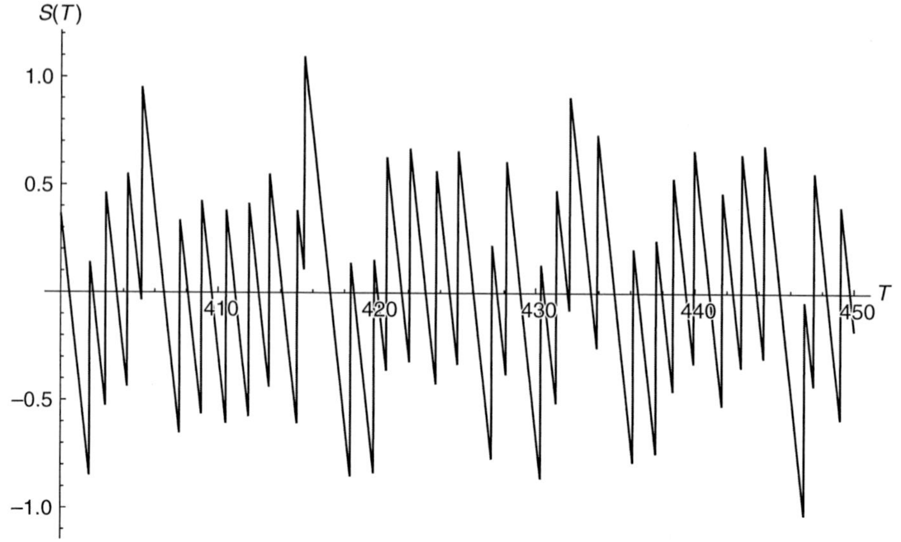

Figure 10.5 The function $S(T)$ on the domain $[400,450]$.

Proof Firstly

$$\vartheta'(t) = -\frac{1}{2}\log\pi + \frac{1}{4}\left(\frac{\Gamma'(1+2it)}{\Gamma(1+2it)} + \frac{\Gamma'(1-2it)}{\Gamma(1-2it)}\right).$$

Now use the estimate given in Lemma 10.13, namely

$$\frac{\Gamma'(s)}{\Gamma(s)} = \log s - \frac{1}{2s} + R_1, \quad \text{where } |R_1| \le \frac{1}{10|s|^2},\ |s| > 2,\ \Re s > 0,$$

to get

$$\vartheta'(t) = \frac{1}{2}\log\left(\frac{2t}{\pi}\right) + R_2, \quad \text{where } |R_2| \le \frac{21}{160t^2},\ t > 2,$$

completing the proof. □

Lemma 10.31 *For* $168\pi \le t < t+h$, *if* $\zeta(s)$ *has no zero with imaginary part in* $[t,\ t+h]$ *and* $S(t)$ *has no zero in* $(t,\ t+h)$ *then* $h < 1.5$ *uniformly in* t.

Proof Assume, to get a contradiction, that the given conditions hold with $h = 1.5$. If the sign of $S(t)$ is uniformly negative on $[t,\ t+1.5]$, because $\vartheta(t)$ is concave up, considering the area of a triangle we can write, using Lemma 10.30 and (10.14),

$$\left|\int_t^{t+h} S(u)\,du\right| \ge \frac{h^2}{2\pi}\vartheta'(t).$$

If the sign of $S(t)$ is positive then, since $\vartheta'(t)$ is increasing,

$$\left|\int_t^{t+h} S(u)\,du\right| \geq \frac{1}{2}h^2(S(t) - S(t+h))$$
$$= \frac{h}{2\pi}(\vartheta(t+h) - \vartheta(t))$$
$$\geq \frac{h^2}{2\pi}\vartheta'(t).$$

Hence in both cases, by Lemma 10.30 we get

$$\left|\int_t^{t+h} S(u)\,du\right| \geq \frac{h^2}{2\pi}\log\left(\frac{2t}{\pi}\right) - \frac{21h^2}{320\pi t^2}.$$

Therefore, by Lemma 10.29

$$2.067 + 0.059\frac{h}{t} + \frac{21h^2}{320\pi t^2} + \frac{h^2}{2\pi}\log\frac{\pi}{2} \geq \left(\frac{h^2}{2\pi} - 0.059\right)\log t.$$

Substituting $h = 1.5$, this inequality implies $168\pi < t < 1724$. In the range $t \in [168\pi, 1724]$ the maximum separation of the imaginary parts of zeta zeros is less than 2.981 (with the maximum being $\gamma_{380} - \gamma_{379}$). □

It follows from Lemma 10.31 that for $t \geq 168\pi$, if $\zeta(s)$ has no zero with imaginary part in $[t,\ t+1.5]$ then $S(t)$ cuts $(t,\ t+1.5)$ at a point of continuity.

Theorem 10.32 *If $t \geq 168\pi$ then $\zeta(s)$ has a zero with imaginary part satisfying $|t - \gamma| \leq 1.5$.*

Proof Assume this is false for some t so $\zeta(s)$ has no zero with imaginary part in $[t - 1.5,\ t]$ or in $[t,\ t + 1.5]$. By Lemma 10.31, $S(t)$ changes sign on $(t - 1.5,\ t)$ and on $(t,\ t + 1.5)$. These points of sign change must be separated by at least one point of discontinuity of $S(t)$, which are in one-to-one correspondence with the imaginary parts γ of the zeros of $\zeta(s)$, so there is at least one such γ with $t - 1.5 < \gamma < t + 1.5$. □

Corollary 10.33 *[98, lemma 3.5] For each t with $|t| \geq 21$ at least one of the conditions (i) or (ii) of Lemma 10.28 is true.*

Remark The study of distances ("gaps") between zeros of the Riemann zeta function has received a great deal of attention: for the initial papers see [119] and [129], but also the very large body of work which is random matrix theory – see for example [89]. A theorem of Littlewood has been made explicit, at least in part, by Hall and Hayman [72, theorem 1]: there exists a T_0 such that for $T > T_0$ there exists a zeta zero $\rho = \beta + i\gamma$ such that

$$|\gamma - T| \leq \frac{\pi/4}{\log\log\log T - \log\log\log\log T - 32}.$$

One could find the explicit value of T_0 and relate this result to that of Theorem 10.32 for example.

Lemma 10.34 *[98, lemma 3.6] If $\sigma > 9/2$ and $|t| \geq 21$ then*

$$\Re\left(\frac{\xi'(\sigma+it)}{\xi(\sigma+it)}\right) \geq \frac{\xi'(\sigma)}{\xi(\sigma)}.$$

Proof Let (σ, t) be in the given ranges, so in particular $\sigma > 1$. From the definition of $\xi(s)$, taking the logarithmic derivative and using the representation of the logarithmic derivative of $\Gamma(z)$ (see Volume Two [32, Theorem H.1])

$$\frac{1}{2}\frac{\Gamma'(s/2)}{\Gamma(s/2)} = -\frac{\gamma}{2} - \frac{1}{s} + \sum_{n=1}^{\infty}\left(\frac{1}{2n} - \frac{1}{2n+s}\right),$$

where γ is Euler's constant, gives

$$\begin{aligned}
\Re\frac{\xi'(s)}{\xi(s)} &= \Re\left(\frac{1}{s}\right) + \Re\left(\frac{1}{s-1}\right) + \frac{1}{2}\Re\frac{\Gamma'(s/2)}{\Gamma(s/2)} + \Re\frac{\zeta'(s)}{\zeta(s)} - \tfrac{1}{2}\log\pi \\
&= \frac{\sigma}{\sigma^2+t^2} + \frac{\sigma-1}{(\sigma-1)^2+t^2} - \frac{\gamma}{2} - \frac{\sigma}{\sigma^2+t^2} \\
&\quad + \sum_{n=1}^{\infty}\left(\frac{1}{2n} - \frac{\sigma+2n}{(\sigma+2n)^2+t^2}\right) + \Re\frac{\zeta'(s)}{\zeta(s)} - \tfrac{1}{2}\log\pi
\end{aligned}$$

and

$$\Re\frac{\zeta'(s)}{\zeta(s)} = -\sum_{n=1}^{\infty}\frac{\Lambda(n)}{n^\sigma}\cos(t\log n).$$

From this we can conclude the minimum of this latter sum is attained at $t = 0$, since each cosine term is bounded above by 1 and each coefficient is positive.

Now let

$$g(s) := \frac{\xi'(s)}{\xi(s)} - \frac{\zeta'(s)}{\zeta(s)} = \frac{1}{s} + \frac{1}{s-1} + \frac{1}{2}\frac{\Gamma'(s/2)}{\Gamma(s/2)} - \tfrac{1}{2}\log\pi.$$

Then

$$\begin{aligned}
&\Re(g(\sigma+it) - g(\sigma)) \\
&= \left(\frac{\sigma-1}{(\sigma-1)^2+t^2} - \frac{1}{\sigma-1}\right) + \sum_{n=1}^{\infty}\left(\frac{1}{\sigma+2n} - \frac{\sigma+2n}{(\sigma+2n)^2+t^2}\right) \\
&\geq t^2\left(\sum_{n=1}^{\infty}\frac{1}{(\sigma+2n)((\sigma+2n)^2+t^2)}\right) - \frac{t^2}{(\sigma-1)((\sigma-1)^2+t^2)}.
\end{aligned}$$

Therefore all we need to show is

$$\sum_{n=1}^{\infty} \frac{1}{(\sigma+2n)((\sigma+2n)^2+t^2)} > \frac{1}{(\sigma-1)((\sigma-1)^2+t^2)}. \tag{10.15}$$

To this end let $k := 1+2n$ and $\sigma_1 = \sigma - 1$. If we consider the ratio of each fixed term on the left with the term on the right

$$R(\sigma_1, t) := \frac{\sigma_1 t^2 + \sigma_1^3}{(\sigma_1+k)t^2 + (\sigma_1+k)^3},$$

then for $t > 0$ we have

$$\frac{\partial R}{\partial t} = \frac{2t((\sigma_1+k)^3 - \sigma_1^3(\sigma_1+k))}{((\sigma_1+k)t^2+(\sigma_1+k)^3)^2} > 0,$$

so for fixed σ_1 the minimum of $R(\sigma_1, t)$ is at $t = 21$. In addition

$$\frac{\partial R}{\partial \sigma_1} = \frac{kt^4 + k^3t^2 + 3k\sigma_1(\sigma_1+k)^2}{((\sigma_1+k)t^2+(\sigma_1+k)^3t^2)^2} > 0,$$

so for fixed $t > 0$ the minimum occurs at $\sigma_1 = 7/2$. Therefore we need only check the result of the theorem at $\sigma = 9/2$, $t = 21$. At this point we find

$$\sum_{n=1}^{3} \frac{1}{(\sigma+2n)((\sigma+2n)^2+t^2)} > \frac{1}{(\sigma-1)((\sigma-1)^2+t^2)}.$$

This completes the proof since all the terms in the sum are positive. □

Theorem 10.35 *[66, p. 2] Let a satisfy $\frac{1}{2} \le a \le 1$. If $\zeta(s)$ has no zeros for $\sigma > a$ then for $s > a$ we have*

$$\frac{\xi'(\sigma)}{\xi(\sigma)} = \inf\left\{\Re\frac{\xi'(\sigma+it)}{\xi(\sigma+it)} : t \in \mathbb{R}\right\}.$$

Proof This result follows directly from Lemmas 10.25, 10.27, 10.28 and 10.34, together with Corollary 10.33. □

10.10 The Order of Elements of the Symmetric Group

In this section we give a summary of the equivalence to RH arising out of the work of Landau and Shah, and developed by Massias, Nicolas and Robin, namely the **symmetric group criterion**. We give some of the introductory proofs, but leave the interested reader to consult the references for details of the main derivations, especially those references given in [112].

Let $n \in \mathbb{N}$. Denote by S_n the group of all permutations of n unlike symbols, so the size $|S_n| = n!$. The group product is composition of functions, where a permutation which sends $i \to \sigma(i)$ is regarded as a function σ. These groups are central to any study of finite groups. This is because of Caley's theorem, which states that any group of order n is isomorphic to a subgroup of S_n. Indeed S_n has many subgroups; for example S_4 has 30 and S_5 over 100.

These groups appear in many parts of mathematics and its applications, including Galois theory, the definition of the determinant, symmetries of objects including molecules, cryptography, etc. For low values of n these groups have concrete isomorphic realizations: for example S_1 consists only of the identity, S_2 is the largest Abelian S_n, S_3 is the group of rotational and reflection symmetries of an equilateral triangle, S_4 is the group of rotations about opposite faces, diagonals and edges of a cube, and S_5 is the Galois group of the general polynomial of the 5th degree.

In this section we need only a few simple properties of S_n. A **cycle** is a permutation which can be represented as an m-tuple of distinct elements $(n_1, n_2, \ldots, n_m)$, with $m \leq n$, and such that the corresponding permutation is

$$n_1 \to n_2 \to \cdots \to n_m \to n_1.$$

We say the **length** of such a cycle is m. The **order** of a permutation σ is the least $k \geq 1$ such that $\sigma^k = e$, the identity. Two cycles are equal if the corresponding permutations are equal as functions. The main properties of cycle representations are given by the following:

- Every $\sigma \in S_n$ may be written as a cycle or a product of disjoint cycles.
- If two cycles are disjoint, the corresponding permutations commute.
- The representation of a permutation in terms of disjoint cycles is unique up to commutation.
- The order of a permutation $\sigma \in S_n$ is the least common multiple (LCM) of the lengths of the cycles in the cycle representation of σ.

For an easy-to-read introduction to S_n, see Gallian [65, chapter 5]. More advanced material is in Sagan [149].

Now let $g(n)$ be the maximum order of any permutation in S_n. In 1903 Landau [103] proved that as $n \to \infty$ we have

$$\log g(n) \sim \sqrt{n \log n} \quad \text{or} \quad g(n) \sim e^{\sqrt{n \log n}}.$$

Values of $g(n)$ are given in Table A.6 and in part A000793 of the *Online Encyclopedia of Integer Sequences* [158]. Landau's theorem [103] is included here as Theorem 10.39. We based our presentation of this result on the proof of William Miller [118].

In 1939 Shah [156] improved Landau's asymptotic formula by proving

$$\log g(n) = \sqrt{n \log n}\left(1 + \frac{\log\log n}{2\log n} + O\left(\frac{1}{\log n}\right)\right).$$

In 1980 Szalay [165] made further improvements to Landau's theorem, arriving at the very precise expression

$$\log g(n) = \sqrt{n \log n \left(1 + \frac{\log\log n - 1}{\log n} + \frac{\log\log n - 2 + o(1)}{\log^2 n}\right)}.$$

In this section we summarize the properties of $g(n)$ and prove some preliminary lemmas, including a simpler proof, as Theorem 10.40, of a result of Shah which is essential to the development of the equivalence. Then we state two theorems. The second spells out the equivalence to RH, but it is really a corollary of the first more detailed result. This is based on the work of Massias, Nicolas and Robin [112]. Here is the statement of the second theorem:

Theorem 10.42 *The Riemann hypothesis is equivalent to the statement that for all n sufficiently large we have*

$$\log g(n) < \sqrt{\mathrm{li}^{-1}(n)}$$

where $\mathrm{li}(x)$ *for* $x > 0$ *is the Cauchy principal value of the logarithmic integral*

$$\mathrm{li}(x) := \int_0^x \frac{dt}{\log t},$$

and where $\mathrm{li}^{-1}(x)$ *is the functional inverse.*

Finally in this section, we summarize the elements of a statistical study of S_n made by Erdős and Turan [55]. This is in order that the reader might see that elements having the maximum order in S_n are exceptional, and that the order of an "average" element is significantly less.

To begin, we define the arithmetic function

$$h(n) := \sum_{1 \le i \le k} p_i^{a_i}, \quad \text{where } n = \prod_{1 \le i \le k} p_i^{a_i}$$

is the standard prime factorization. Also set $h(1) = 0$ so $h(n)$ is an additive function. A plot of $h(n)$ for $1 \le n \le 10^3$ is given in Figure 10.6.

Lemma 10.36 *[118] For all* $n \in \mathbb{N}$ *we have*

$$g(n) = \max_{h(m) \le n} m.$$

Proof If $n = 1$ the result is immediate since $h(1) = 0$ and $h(2) = 2$, so we can assume $n > 1$.

(1) Let $a_1, a_2, \ldots, a_k$ be a set of positive integers and $m = [a_1, \ldots, a_k]$ their LCM. Then we **claim** $h(m) \le \sum_{1 \le i \le k} a_i$. To see this, suppose first $m = 1$ so $m = [1, 1, \ldots, 1]$ and the inequality is immediate, and we can assume $m > 1$. Suppose we have a set of positive integers with $m = [a_1, \ldots, a_k]$ but $h(m) >$

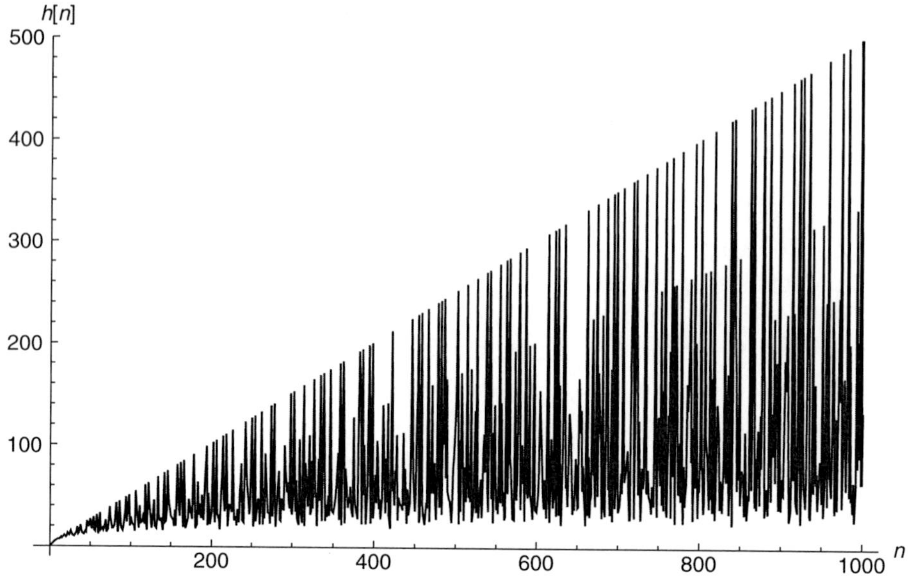

Figure 10.6 Values of the function $h(n)$ excluding prime powers.

$\sum_{1\leq i\leq k} a_i$. Let $\{a_i\}$ be a set with the smallest such sum. Since removing a term with $a_i = 1$ reduces the sum, but preserves the value of the LCM, we can assume each $a_i \geq 2$. If we could write one of the a_i as $a_i = xy$ with $(x, y) = 1$ and both $x, y \geq 2$ then since $x+y < xy = a_i$ we could replace a_i with the two integers x and y, and again reduce the sum. Hence we can assume each of the a_i is a power of a prime. Finally if two of the a_i were powers of the same prime we could delete the one with the smaller or equal power, and get a smaller sum, but the same LCM and so satisfying $h(m) > \sum_{1\leq i\leq k} a_i$. But in case the a_i are powers of distinct primes we have $h(m) = \sum_i^k a_i$, a contradiction, which proves the claim.

(2) We will now show that there exists a permutation on n elements of order m if and only if $h(m) \leq n$. To see this, first suppose $h(m) \leq n$. Let $m = \prod_{1\leq i\leq k} p_i^{\alpha_i}$ and define a permutation $\sigma \in S_n$ with cycle lengths $(p_i^{\alpha_i})$ and $n - h(m)$ fixed points. Then the order of this permutation is m. If such a permutation of order m exists let (a_i) be its cycle lengths so their sum is n and LCM is m. Hence, by Step (1) we have $h(m) \leq n$.

(3) Now we **claim** that among all of the permutations on n symbols with maximum order there exists at least one with non-trivial cycle orders being powers of distinct primes: let m be the maximum order of a permutation so we can write $g(n) = m$. Then by Step (2) we have $h(g(n)) \leq n$. Use the construction of Step (2), with m having this value $g(n)$, to construct the required permutation.

(4) We can now complete the proof. Let $\theta := \max_{h(m)\le n} m$. Now by Step (3) using a permutation of maximum order, $h(g(n)) \le n$ so $g(n) \le \theta$. If on the other hand m is such that $h(m) \le n$, by Step (2) there is a permutation of order m so $m \le g(n)$. Hence $\theta \le g(n)$ so $\theta = g(n)$ and the proof of the lemma is complete. □

This section includes a version of Landau's theorem $\log g(n) \sim \sqrt{n \log n}$ [105, pp. 222–229] based on the proof in Miller's paper [118], which on the face of it looks very similar to Landau's. In Table A.6 the first 160 values of $g(n)$ are given. Extensive evaluations were carried out by Morain up to $n = 32\,000$ with some details in [120].

For $x > 0$ let $A(x) := \sum_{p \le x} p$ be the sum of all primes up to x.

Lemma 10.37 *As $x \to \infty$ we have*

$$A(x) = \frac{x^2}{2\log x} + O\left(\frac{x^2}{\log^2 x}\right).$$

Proof This is a simple application of Abel's identity [6, theorem 4.2], splitting the integral for the error at $\sqrt{x}$. □

The reader may wish to compare the use of Lemma 10.37 with the method of Shah [156] who takes $A(x)$ to the fourth order. This is not needed in the proof, but deriving it could be good practice for the energetic reader.

Lemma 10.38 *As $x \to \infty$ we have*

$$A(x) = \frac{x^2}{2\log x} + \frac{1}{4}\frac{x^2}{\log^2 x} + \frac{1}{4}\frac{x^2}{\log^3 x} + O\left(\frac{x^2}{\log^4 x}\right).$$

Proof First use induction to show that

$$\sum_{p \le x} p = -\sum_{n=1}^{x} \pi(n) + \pi(x)(\lfloor x \rfloor + 1).$$

In addition, integrating the definition of $\mathrm{Li}(x)$ successively by parts leads to the asymptotic expansion, for $x \to \infty$,

$$\mathrm{Li}(x) = \frac{x}{\log x} \sum_{n=0}^{\infty} \frac{n!}{\log^n x}.$$

Using these expressions and the prime number theorem in the form

$$\pi(n) = \mathrm{Li}(n) + O(x \exp(-c\sqrt{\log x}))$$

enables us to derive

$$\sum_{n=1}^{x} \pi(n) = \sum_{n=2}^{x} \left(\frac{n}{\log n} + \frac{n}{\log^2 n} + \frac{2n}{\log^3 n} + O\left(\frac{n}{\log^4 n}\right)\right)$$

$$= \int_2^x u\left(\frac{1}{\log u} + \frac{1}{\log^2 u} + \frac{2}{\log^3 u}\right) du + O\left(\int_2^x \frac{u\,du}{\log^4 u}\right)$$
$$= \frac{x^2}{2\log x} + \frac{3}{4}\frac{x^2}{\log^2 x} + \frac{7}{4}\frac{x^2}{\log^3 x} + O\left(\frac{x^2}{\log^4 x}\right).$$

Therefore

$$A(x) = \sum_{p\le x} p = \frac{x^2}{2\log x} + \frac{1}{4}\frac{x^2}{\log^2 x} + \frac{1}{4}\frac{x^2}{\log^3 x} + O\left(\frac{x^2}{\log^4 x}\right).$$

This completes the proof. □

Theorem 10.39 *[105, 118] As $n \to \infty$ we have $\log g(n) \sim \sqrt{n\log n}$.*

Proof (1) Let n be a positive integer and let $P = P(n)$ be the largest prime such that the sum of the primes less than P does not exceed n and define $F(n) := \prod_{p<P} p$. Then because $\log F(n) = \theta(P-1)$ and $P \sim P-1$ we have, by the prime number theorem, $\log F(n) \sim P$.

(2) Now for $x > 0$ let $A(x) := \sum_{p\le x} p$. By the definitions of $A(x)$ and P we have $A(P-1) \le n < A(P)$. But for all $x > 0$ we have $A(x-1) \sim A(x)$ and therefore, by Lemma 10.37, we have $P^2/(2\log P) \sim n$, which implies that $P^2/(2n\log P) \to 1$.

If P is not asymptotic to $\sqrt{n\log n}$ then there is an $\epsilon > 0$ such that for an infinite number of values of n one of the inequalities

$$P \le (1-\epsilon)\sqrt{n\log n} \quad \text{or} \quad P \ge (1+\epsilon)\sqrt{n\log n}$$

holds. Because $x \to x^2/\log x$ is increasing for $x > \sqrt{e}$, if the first inequality is true then

$$\frac{P^2}{2n\log P} \le \frac{(1-\epsilon)^2\log n}{\log n + \log\log n + 2\log(1-\epsilon)}.$$

However, as $n \to \infty$ the right-hand side approaches $(1-\epsilon)^2$ but the left-hand side tends to 1. Similarly the second inequality does not hold for infinitely many n. Hence $P \sim \sqrt{n\log n}$ and so, by Step (1) we have $\log F(n) \sim \sqrt{n\log n}$.

(3) Because, by Step (2) of Lemma 10.36, there is a permutation on n letters of order $F(n)$ we have $F(n) \le g(n)$.

(4) Let $q_1, \dots, q_s$ be all of the distinct primes dividing $g(n)$ in increasing order. Then we **claim** that

$$\sum_{j=1}^{s} \log q_j < 2 + \log F(n) + \log P.$$

Note that we can assume $n \ge 2$ so $P \ge 3$. Let $p_1, \dots, p_r$ be the *odd* primes with $p_j \le P$ and such that $p_j \nmid g(n)$ for $1 \le j \le r$. Hence the set of primes not exceeding P, other than possibly 2, is the list $p_1, \dots, p_r, q_1, \dots, q_{t-1}$ where these q are the $q_i \le P$ which divide $g(n)$. Now because $h(m) \ge q_1 + \cdots + q_s$, by Lemma 10.36 we get

$$\sum_{j=1}^{s} q_j \le h(g(n)) \le n < \sum_{p \le P} p.$$

Cancelling like terms we get

$$\sum_{j=t}^{s} q_j \le 2 + \sum_{j=1}^{r} p_j.$$

But for $t \le j \le s$ we have $3 \le P \le q_j$ and for $1 \le i \le r$ we have $3 \le p_i \le P$. Therefore since for $x \ge 3$ the function $x \to x/\log x$ is increasing, we can write

$$\frac{P}{\log P}\log q_j \le q_j \quad \text{and} \quad p_i \le \frac{P}{\log P}\log p_i.$$

Therefore

$$\sum_{j=t}^{s} \log q_j \le \frac{2\log P}{P} + \sum_{j=1}^{r} \log p_j \le 2 + \sum_{j=1}^{r} p_j.$$

Finally add the sum $\sum_{1 \le j \le t-1} \log q_j$ to each side, and use the fact that $p_1, \ldots, p_r, q_1, \ldots, q_{t-1}$ is a list of all of the primes not greater than P to complete the proof of the claim.

(5) If q is a prime and $e > 1$ an integer being the maximum power for which $q^e \mid g(n)$, we **claim** $q^e \le 2P$. To see this let Q be the smallest prime not dividing $g(n)$ so each of the primes less than Q divides $g(n)$. Thus their sum is at most $h(g(n)) \le n$. Also the sum of the primes $p \le P$ is greater than n. Hence $Q \le P$. To complete the proof we will show that $q^e \le 2Q$. Assume, to get a contradiction, that $q^e > 2Q$.

Now let $t \in \mathbb{N}$ be the unique positive integer such that $Q^{t-1} < q < Q^t$ and let $m := (Q^t/q)g(n)$. Then $m > g(n)$ and

$$h(m) = h(g(n)) + (Q^t - q^e + q^{e-1}). \tag{10.16}$$

If $q < Q$, so $t = 1$, then since $q^e > 2Q$ we have

$$-q^e + q^{e-1} \le -\frac{q^e}{2} < -Q$$

so the second term on the right-hand side of (10.16) is negative. If $q > Q$ we get

$$Q^t - q^e + q^{e-1} < qQ - q(q-1) \le qQ - qQ = 0.$$

Therefore, in each case we have $h(m) < h(g(n)) \le n$. But $m > g(n)$, which is a contradiction. Therefore $q^e \le 2Q$, from which the claim follows directly.

(6) Now let $g(n) = \prod_{1\le j\le s} q_j^{e_j}$ be the prime factorization. Write $\log g(n)$ as the sum of the terms $\log q_j$ where $e_j = 1$ and $e_j \log q_j$ where $e_j > 1$. By Step (2) the first sum is at most $2 + \log F(n) + \log P$. By Step (5) each term in the second sum is at most $\log 2P$ and there at most $\sqrt{2P}$ terms. Therefore

$$\log F(n) \le \log g(n) \le 2 + \log F(n) + \log P + \sqrt{2P}\log 2P$$
$$\implies \quad \lim_{n\to\infty} \frac{\log g(n)}{P} = 1,$$

and so by Steps (1) and (2) we have

$$\log g(n) \sim P \sim \log F(n) \sim \sqrt{n\log n},$$

which completes the proof. □

The present author found parts of the proof of the theorem of Shah [156] mysterious, so here we present a different derivation of the same result, based on the proof of Landau's theorem given by William Miller, Theorem 10.39.

Theorem 10.40 *As $n \to \infty$,*

$$\log g(n) = \sqrt{n\log n}\left(1 + \frac{\log\log n}{2\log n} + O\left(\frac{1}{\log n}\right)\right).$$

Proof **(1)** By part (2) of the proof of Theorem 10.39 we have $P \sim \sqrt{n\log n}$, and we can write $P = \sqrt{n\log n}(1+\alpha)$, so that $\alpha \to 0$ as $n \to \infty$. Also we have $A(P-1) \le n < A(P)$, so by Lemma 10.37 we have

$$\begin{aligned} n &= \frac{P^2}{2\log P} + O\left(\frac{P^2}{\log^2 P}\right) \\ &= \frac{n\log n(1+\alpha)^2}{\log n + \log\log n + 2\log(1+\alpha)} + O\left(\frac{n}{\log n}\right) \\ \implies \quad 1 &= \frac{1+2\alpha+\alpha^2}{1 + \dfrac{\log\log n}{\log n} + O\left(\dfrac{\alpha}{\log n}\right)} + O\left(\frac{1}{\log n}\right). \end{aligned}$$

Therefore

$$\begin{aligned} \alpha &= \frac{\log\log n}{2\log n} + O(\alpha^2) + O\left(\frac{1}{\log n}\right) \\ (\alpha + O(\alpha^2))^2 &= \left(\frac{\log\log n}{2\log n}\right)^2 + O\left(\frac{\log\log n}{\log^2 n}\right) \\ \alpha^2(1+O(\alpha)) &= O\left(\left(\frac{\log\log n}{\log n}\right)^2\right) = O\left(\frac{1}{\log n}\right). \end{aligned}$$

Therefore

$$\alpha = \frac{\log\log n}{2\log n} + O\left(\frac{1}{\log n}\right).$$

(2) By Steps (1) and (6) of the proof of Theorem 10.39 we have

$$\theta(P-1) \le \log g(n) \le 2 + \theta(P-1) + \log P + \sqrt{2P}\log(2P).$$

Therefore, using the result of Step (1),

$$\begin{aligned}
\log g(n) &= \theta(P-1)\left(1 + O\left(\frac{\sqrt{P}\log P}{\theta(P)}\right)\right) \\
&= (P-1)\left(1 + O\left(\frac{1}{\log P}\right)\right)\left(1 + O\left(\frac{\log P}{\sqrt{P}}\right)\right) \\
&= P\left(1 + O\left(\frac{1}{P}\right)\right)\left(1 + O\left(\frac{1}{\log P}\right)\right)\left(1 + O\left(\frac{\log P}{\sqrt{P}}\right)\right) \\
&= P\left(1 + O\left(\frac{1}{\log P}\right)\right) \\
&= \sqrt{n\log n}\left(1 + \frac{\log\log n}{2\log n} + O\left(\frac{1}{\log n}\right)\right)\left(1 + O\left(\frac{1}{\log n}\right)\right) \\
&= \sqrt{n\log n}\left(1 + \frac{\log\log n}{2\log n} + O\left(\frac{1}{\log n}\right)\right)
\end{aligned}$$

and this completes the proof. □

Figure 10.7 is a plot of $\log g(n)/\sqrt{n\log n}$.

Nicolas studied the properties of $g(n)$ in detail in the 1960s, and Massias continued this study, publishing related work in his Limoges doctoral thesis in 1985. Several papers were also published on the topic by these two authors together with Robin. See the references in [112] and [113]. The first set of results stated below are for the main part proved in detail in Nicolas' paper [125], where the methods are similar to those used in Chapter 6. There is also a summary with sketched proofs in [113]. The RH equivalence is proved in [112]. The introductory results we have given above should be sufficient background for the reader to read these papers.

Recall the definition: $h(1) = 0$ and, if $n = \prod_{p|n} p^{\nu_p(n)}$ is the standard prime factorization set,

$$h(n) := \sum_{p|n} p^{\nu_p(n)}.$$

We have seen in Lemma 10.36 that

$$g(n) = \max_{h(m)\le n} m.$$

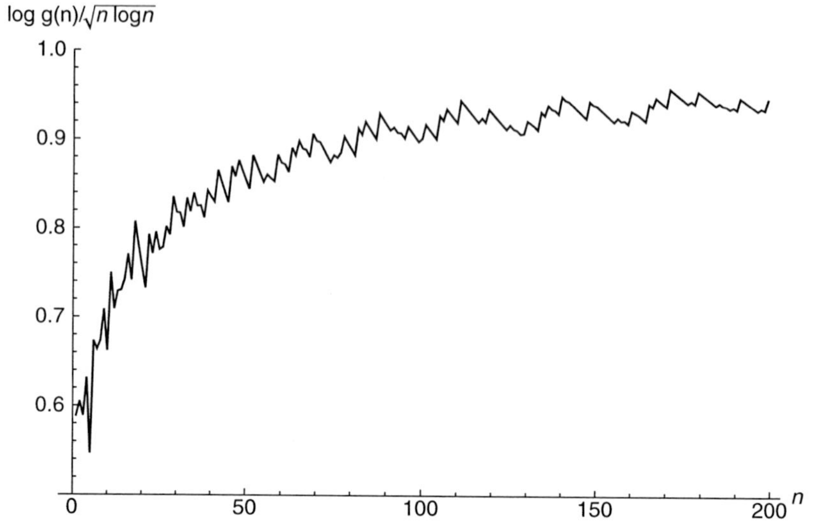

Figure 10.7 Values of the function $\log g(n)/\sqrt{n\log n}$ for $3 \leq n \leq 1000$.

It follows from this expression that we have $m = g(n)$ for some n if and only if $m' > m$ implies $h(m') > h(m)$.

Definition of the set G: we say that $n \in G$ if there exists $\rho > 0$ such that

$$\forall\, m \geq 1, \quad g(m) - \rho \log m \geq g(n) - \rho \log n. \tag{10.17}$$

It follows, using a similar method to that used in the introduction to Chapter 6 to show that colossally abundant numbers are superabundant, that $G \subset g(\mathbb{N})$. To each number $\rho > 2/\log 2$ there corresponds an $n_\rho \in G$ defined in the following manner:

$$n_\rho = \prod_{p \leq x_1} p^{\alpha_p} \tag{10.18}$$

with $x_1/\log x_1 = \rho$ so $x_1 > 4$. We also introduce the real numbers x_i, $i \geq 2$, by

$$\frac{x_i^i - x_i^{i-1}}{\log x_i} = \rho,$$

which enables us to write

$$n_\rho = \prod_{i=1}^{\infty} \prod_{p \leq x_i} p. \tag{10.19}$$

We have relationships such as

$$x_2 = \sqrt{\frac{x_1}{2}}\left(1 + O\left(\frac{1}{\log x_1}\right)\right). \tag{10.20}$$

Recall that $\theta := \sup\{\Re\zeta(\rho) : \zeta(\rho) = 0\}$, and that $f(n) = \Omega_\pm(g(n))$ as $n \to \infty$ provided there are two unbounded positive integer sequences (n_i) and (m_i) such that for all $i \in \mathbb{N}$ we have $f(n_i) \geq g(n_i)$ and $f(m_i) \leq -g(m_i)$.

The symmetric group criterion is a consequence of a more refined result than Theorem 10.40, and is Theorem 10.41 stated below:

Theorem 10.41 *[112, theorem 1]*

(i) If $\theta = 1$ *there is an* $a > 0$ *such that*

$$\log g(n) = \sqrt{\mathrm{li}^{-1}(n)} + O\left(\sqrt{n}e^{-a\sqrt{\log n}}\right).$$

If $\theta < 1$ *then*

$$\log g(n) = \sqrt{\mathrm{li}^{-1}(n)} + O\left((n\log n)^{\theta/2}\right).$$

(ii) If $\theta > \frac{1}{2}$ *then for all* $\xi < \theta$

$$\log g(n) = \sqrt{\mathrm{li}^{-1}(n)} + \Omega_\pm\left((n\log n)^{\xi/2}\right).$$

(iii) If there exists a zero of $\zeta(s)$ *with real part* $\theta > \frac{1}{2}$ *then*

$$\log g(n) = \sqrt{\mathrm{li}^{-1}(n)} + \Omega_\pm\left((n\log n)^{\theta/2}\right).$$

(iv) If the Riemann hypothesis is true then for all n sufficiently large

$$\log g(n) < \sqrt{\mathrm{li}^{-1}(n)}.$$

(v) If the Riemann hypothesis is true then as $n \to \infty$

$$\log g(n) = \sqrt{\mathrm{li}^{-1}(n)} - \left(\frac{2-\sqrt{2}}{3}\right)(n\log n)^{1/4} + \Omega_\pm\left((n\log n)^{1/4}\right).$$

Theorem 10.42 (Symmetric group criterion) *The Riemann hypothesis is equivalent to the statement that for all n sufficiently large we have*

$$\log g(n) < \sqrt{\mathrm{li}^{-1}(n)},$$

where li(n) *is the Cauchy principal value of the logarithmic integral*

$$\mathrm{li}(n) := \int_0^n \frac{dt}{\log t}.$$

Of course, knowing the maximal order of any element of S_n, or any group, does not give insight into the way the orders of elements are distributed. In a series of seven papers published in the 1960s and 1970s, commencing with [55], Erdős and Turan revealed many facts regarding the distribution of orders, including that few elements $\sigma \in S_n$ have orders comparable with the maximal asymptotic order, $g(n) \sim \exp(\sqrt{n \log n})$. They noted for example that there were $(n-1)!$ elements σ, consisting of a single cycle, having order n. They proved that the order of a "generic" element is much smaller than the maximum. Here we simply state two of their main results.

Theorem 10.43 *Given $\epsilon, \delta > 0$ and n sufficiently large, depending on ϵ and δ, we have, other than for at most $\delta n!$ permutations σ:*

$$e^{(1/2-\epsilon)\log^2 n} \leq \operatorname{ord}(\sigma) \leq e^{(1/2+\epsilon)\log^2 n}.$$

They indicated they could use the method they developed to show also that:

Theorem 10.44 *Provided $n \to \infty$ through integers such that $\omega(n) \to \infty$ we have for $\sigma \in S_n$*

$$\left|\log \operatorname{ord}(\sigma) - \frac{1}{2}\log^2 n\right| \leq \omega(n) \log^{3/2} n$$

with at most $o(n!)$ exceptions.

10.11 Hilbert–Pólya Conjecture

To describe this conjecture and equivalence we first need some definitions. Let H be a complex Hilbert space and let Δ be a dense linear subspace and $T, \Delta \to H$ a linear transformation. Let

$$\Delta_* := \{x \in H : y \to \langle x, Ty \rangle \text{ is a continuous linear functional on all of } H\}.$$

By the Riesz representation theorem (Volume Two [32, Theorem J.1]), for each $x \in \Delta_*$ there is a unique $z \in H$ such that $\langle x, Ty \rangle = \langle z, y \rangle$ for all $y \in \Delta$. We define $T^*(x) = z$, and call the linear transformation T^* the **adjoint** of T. The map T is called **self-adjoint** if the domain of its adjoint transformation T^* is also Δ, so $\Delta_* = \Delta$, and on this domain we have $T = T^*$, so for all $f, g \in \Delta$ we have $\langle Tf, g \rangle = \langle f, Tg \rangle$. Thus T is symmetric, a weaker condition than being self-adjoint.

The Hilbert–Pólya conjecture, as formulated by Hugh Montgomery in 1973 [119] possibly for the first time, is that if we write each non-trivial zero of $\zeta(s)$ with positive imaginary part as $\frac{1}{2} + i\gamma_n$, then the numbers γ_n correspond to the eigenvalues of an unbounded self-adjoint operator T. This of course would force the γ_n to be real and solve RH.

Since (see Lorch [110, theorem 4-1]) a self-adjoint transformation defined on all of H is necessarily bounded, we must have $\Delta \neq H$.

It appears that the conjecture goes back to the very conception of RH. In considering what was left of Riemann's handwritten working paper collection in the Göttingen University library, John Keating found a note relating to the stability of a rotating fluid on the same page as notes on the zeros of $\zeta(s)$ (see du Sautoy [151, p. 286]). The condition for stability of the fluid subject to perturbation was that a set of eigenvalues should be on a straight line! An implicit connection between RH and something physical!

Andrew Odlyzko attempted, in the early 1980s while Pólya was still alive, to trace the formulation of the conjecture back to Pólya and Hilbert by corresponding with Pólya (see Volume Two [32, Chapter 5]). Pólya recalled answering a question of Landau as to whether or not there was a physical reason why RH should be true. Pólya's answer was that it would be the case if the nontrivial zeros of the function $\xi(s)$ were so connected with the physical problem that RH would be equivalent to all of the eigenvalues of that problem being real. Odlyzko was not able to find any reported statement by Hilbert, but the name of the conjecture is now well established by use.

While up to the time of writing Hilbert–Pólya has failed to provide a proof of RH, the conjecture has inspired a great deal of work and progress in the intersection of mathematics and physics. The main focus has been to design the Hilbert space H, domain Δ and transformation T. Macroscopic classical mechanical systems, quantum mechanical and quite a few others have been investigated, but no one seems to have come close to providing what is needed. Some believe asking for the transformation to be self-adjoint is too much, or that the structures should be constructed out of mathematics alone, i.e. starting with the integers. Sir Michael Berry and his co-workers have given quite precise specifications on what a successful structure might need to satisfy [11].

As part of this story of successful synergies between mathematics and physics, there is the well-known and often retold interaction between Hugh Montgomery and Freeman Dyson over tea in the Institute for Advanced Study, Princeton [151, p. 262]. This became related to the great body of work on random matrices, which was originally developed as a way to study, statistically, the energy levels of heavy nuclei, i.e. those with large numbers of neutrons and protons. The relationships with the results of computations of zeta zero imaginary parts, especially those carried out by Andrew Odlyzko [79, 130], have been startling.

Background reading might include Conrey [41], Katz and Sarnak [90] and Rudnick and Sarnak [147]. On the physics side, there is the survey preprint of Schumayer and Hutchinson [155].

One of the remarkable particular discoveries of random matrix theory has been very precise expressions for moments of distribution functions, providing corresponding conjectures for the moments of $\zeta(s)$. Here are the conjectured moments of John Keating and Nina Snaith [91] for $\zeta(s)$ for all $k \in \mathbb{N}$ and s on the critical line:

$$\frac{1}{T}\int_0^T |\zeta(\tfrac{1}{2}+it)|^{2k}\,dt \sim g_k a(k)\left(\log \tfrac{T}{2\pi}\right)^{k^2},$$

where

$$a(k) := \prod_p \left(1 - \frac{1}{p}\right)^{k^2} \sum_{m=0}^{\infty} \left(\frac{\Gamma(m+k)}{m!\Gamma(k)}\right)^2 \frac{1}{p^m}$$

and

$$g_1 = 1, \quad g_2 = 1/12, \quad g_3 = 42/9!, \quad g_4 = 2024/16!,$$

and if

$$G(n) := \prod_{i=0}^{n-2} i! \quad \Longrightarrow \quad g_k = \frac{G(k+1)^2}{G(2k+1)}.$$

Estimates for moments of zeta have been very significant, and have been so for a long time. For example Landau in his treatise of 1908 included the following theorem for the case $k = 1$ [51, section 9.7]: Given $\epsilon > 0$ and $\sigma_0 > \frac{1}{2}$ there is a T_0 such that for all $\sigma \geq \sigma_0$ and $T \geq T_0$ we have

$$\left| \frac{1}{T-1} \int_1^T |\zeta(\sigma + it)|^2 \, dt - \zeta(2\sigma) \right| < \epsilon.$$

This estimate was used by Bohr and Landau [51, section 9.6], together with Jensen's formula (Volume Two [32, Theorem B.5]), to show that the number of zeta zeros in any rectangle $R := [\delta, 1] \times [0, T]$ for any $\delta > \frac{1}{2}$ is bounded above by $K_\delta T$. Since the total number of zeta zeros $N(T)$ in $[0, 1] \times [0, T]$ satisfies $N(T) \sim T/(2\pi)\log(T/(2\pi e))$, the proportion of roots in R divided by the total number up to T tends to zero as $T \to \infty$. This result is still one of the best pieces of analytic evidence for the truth of RH.

For more recent significant developments on the topic of estimates for moments and references, there is the article of Soundararajan [160].

Searching for a single self-adjoint transformation which would resolve RH could be asking too much. For example associating each zero with an individual transformation depending on the zero would give the same result. The work of Ross Barnett and the author leading to [21], consistent with random matrix theory, seems to indicate a relationship between rotations and zeta zeros, which has yet to be fully developed.

Theorem 10.45 (Hilbert–Pólya criterion) *The Riemann hypothesis is equivalent to the following property: for each critical zero ρ of $\zeta(s)$, when written in the form $\rho = \frac{1}{2} + i\eta$, there is a self-adjoint transformation T_ρ of a Hilbert space H_ρ such that η is an eigenvalue of T_ρ.*

Proof If RH is true let $H = \ell_2(\mathbb{N}, \mathbb{C})$ with Euclidean inner product, and for $n \in \mathbb{N}$ let $e_n := (0, 0, \ldots, 1, 0, \ldots)$ be the nth standard basis element. If $\rho_n = \frac{1}{2} + i\eta_n$

is the nth complex zero with positive imaginary part, define $T : \Delta \to H$ by setting $T(e_n) = \eta_n e_n$, so

$$T(x) = \sum_{n \in \mathbb{N}} \eta_n x_n e_n.$$

Let $\Delta := \{x \in H : \|T(x)\| < \infty\}$. Since the linear subspace generated by any finite subset of the (e_n) is in Δ, Δ is dense in H. Since by RH each η_n is real, T is self-adjoint. Setting $T_\rho = T$ completes the derivation. □

10.12 Epilogue

Looking back over the equivalences to RH described in Chapters 4 through 10, the reader might be surprised at their variety, but also at the few properties of $\zeta(s)$ which have been used to derive them: von Mangoldt's theorem, a simple zero-free region, both known for over 100 years, and a better critical zero line height H – but even deriving improvements to H often uses methods based on those dating back to Turing. The functional equation barely appears, neither does the lower bound for the proportion of zeros on the critical line, nor upper bounds for zeros possibly off the line.

In addition, the properties of primes that are used are slight and based on not much more than the prime number theorem – no quadratic reciprocity, no modular restrictions. It is some explicit estimates for arithmetic quantities which are used to derive the simple inequalities which have been shown to be equivalent to RH. Work on these estimates has thus been shown to be invaluable.

Another observation is that when RH fails, the failure has dramatic consequences. Just one off-critical-line zero will result in an infinite number of failures for an equivalent inequality.

One approach to resolving RH, begun as reported in Chapter 8, would be to further explore properties of potential counterexamples, or sets of counterexamples, i.e. integers which do not satisfy an equivalent inequality. Finding a set of restrictions which are inconsistent would be the goal. Another approach would be to work at weakening a given inequality and strengthening a related unconditional form, until they merge.

Bringing to bear on equivalences recent better understanding of prime properties, such as primes in progressions and infinite subsequences of primes with $p_{n+1} - p_n \leq K$ for some explicit (and small) value of K, as well as more classical properties, could also be fruitful. In this way it is properties of primes which could resolve RH, rather than the reverse!

In addition to the equivalences described here, there are a host of other equivalences which have been demonstrated. These are often fascinating and, on the face of it, it is difficult to detect any connection to RH. Many of these are described in Volume Two [32].

One thing is for certain – when it comes to equivalences to RH, the last word has not been spoken.

10.13 Unsolved Problems

(1) From Section 10.3, the trivial bound is

$$\sum_{j=1}^{m} \left| f_j^n - \frac{j}{m} \right| \ll n^2.$$

Improve this bound by finding a lower power of n.

(2) Again from Section 10.3, show for all j with $1 \leq j \leq m$ that

$$\left| f_j^n - \frac{j}{m} \right| \leq \frac{1}{n}.$$

To this end, using telescoping sums from 0 to 1 and the bound $f_{j+1}^n - f_j^n \geq 1/n^2$ we should be able to get upper and lower bounds for f_j^n in terms of j.

(3) Prove that if the Riemann hypothesis is false there exists a zero $\rho = \beta + i\gamma$ of $\zeta'(s)$ with $0 < \beta < \frac{1}{2}$.

(4) Find an explicit lower bound for n in the RH equivalence of Section 10.10.

(5) In Section 10.9 find the relationship between the result of Hall and Hayman and Theorem 10.32.

Appendix A
Tables

The following RHpack (see Appendix B) variables and functions relate to the material in this appendix: GenerateHC, GenerateHR, GenerateSA, GenerateCA, GenerateLandau, FindNextXA, CAPrimes, InitialNR, InitialHC, InitialSA, InitialSA1000, InitialCA, InitialCA150, InitialXA, InitialLandau, InitialCAPrimes, HardyRamanujanToPrimorialForm and PrimorialFormToInteger.

A.1 Extremely Abundant Numbers

See Chapter 9 and the item A217867 in the *Online Encyclopedia of Integer Sequences* [158]. Recall that

$$G(n) := \frac{\sigma(n)}{n \log\log n}, \quad n \geq 2.$$

Table A.1 The first seven extremely abundant numbers.

Rank	Factors	$e^\gamma - G(n)$
1	$2^5 \cdot 3^2 \cdot 5 \cdot 7$	0.0252580790
2	$2^8 \cdot 3^5 \cdot 5^3 \cdot 7^2 \cdot 11^2 \cdot 13^2 \cdot 17^1 \cdots 113^1$	0.0238889777
3	$2^8 \cdot 3^5 \cdot 5^3 \cdot 7^2 \cdot 11^2 \cdot 13^2 \cdot 17^1 \cdots 127^1$	0.0237040357
4	$2^8 \cdot 3^5 \cdot 5^3 \cdot 7^2 \cdot 11^2 \cdot 13^2 \cdot 17^1 \cdots 131^1$	0.02342819838
5	$2^8 \cdot 3^5 \cdot 5^3 \cdot 7^2 \cdot 11^2 \cdot 13^2 \cdot 17^1 \cdots 137^1$	0.0232960582
6	$2^8 \cdot 3^5 \cdot 5^3 \cdot 7^2 \cdot 11^2 \cdot 13^2 \cdot 17^1 \cdots 139^1$	0.0228625950
7	$2^9 \cdot 3^5 \cdot 5^3 \cdot 7^2 \cdot 11^2 \cdot 13^2 \cdot 17^1 \cdots 139^1$	0.0228130145

A.2 Small Numbers Not Satisfying Robin's Inequality

See Chapter 8. Numbers not satisfying Robin's inequality satisfy

$$\frac{\sigma(n)}{ne^{\gamma}\log\log n} > 1.$$

Table A.2 Numbers $n \le 5041$ with $\sigma(n)/n \ge e^{\gamma}\log\log n$.

Rank	n	Factors	$\sigma(n)/(ne^{\gamma}\log\log n)$
1	1	1	—
2	2	2	−2.29784
3	3	3	7.95991
4	4	2	3.00812
5	5	5	1.41579
6	6	$2\cdot 3$	1.92545
7	8	2^3	1.43797
8	9	3^2	1.03024
9	10	$2\cdot 5$	1.21174
10	12	$2^2\cdot 3$	1.43927
11	16	2^4	1.06673
12	18	$2\cdot 3^2$	1.14614
13	20	$2^2\cdot 5$	1.07462
14	24	$2^3\cdot 3$	1.21395
15	30	$2\cdot 3\cdot 5$	1.10079
16	36	$2^2\cdot 3^2$	1.1196
17	48	$2^4\cdot 3$	1.07517
18	60	$2^2\cdot 3\cdot 5$	1.11527
19	72	$2^3\cdot 3^2$	1.04641
20	84	$2^2\cdot 3\cdot 7$	1.00581
21	120	$2^3\cdot 3\cdot 5$	1.07559
22	180	$2^2\cdot 3^2\cdot 5$	1.03387
23	240	$2^4\cdot 3\cdot 5$	1.0231
24	360	$2^3\cdot 3^2\cdot 5$	1.02942
25	720	$2^4\cdot 3^2\cdot 5$	1.00087
26	840	$2^3\cdot 3\cdot 5\cdot 7$	1.0094
27	2520	$2^3\cdot 3^2\cdot 5\cdot 7$	1.01322
28	5040	$2^4\cdot 3^2\cdot 5\cdot 7$	1.00556

A.3 Superabundant Numbers

Recall that a superabundant number n is a positive integer for which $\sigma(m)/m < \sigma(n)/n$ for all $1 \leq m < n$. See the item A004394 in the *Online Encyclopedia of Integer Sequences* [158].

Table A.3 The first 31 superabundant numbers.

Rank	n	Factors	$\sigma(n)/n$
1	1	1	—
2	2	2	−2.29784
3	4	2^2	1.750
4	6	$2 \cdot 3$	2.000
5	12	$2^2 \cdot 3$	2.333
6	24	$2^3 \cdot 3$	2.500
7	36	$2^2 \cdot 3^2$	2.528
8	48	$2^4 \cdot 3$	2.583
9	60	$2^2 \cdot 3 \cdot 5$	2.800
10	120	$2^3 \cdot 3 \cdot 5$	3.000
11	180	$2^2 \cdot 3^2 \cdot 5$	3.033
12	240	$2^4 \cdot 3 \cdot 5$	3.100
13	360	$2^3 \cdot 3^2 \cdot 5$	3.250
14	720	$2^4 \cdot 3^2 \cdot 5$	3.385
15	840	$2^3 \cdot 3 \cdot 5 \cdot 7$	3.429
16	1260	$2^2 \cdot 3^2 \cdot 5 \cdot 7$	3.467
17	1680	$2^4 \cdot 3 \cdot 5 \cdot 7$	3.543
18	2520	$2^3 \cdot 3^2 \cdot 5 \cdot 7$	3.714
19	5040	$2^4 \cdot 3^2 \cdot 5 \cdot 7$	3.838
20	10080	$2^5 \cdot 3^2 \cdot 5 \cdot 7$	3.900
21	15120	$2^4 \cdot 3^3 \cdot 5 \cdot 7$	3.937
22	25200	$2^4 \cdot 3^2 \cdot 5^2 \cdot 7$	3.966
23	27720	$2^3 \cdot 3^2 \cdot 5 \cdot 7 \cdot 11$	4.052
24	55440	$2^4 \cdot 3^2 \cdot 5 \cdot 7 \cdot 11$	4.187
25	110880	$2^5 \cdot 3^2 \cdot 5 \cdot 7 \cdot 11$	4.255
26	166320	$2^4 \cdot 3^3 \cdot 5 \cdot 7 \cdot 11$	4.294
27	277200	$2^4 \cdot 3^2 \cdot 5^2 \cdot 7 \cdot 11$	4.327
28	332640	$2^5 \cdot 3^3 \cdot 5 \cdot 7 \cdot 11$	4.364
29	554400	$2^5 \cdot 3^2 \cdot 5^2 \cdot 7 \cdot 11$	4.396
30	665280	$2^6 \cdot 3^3 \cdot 5 \cdot 7 \cdot 11$	4.398
31	720720	$2^4 \cdot 3^2 \cdot 5 \cdot 7 \cdot 11 \cdot 13$	4.509

A.4 Colossally Abundant Numbers

See the item A004490 in the *Online Encyclopedia of Integer Sequences* [158]. Recall that n is colossally abundant if there is an $\epsilon > 0$ with, for all $m > 1$, $\sigma(n)/n^{1+\epsilon} \geq \sigma(m)/m^{1+\epsilon}$. See Chapter 6 for some larger examples.

Table A.4 The first 30 colossally abundant numbers.

Rank	n	Factors	Prime
1	2	2	2
2	6	$2 \cdot 3$	3
3	12	$2^2 \cdot 3$	2
4	60	$2^2 \cdot 3 \cdot 5$	5
5	120	$2^3 \cdot 3 \cdot 5$	2
6	360	$2^3 \cdot 3^2 \cdot 5$	3
7	2 520	$2^3 \cdot 3^2 \cdot 5 \cdot 7$	7
8	5 040	$2^4 \cdot 3^2 \cdot 5 \cdot 7$	2
9	55 440	$2^4 \cdot 3^2 \cdot 5 \cdot 7 \cdot 11$	11
10	720 720	$2^4 \cdot 3^2 \cdot 5 \cdot 7 \cdot 11 \cdot 13$	13
11	1 441 440	$2^5 \cdot 3^2 \cdot 5 \cdot 7 \cdot 11 \cdot 13$	2
12	4 324 320	$2^5 \cdot 3^3 \cdot 5 \cdot 7 \cdot 11 \cdot 13$	3
13	21 621 600	$2^5 \cdot 3^3 \cdot 5^2 \cdot 7 \cdot 11 \cdot 13$	5
14	367 567 200	$2^5 \cdot 3^3 \cdot 5^2 \cdot 7 \cdot 11 \cdot 13 \cdot 17$	17
15	6 983 776 800	$2^5 \cdot 3^3 \cdot 5^2 \cdot 7 \cdot 11 \cdot 13 \cdot 17 \cdot 19$	19
16	160 626 866 400	$2^5 \cdot 3^3 \cdot 5^2 \cdot 7 \cdot 11 \cdot 13 \cdot 17 \cdot 19 \cdot 23$	23
17	321 253 732 800	$2^6 \cdot 3^3 \cdot 5^2 \cdot 7 \cdot 11 \cdot 13 \cdot 17 \cdot 19 \cdot 23$	2
18	9 316 358 251 200	$2^6 \cdot 3^3 \cdot 5^2 \cdot 7 \cdot 11 \cdot 13 \cdot 17 \cdot 19 \cdot 23 \cdot 29$	29
19	288 807 105 787 200	$2^6 \cdot 3^3 \cdot 5^2 \cdot 7 \cdot 11 \cdot 13 \cdot 17 \cdot 19 \cdot 23 \cdot 29 \cdot 31$	31
20	2 021 649 740 510 400	$2^6 \cdot 3^3 \cdot 5^2 \cdot 7^2 \cdot 11 \cdot 13 \cdot 17 \cdot 19 \cdot 23 \cdot 29 \cdot 31$	7
21	6 064 949 221 531 200	$2^6 \cdot 3^4 \cdot 5^2 \cdot 7^2 \cdot 11 \cdot 13 \cdot 17 \cdot 19 \cdot 23 \cdot 29 \cdot 31$	3
22	224 403 121 196 654 400	$2^6 \cdot 3^4 \cdot 5^2 \cdot 7^2 \cdot 11 \cdot 13 \cdot 17 \cdot 19 \cdot 23 \cdot 29 \cdot 31 \cdot 37$	37
23	9 200 527 969 062 830 400	$2^6 \cdot 3^4 \cdot 5^2 \cdot 7^2 \cdot 11 \cdot 13 \cdot 17 \cdot 19 \cdot 23 \cdot 29 \cdot 31 \cdot 37 \cdot 41$	41
24	c_{24}	$2^6 \cdot 3^4 \cdot 5^2 \cdot 7^2 \cdot p_5 \cdots p_{14}$	43
25	c_{25}	$2^7 \cdot 3^4 \cdot 5^2 \cdot 7^2 \cdot p_5 \cdots p_{14}$	2
26	c_{26}	$2^7 \cdot 3^4 \cdot 5^2 \cdot 7^2 \cdot p_5 \cdots p_{15}$	47
27	c_{27}	$2^7 \cdot 3^4 \cdot 5^2 \cdot 7^2 \cdot p_5 \cdots p_{16}$	53
28	c_{28}	$2^7 \cdot 3^4 \cdot 5^2 \cdot 7^2 \cdot p_5 \cdots p_{17}$	59
29	c_{29}	$2^7 \cdot 3^4 \cdot 5^3 \cdot 7^2 \cdot p_5 \cdots p_{17}$	5
30	c_{30}	$2^7 \cdot 3^4 \cdot 5^3 \cdot 7^2 \cdot p_5 \cdots p_{18}$	61

A.5 Primes to Make Colossally Abundant Numbers

See Chapter 6 and the item A073751 in the *Online Encyclopedia of Integer Sequences* [158].

$$\{2,3,2,5,2,3,7,2,11,13,2,3,5,17,19,23,2,29,31,7,$$
$$3,37,41,43,2,47,53,59,5,61,67,71,73,11,79,2,83,3,$$
$$89,97,13,101,103,107,109,113,127,131,137,139,2,149,151,$$
$$7,157,163,167,17,173,179,181,191,193,197,199,19,211,3,$$
$$223,227,229,5,233,239,241,251,2,257,263,269,271,277,$$
$$281,283,293,23,307,311,313,317,331,337,347,349,353,359,$$
$$367,373,379,383,389,397,401,409,419,421,431,433,439,443,$$
$$449,457,461,2,463,467,29,479,487,491,499,503,509,521,$$
$$523,541,31,547,11,557,563,3,569,571,577,587,593,599,$$
$$601,607,613,617,619,631,641,643,647,653,659,661,673,677,$$
$$683,691,701,709,719,727,733,739,743,751,757,761,37,769,$$
$$773,787,797,809,811,7,821,823,827,829,839,2,853,857,$$
$$859,863,877,881,883,887,13,907,911,919,5,929,41,937,$$
$$941,947,953,967,971,977,983,991,997,1009,1013,1019,1021,$$
$$43,1031,1033,1039,1049,1051,1061,1063,1069,1087,1091,1093,$$
$$1097,1103,1109,1117,1123,1129,1151,1153,1163,1171,1181,$$
$$1187,1193,1201,1213,1217,47,1223,1229,1231,1237,1249,1259,$$
$$1277,1279,1283,1289,1291,1297,1301,1303,1307,1319,1321,$$
$$1327,1361,1367,1373,1381,1399,1409,1423,1427,1429,1433,$$
$$1439,1447,1451,1453,1459,1471,3,1481,1483,1487,1489,1493,$$
$$1499,1511,1523,1531,1543,2,53,1549,1553,1559,1567,1571,$$
$$1579,1583,1597,1601,1607,1609,1613,1619,1621,1627,1637,$$
$$1657,1663,1667,1669,1693,1697,1699,1709,1721,1723,1733,$$
$$1741,1747,1753,1759,1777,1783,1787,1789,1801,1811,1823,$$
$$1831,1847,1861,1867,1871,1873,1877,1879,1889,1901,1907,59,$$
$$1913,1931,1933,1949,1951,17,1973,1979,1987,1993,1997,1999,$$
$$2003,2011,2017,2027,2029,2039,61,2053,2063,2069,2081,2083,$$
$$2087,2089,2099,2111,2113,2129,2131,2137,2141,2143,2153,$$
$$2161,2179,2203,2207,2213,2221,2237,2239,2243,2251,2267,$$
$$2269,2273,2281,2287,2293,2297,2309,2311,2333,2339,2341\}$$

A.6 Small Numbers Satisfying Nicolas' Reversed Inequality

The numbers in **bold** are primorials and those marked with an asterisk (*) are multiples of a primorial which are less than the next primorial.

{1, **2**, 3, 4*, 5, 6*, 8*, 9, 10*, 12*, 14*, 15, 16*, 18*, 20*, 22*, 24*, 26*, 28*, **30**, 36, 40, 42, 48, 50, 54, 60*, 66, 70, 72, 78, 84, 90*, 96, 102, 108, 114, 120*, 126, 132, 138, 140, 144, 150*, 156, 162, 168, 174, 180*, 186, 192, 198, 204, **210**, 216, 222, 228, 234, 240, 246, 252, 258, 264, 270, 276, 294, 300, 306, 312, 330, 336, 342, 360, 378, 390, 396, 420*, 450, 462, 468, 480, 504, 510, 528, 540, 546, 570, 588, 600, 630*, 660, 672, 690, 714, 720, 750, 756, 780, 798, 810, 840*, 858, 870, 882, 900, 924, 930, 960, 966, 990, 1008, 1020, 1050*, 1080, 1092, 1110, 1122, 1134, 1140, 1170, 1176, 1200, 1218, 1230, 1260*, 1290, 1302, 1320, 1350, 1380, 1386, 1410, 1428, 1440, 1470*, 1500, 1530, 1554, 1560, 1590, 1596, 1620, 1638, 1650, 1680*, 1710, 1722, 1740, 1770, 1800, 1830, 1848, 1860, 1890*, 1920, 1932, 1950, 1980, 2010, 2040, 2070, 2100*, 2130, 2142, 2160, 2184, 2190, 2220, 2250, 2280, **2310**, 2340, 2370, 2394, 2400, 2430, 2460, 2490, 2520, 2550, 2580, 2610, 2640, 2670, 2700, 2730, 2760, 2772, 2790, 2820, 2850, 2856, 2880, 2910, 2940, 2970, 3000, 3030, 3060, 3090, 3120, 3150, 3180, 3210, 3234, 3240, 3270, 3276, 3300, 3330, 3360, 3390, 3420, 3450, 3480, 3510, 3540, 3570, 3600, 3630, 3660, 3690, 3696, 3720, 3780, 3810, 3822, 3870, 3900, 3930, 3960, 3990, 4020, 4080, 4110, 4140, 4158, 4170, 4200, 4230, 4260, 4290, 4350, 4368, 4380, 4410, 4440, 4560, 4590, 4620*, 4650, 4680, 4770, 4830, 4920, 4950, 5040, 5070, 5082, 5100, 5130, 5160, 5220, 5250, 5280, 5460, 5520, 5544, 5550, 5580, 5610, 5670, 5700, 5850, 5880, 5940, 6006, 6090, 6120, 6210, 6240, 6270, 6300, 6510, 6600, 6630, 6720, 6840, 6900, 6930*, 6960, 7020, 7140, 7260, 7350, 7410, 7560, 7590, 7650, 7770, 7800, 7854, 7920, 7980, 8160, 8190, 8250, 8280, 8400, 8550, 8580, 8610, 8670, 8778, 8820, 8910, 8970, 9030, 9120, 9180, 9240*, 9282, 9360, 9450, 9570, 9660, 9690, 9750, 9870, 9900}

A.7 Heights of Integers

The height $h(n)$ of an integer $n = p_1^{\alpha_1} \cdots p_m^{\alpha_m}$ is $\sum_{i=1}^{n} p_i^{\alpha_i}$.

Table A.5 Initial values of $h(n)$; compare with item A008475 in OEIS [158].

n	$h(n)$	n	$h(n)$	n	$h(n)$	n	$h(n)$
1	0	13	13	25	25	37	37
2	2	14	9	26	15	38	21
3	3	15	8	27	27	39	16
4	4	16	16	28	11	40	13
5	5	17	17	29	29	41	41
6	5	18	11	30	10	42	12
7	7	19	19	31	31	43	43
8	8	20	9	32	32	44	15
9	9	21	10	33	14	45	14
10	7	22	13	34	19	46	25
11	11	23	23	35	12	47	47
12	7	24	11	36	13	48	19

A.8 Maximum Order of an Element of the Symmetric Group

Table A.6 Initial values of $g(n)$; compare with item A000793 in OEIS [158].

n	$g(n)$	n	$g(n)$	n	$g(n)$	n	$g(n)$
1	1	13	60	25	1260	37	13860
2	2	14	84	26	1260	38	16380
3	3	15	105	27	1540	39	27720
4	4	16	140	28	2310	40	30030
5	6	17	210	29	2520	41	32760
6	6	18	210	30	4620	42	60060
7	12	19	420	31	4620	43	60060
8	15	20	420	32	5460	44	60060
9	20	21	420	33	5460	45	60060
10	30	22	420	34	9240	46	60060
11	30	23	840	35	9240	47	120120
12	60	24	840	36	13860	48	120120

Appendix B
RHpack Mini-Manual

B.1 Introduction

This appendix is the manual for a set of functions written to assist the reader to reproduce, and possibly extend, the calculations mentioned in the main part of the book. The software for the package is provided over the world wide web at the webpage for the book linked to the author's homepage:

www.math.waikato.ac.nz/~kab

and is in the form of a standard *Mathematica* add-on package. To use the functions in the package you will need to have a version of *Mathematica* at level 7.0 or higher.

B.1.1 Installation

First connect to the website given in the paragraph above and click on the link for RHpack listed under the heading "Software" to get to the RHpack homepage. Instructions on downloading the files for the package will be given on the homepage. If you have an earlier version of RHpack, first delete the file `RHpack.m`. The name of the package file is `RHpack.m`. To install the package, if you have access to the file system for programs on your computer, place a copy of the file in the standard repository for *Mathematica* packages or any other directory, which will be listed by evaluating $Path in *Mathematica*, to which you have access. You can then type `<<RHpack.m` and then press the Shift and Enter keys to load the package. You may need administrator or super-user status to complete this installation. Alternatively, place the package file `RHpack.m` anywhere in your own file system where it is safe and accessible.

Instructions for Windows systems The package can be loaded by typing

```
SetDirectory["c:\\your\\directory\\path"]<Shift/Enter>
<<RHpack.m;ResetDirectory[]<Shift/Enter>
```

or

```
Get["RHpack.m",Path->{"c:\\your\\directory\\path"}]
<Shift/Enter>
```

where the path-name in quotes should be replaced by the actual path-name of directories and subdirectories which specify where the package has been placed on the computer you are using. If the double backslash does not work, try a single backslash.

Instructions for Unix/Linux systems These are the same as for Windows, but the path-name syntax style should be like

```
/usr/home/your/subdirectory
```

Instructions for Macintosh systems These are the same as for Windows, but the path-name syntax style should be like

```
HD:Users:ham:Documents:
```

All systems The package should load printing a message. The functions of RHpack are then available to any *Mathematica* notebook you subsequently open.

B.1.2 About This Mini-Manual

This appendix contains a list of all of the functions available in the package RHpack followed by a manual entry for each function in alphabetical order. To obtain information about bug fixes and updates to RHpack, consult the website for the text linked to the author's homepage (see Section B.1).

There are many issues to do with computer algebra and mathematical software that will arise in any serious evaluation or use of *Mathematica* and RHpack. Some notes follow:

RHpack function arguments are first evaluated and then checked for correct data type. If the user calls a function with an incorrect number of arguments or an argument of incorrect type, rather than issue a warning and proceeding to compute (default *Mathematica* style), RHpack prints an error message, aborts the evaluation and returns the user to the top level, no matter how deeply nested the function which makes the erroneous call happens to be placed. This is to assist users to debug programs which include calls to this package.

Every function in RHpack has a short name which can be used in place of the *Mathematica* style long name. Functions which include numerical floating-point evaluations frequently have an integer argument "d". This represents the precision which is used for internal computations, and not the accuracy of the resulting output.

The time that it takes for functions to evaluate is variable. For some argument ranges, stored values have been used to speed up the computations. This applies for example to $M(x)$ and $\theta(x)$.

Appendix A gives tables of numbers such as superabundant and colossally abundant. RHpack includes in some instances larger sets of numbers: for example 1000 superabundant and 20 extremely abundant numbers are stored explicitly.

B.2 RHpack Functions

■ AllPrimesFactor (apf)

This function returns the prime factors of a natural number with all primes up to the maximum prime divisor appearing.

AllPrimesFactor[n] ⟶ L

n is a natural number,
L is a list of lists of the form $\{\{2,a_2\},\{3,a_3\},\ldots,\{p,a_p\}\}$ where p is the maximum prime dividing n and, for each sublist $\{q,a_q\}$, a_q is the maximum power to which q divides n.

■ BacklundB (blb)

This function computes an upper bound for the error in the approximation to the number of critical zeta zeros up to height T, given by

$$|N(T) - F(T)| \leq B(T).$$

We have $B(T) := a_1 \log T + a_2 \operatorname{loglog} T + a_3$ where the values a_i are stored in the global variables $\$a_1, \$a_2, \$a_3$, and are best possible at the time of writing. See Section 2.2.

BacklundB[T, d] ⟶ r

T is a positive real number representing the height up to which zeta zeros are being counted,
d is a precision to which evaluations to compute $B(T)$ will be made internally,
r is a floating-point number being the value of $B(T)$.

■ CAPrimes (cap)

This function computes a list of primes for generating colossally abundant numbers – see Table A.5. This has been shown to work up to $n = 10^{11}$.

CAPrimes[n] ⟶ L

n is a natural number being the number of primes which are to be generated,
L is an ordered list of primes such that the product of the first j primes is the jth colossally abundant number.

■ CheckCNk (cnk)

This function computes the $k \le k_0$ such that $c(N_1) \le c(N_k) \le c(N_{66})$ fails, to precision d – see Theorem 5.34.

CheckCNk[k0, d] ⟶ L

k0 is an integer being the upper bound for which the values of $c(N_k)$ are checked,

d is a natural number being the precision to which internal calculations are made,

L is a list of values of k for which either $c(N_k) < c(N_1)$ or $c(N_k) > c(N_{66})$.

■ ColossallyAbundantF (caf)

This function evaluates the function $F(x,\alpha)$ defined in Section 6.3, to precision d.

ColossallyAbundantF[x, a, d] ⟶ r

a is a natural number,

x is a real number greater than 1,

d is a natural number being the precision for internal computations,

r is a floating-point number being the value of

$$\log_x\left(1 + \frac{1}{x^a + x^{a-1} + \cdots + x}\right).$$

■ ColossallyAbundantInteger (cai)

This function returns the colossally abundant integer corresponding to a given $e > 0$ – see Section 6.3.

ColossallyAbundantInteger[e] ⟶ L

e is a positive real number,

L is a list, the first element being a colossally abundant number, the second being a list of prime powers for the prime divisors of n, in ascending order.

■ ColossallyAbundantQ (caq)

This function determines whether n is a colossally abundant integer – see Table A.4.

ColossallyAbundantQ[n] ⟶ P

n is a natural number,

P is True if n is colossally abundant and False otherwise.

■ **CompositeQ (coq)**

This function determines whether or not n is a composite integer.

CompositeQ[n] ⟶ P

n is a natural number,
P is True if n is composite, False otherwise.

■ **ComputeC (coc)**

This function evaluates $c(n) = (n/\varphi(n) - e^{\gamma} \log\log n)\sqrt{\log n}$, to precision d – see Section 5.4.

ComputeC[n, d] ⟶ r

n is a natural number 2 or more,
d is the precision to which evaluations to compute $c(n)$will be made internally,
r is a floating-point number.

■ **ComputeDeltas (cod)**

This function computes the values of

$$\delta_i := \frac{\psi(q_i) - \theta(q_i)}{\sqrt{q_{i+1}}}$$

which are used in the proof of Lemma 5.18. The q_i are the ordered squarefull powers of primes up to 599^3. It also prints a table of the first 11 δ_i.

ComputeDeltas[d] ⟶ min

d is a natural number being the precision for internal computations,
min is the minimum value of the δ_i.

■ **Computef (cof)**

This function evaluates $f(x) = e^{\gamma} \log\theta(x) \prod_{p \le x}(1 - 1/p)$, to precision d – see Chapters 5 and 7.

Computef[x, d] ⟶ r

x is a positive real number,
d is a natural number being the precision for internal computations,
r is a floating-point number representing the value of $f(x)$.

■ **ComputeH (coh)**

This function computes the value of

$$H(x) := 1 + \sum_{n=4}^{\log x/\log 2} x^{1/n-1/3}.$$

See the proof of Lemma 5.17.

ComputeH[x, d] ⟶ r

x is a positive real number not less than 16,
d is a natural number being the precision for internal computations,
r is a floating-point number being the value of $H(x)$.

■ ComputeKm (ckm)

Note the definition

$$k_m(h) = \sum_{\gamma>h} \frac{1}{\gamma^{m+1}},$$

where $\rho=\beta+i\gamma$ are complex zeta zeros. This function finds upper bounds for the values $k_m(0)$ recursively, to precision d, starting with the bound $k_1(0) \leq 0.0463$ – see Table 3.1.

ComputeKm[m, d] ⟶ L

m is a natural number,
d is a natural number being the precision for internal computations,
L is a list of floating-point numbers representing computed upper bound for $\{k_1(0),\ k_2(0),\ \ldots, k_m(0)\}$.

■ ComputeTheta (cth)

This function evaluates $\theta(x) = \sum_{p\leq x} \log p$.

ComputeTheta[x, d] ⟶ r

x is a real number,
d is a natural number being the precision for internal computations,
r is a floating-point number which is an approximation to $\theta(x)$.

■ ComputeThetaValues (ctv)

This function computes b values, at integer points commencing at 1, of a list of $\theta(x)$ values, to precision d.

ComputeThetaValues[b, d] ⟶ L

b is a natural number being the number of $\theta(x)$ values to be computed,
d is a natural number being the precision for internal computations,
L is a list of b $\theta(x)$ values.

■ CriticalEpsilonValues (cev)

This function computes a list of critical values at primes up to b and powers up to *jmax*, in decreasing order – see Section 6.3.

CriticalEpsilonValues[b, jmax] ⟶ L

b is a natural number,
jmax is a natural number,
L is a list of critical epsilon values in decreasing order in symbolic form for primes up p_b and exponents j up to $jmax$.

■ DivisibilityGraph (dvg)

This function returns the $n \times n$ divisibility graph of Section 10.5.

DivisibilityGraph[n] ⟶ L

n is a natural number,
L is a list of terms $i \to j$ where $1 \le i, j \le n$ is a subset of pairs, representing the divisibility graph.

■ ExtraordinaryQ (exq)

This function determines whether n is an extraordinary number up to a finite number of multiples of n – see Chapter 9.

ExtraordinaryQ[n, b, d] ⟶ P

n is a natural number,
b is a natural number,
d is a natural number being the precision for internal computations,
P is True if n is left abundant and right abundant up to multiples not greater than b, both to precision d, otherwise False.

■ ExtremelyAbundantQ (eaq)

This function determines whether n is an extremely abundant number – see Table A.1

ExtremelyAbundantQ[n, d] ⟶ P

n is a natural number with $n \ge 10080$,
d is a natural number being the precision for internal computations,
P is True if $G(m) < G(n)$, to precision d, for $10080 \le m < n$, otherwise False.

■ FareyFractions (ffr)

This function computes the list of Farey fractions for denominator n, in increasing order – see Section 10.3.

FareyFractions[n] ⟶ L

n is a natural number ,
L is a list of the Farey fractions $\{0, 1/n, \ldots, 1\}$ of order n.

■ FHalf (fha)

This function computes $F_{1/2}(z)$ – see Lemma 5.16.

FHalf[x, d] ⟶ r

x is a positive real number,
d is a natural number being the precision for internal computations,
r is a floating-point number representing the value of $F_{1/2}(x)$.

■ **FindNextXA (fnx)**

This function finds an extremely abundant number larger than a given extremely abundant number – see Chapter 9. This function does not check that its argument is extremely abundant, and simply checks the size of Grönwall's function $G(n)$ against its values at multiples by primes which preserve its Hardy–Ramanujan status. Because of this, some extremely abundant numbers will be missed.

FindNextXA[n] ⟶ m

n is a natural number which should be a Hardy–Ramanujan number,
m is a natural number with $G(m) > G(n)$ or False if it fails to find such an m.

■ **GenerateCA (gca)**

This function generates the b smallest colossally abundant numbers, with some restriction on the size of b – see Table A.4.

GenerateCA[b] ⟶ L

b is a natural number,
L is a list of b colossally abundant numbers in increasing order.

■ **GenerateHC (ghc)**

This function generates all highly composite numbers up to b – see Section 8.2.

GenerateHC[b] ⟶ L

b is a natural number,
L is a list of the highly composite numbers up to b.

■ **GenerateHR (ghr)**

This function generates a list of all Hardy–Ramanujan numbers up to b – see Section 8.2.

GenerateHR[b] ⟶ L

b is a natural number,
L is a sorted list of all Hardy–Ramanujan numbers less than or equal to b.

■ **GenerateLandau (gla)**

This function generates values of Landau's function $g(n)$ up to $n = b$ – see Section 10.10. It uses the algorithm of Heinz and Alcover.

GenerateLandau[b] ⟶ LL

b is a natural number,
LL is a list of lists, each sublist being a pair $\{n,\ g(n)\}$.

■ **GenerateSA (gsa)**

This function generates all superabundant numbers up to b – see Section 6.2.

GenerateSA[b] ⟶ L

b is a natural number,
L is a list of all superabundant numbers up to b in increasing order.

■ **GronwallG (grg)**

This function evaluates Grönwall's function $G(n) = \sigma(n)/(n \log\log n)$, to precision d.

GronwallG[n, d] ⟶ r

n is a natural number,
d is a natural number being the precision for internal computations,
r is a floating-point number representing the value of $G(n)$.

■ **HardyRamanujanQ (hrq)**

This function determines whether n is a Hardy–Ramanujan integer – see Section 8.2.

HardyRamanujanQ[n] ⟶ P

n is a natural number,
P is True if n is a Hardy–Ramanujan number, otherwise False.

■ **HardyRamanujanToPrimorialForm (hr2pf)**

This function converts a HardyRamanujan integer to primorial form – see Section 6.1.

HardyRamanujanToPrimorialForm[n] ⟶ L

n is a natural number which is Hardy–Ramanujan,
L is a list of primes in non-increasing order of prime value such that n is the product of the primorials corresponding to those primes to the given powers such that each prime is the largest in an associated primorial.

■ **HighlyCompositeQ (hcq)**

This function determines whether n is a highly composite integer – see Section 6.1.

HighlyCompositeQ[n] ⟶ P

n is a natural number,
P is True if n is highly composite, otherwise False.

■ Initial Set Elements

The following symbols evaluate to sorted lists of the smallest types of numbers given by

InitialCA (ica) colossally abundant numbers,
InitialCAPrimes (icap) primes to make colossally abundant numbers,
InitialCA150 (ca1) the first 150 colossally abundant numbers in primorial form,
InitialHC (ihc) the first 41 highly composite numbers ,
InitialLandau (ila) values of Landau's function $g(n)$,
InitialSA (isa) superabundant numbers,
InitialSA1000 (sa1) the first 1000 superabundant numbers in primorial form,
InitialNR (inr) natural numbers which do not satisfy Nicolas' inequality,
InitialXA (ixa) extremely abundant numbers.

Primorial form for a Hardy–Ramanujan integer is the encryption $n \to \{p_1, p_2, \ldots, p_m\}$ where the p_i are primes, which are not necessarily distinct, in increasing order, and which are such that n is the product of the primorials $p_i\#$, which in each case is the product of all of the primes up to and including p_i. See the RHpack functions HardyRamanujanToPrimorialForm, PrimorialFormToInteger and PrimorialFormQ.

■ IntegerHeight (iht)

This function computes the function $h(n)$ of Section 10.10.

IntegerHeight[n] ⟶ m

n is a natural number,
m is a natural number being $h(n)$.

■ KFreeQ (kfq)

This function determines whether the integer n is k-free.

KFreeQ[n, k] ⟶ P

n is a natural number,
k is a natural number,
P is True if n is k-free, otherwise False.

■ KthApproximatePrimorial (app)

This function returns the kth primorial, to precision d – see Chapter 5.

KthApproximatePrimorial[k, d] ⟶ r

k is a natural number,

d is a natural number being the precision for internal computations,
r is a floating-point number representing $N_k = p_1 \cdots p_k$.

■ **KthPrimorial (kpp)**

This function returns the kth primorial N_k as an exact integer.

KthPrimorial[k] ⟶ m

k is a natural number,
m is a natural number being the value of N_k, the kth primorial.

■ **LagariasInequalityQ (liq)**

This function checks $\sigma(n) < H_n + \exp(H_n)\log(H_n)$, to precision d – see Section 7.10.

LagariasInequalityQ[n ,d] ⟶ P

n is a natural number,
d is a natural number being the precision for internal computations,
P is True if the inequality is satisfied at precision d, False otherwise.

■ **Lambda (lam)**

This function evaluates the function $\Lambda(s)$ of Section 10.8 for complex s.

Lambda[s, d] ⟶ r

s is a real or complex number,
d is a natural number being the precision for internal computations,
r is a floating-point complex number.

■ **LargestZetaZero (lzz)**

This function finds the zeta zero $\rho = \frac{1}{2} + i\gamma$ with maximum $\gamma \leq h$, to precision d for $h \leq 10^6$.

LargestZetaZero[h, d] ⟶ r

h is a positive real number with $h \leq 10^6$,
d is a natural number being the precision for internal computations,
r is the value of the y-coordinate of the largest zeta zero with imaginary part less than or equal to h.

■ **LeftAbundantQ (laq)**

This function determines whether n is a left abundant number – see Chapter 9.

LeftAbundantQ[n, d] ⟶ P

n is a natural number,

d is a natural number being the precision for internal computations,
P is True if $G(n/p) \leq G(n)$, to precision d, for all prime divisors $p \mid n$, False otherwise.

■ **LogKthPrimorialBounds (lppb)**

This function gives upper and lower bounds of Lemma 8.18 for $\log N_k$.

LogKthPrimorialBounds[k, d] $\longrightarrow$ **L**

k is a natural number with $k \geq 198$,
d is a natural number being the precision for internal computations,
L is a list of two floating-point numbers $\{l, u\}$ such that $l \leq \log N_k \leq u$.

■ **M**

This function evaluates the sum of the Möbious $\mu(n)$ function up to $x > 0$ – see Chapter 4 and Section 10.3.

M[x] $\longrightarrow$ **s**

x is a positive real number,
s is an integer representing the sum $\sum_{n=1}^{x} \mu(n)$.

■ **MaxPrime (mxp)**

This function finds the maximum prime less than or equal to x.

MaxPrime[x] $\longrightarrow$ **p**

x is a real number,
p is a prime number which is maximal such that $p \leq x$.

■ **NextZetaZero (nzz)**

This function finds the next zeta zero with height larger than h, to precision d. It is suitable only for modest values of h.

NextZetaZero[h, d] $\longrightarrow$ **r**

h is a positive real number,
d is a natural number being the precision for internal computations,
r is a floating-point number representing the y-coordinate of the smallest zeta zero ρ with $\gamma > h$, where $\rho = \frac{1}{2} + i\frac{1}{2}$.

■ **NiceFactorInteger (nfi)**

This function factors a natural number n with primes to the power 1 appearing as singletons.

NiceFactorInteger[n] $\longrightarrow$ **L**

n is a natural number,

L is list of list pairs $\{p,e\}$ where p is prime and $p^e \| n$ and $e \geq 2$ and singletons p where $p^1 \| n$.

■ NicolasInequalityQ (niq)

This function checks $n/\varphi(n) < e^\gamma \log\log n$ to precision d – see Chapter 5.

NicolasInequalityQ[n, d] ⟶ P

n is a natural number,
d is a natural number being the precision for internal computations,
P is True if the inequality is satisfied to precision d and False otherwise.

■ NicolasTwoFailures (n2f)

This function computes the number of failures to Nicolas inequality at primorials

$$\left(\frac{n}{\varphi(n)} - e^\gamma \log\log n\right)\sqrt{\log n} < e^\gamma(4+\gamma-\log\pi-2\log 2)$$

in the range $k1 \leq k \leq k0$ with $n = N_k$ – see the proof of Theorem 5.34 and Corollary 5.35. This function is suitable only for small values of k.

NicolasTwoFailures[k1, k0, d] ⟶ n

k1 is a natural number,
k0 is a natural number with $k1 \leq k0$,
d is a natural number being the precision for internal computations,
n is a non-negative integer representing the number of integer points k in $[k1, k0]$ at which the inequality fails at $n = N_k$.

■ NOverPhi (nop)

This function evaluates $n/\varphi(n)$, to precision d – see Chapter 5.

NOverPhi[n, d] ⟶ r

n is a natural number,
d is a natural number being the precision for internal computations,
r is a floating-point number representing the value $n/\varphi(n)$.

■ NthHarmonicNumber (nha)

This function evaluates the n harmonic number, H_n, to precision d – see Section 7.10.

NthHarmonicNumber[n, d] ⟶ r

n is a natural number,
d is a natural number being the precision for internal computations,
r is a floating-point number representing the value of H_n.

■ **NthPrime (npr)**

This function returns the nth prime p_n where $p_1 = 2$ – see Section 8.5.

NthPrime[n] $\longrightarrow$ **c**

n is a natural number,
d is a natural number being the precision for internal computations,
p is a rational prime being the nth largest prime number.

■ **NthPrimeBounds (prb)**

This function gives the upper and lower bounds of Rosser and Schoenfeld [145] for the nth prime – see Section 8.5.

NthPrimeBounds[n, d] $\longrightarrow$ **L**

n is a natural number being the index for the nth prime p_n with $n > 20$,
d is a natural number being the precision for internal computations,
L is a list of two floating-point numbers, being $\{l, u\}$ where $l \le p_n \le u$.

■ **PlotCNk (pcnk)**

This function prints a plot of the values of $c(N_k)$ for $k1 \le k \le k0$ – see Chapter 5. It returns the value of $c(N_k)$ at $k = k0$.

PlotCNk[k1, k0, d] $\longrightarrow$ **r**

k1 is a natural number,
k0 is a natural number with $k1 \le k0$,
d is a natural number being the precision for internal computations,
r is a floating-point number representing the value of $c(N_{k0})$.

■ **PlotDivisibilityGraph (pdg)**

This function returns a plot of the divisibility graph of Section 10.5. The edges are directed and vertices labelled 1 through n.

PlotDivisibilityGraph[n] $\longrightarrow$ **G**

n is a natural number,
G is a labelled, directed graph.

■ **PlotEulerPhiRatio (pep)**

This function prints a plot of the graph of the proportion of positive integers n for which $\log\log n < n/\varphi(n)$ up to b – see Chapter 5. It returns the value of this proportion at b.

PlotEulerPhiRatio[b] $\longrightarrow$ **r**

b is a positive real number,

r is a floating-point number representing the number of integers which satisfy the inequality divided by b.

■ PlotRobinsInequality (pri)

This function prints a plot of n vs $\sigma(n)/n$ and $e^{\gamma}\log\log n$ on the same axes, for $a \le n \le b$ – see Chapter 7.

PlotRobinsInequality[a, b, d] ⟶ L

a is a natural number,
b is a natural number with $a < b$,
d is a natural number being the precision for internal computations,
L is a two-element list being $\{\sigma(b)/b,\ e^{\gamma}\log\log b\}$.

■ PlotVonMangoldtPsi (pvm)

This function prints a plot of the explicit formula for $\psi_0(x)$ – see Section 2.2. It returns the value of this function at the end point of its domain.

PlotVonMangoldtPsi[a, b, nx, d] ⟶ r

a is a positive real number being the start of the domain to be plotted,
b is a positive real number with $a < b$ being the end point of the domain to be plotted,
r is a floating-point number being the value of $\psi_0(b)$,
nx is a natural number being the number of terms of the form x^{ρ}/ρ used to approximate $\psi_0(x)$,
d is a natural number being the precision for internal computations.

■ PowerfulQ (pfq)

This function determines whether a natural number n is powerful.

PowerfulQ[n] ⟶ P

n is a natural number,
P is True if for each prime $p \mid n$ we have $p^2 \mid n$, otherwise False.

■ PrimePower (prp)

This function returns the maximum power to which p divides n. The number p need not be prime.

PrimePower[p, n] ⟶ e

p is a natural number with $p > 1$,
n is a natural number
e is a non-negative integer being $\nu_p(n)$.

■ PrimorialFormQ (pfq)

This function determines whether a list of integers is in primorial form, i.e. a list of primes in non-decreasing order

PrimorialFormQ[L] ⟶ P

L is a list of integers,
P is True if each element of L is prime, else False.

■ PrimorialFormToInteger (pf2i)

This function computes the integer corresponding to the primorial form.

PrimorialFormToInteger[P] ⟶ n

P is a list of primes in non-decreasing order,
n n is a natural number being the product of the primorials corresponding to the primes in P.

■ Psi

This function computes $\psi(x)$ using the expression

$$\psi(x) = \sum_{n=1}^{\infty} \theta\left(x^{1/n}\right).$$

See Section 2.2.

Psi[x, d] ⟶ r

x is a positive real number being the point at which ψ is to be evaluated,
d is a natural number being the precision for internal computations,
r is a floating-point number representing the value of $\psi(x)$.

■ Psi0

This function computes $\psi_0(x)$ using nz terms – see Section 2.2.

Psi0[x, nz, d] ⟶ r

x is a positive real number being the point at which ψ_0 is to be evaluated,
nz is a natural number being the number of terms in the von Mangoldt sum which will be used,
d is a natural number being the precision for internal computations,
r is a floating-point number representing the value of $\psi_0(x)$.

■ PsiLargeXEpsilon (ple)

This function computes

$$\epsilon(x) = \sqrt{\log x}\exp(-\sqrt{\log x/R}),$$

to precision d, for $R \geq 16$ – see Section 3.4.

PsiLargeXEpsilon[x, R, d] ⟶ r

x is a positive real number,
d is a natural number being the precision for internal computations,
R is a real number with $R \geq 16$,
r is a floating-point number representing the value of $\epsilon(x)$.

■ PsiLargeXEpsilon2 (ple2)

This function computes

$$\epsilon(x) = \sqrt{2\log x}\exp(-\sqrt{\log x/R}),$$

to precision d for $8 \leq R \leq 18$ – see Section 3.4.

PsiLargeXEpsilon2[x, R, d] ⟶ r

x is a positive real number,
R is a real number in the range $8 \leq R \leq 18$,
d is a natural number being the precision for internal computations,
r is a floating-point number representing $\epsilon(x)$.

■ PsiSmallXEpsilon (pse)

This function computes ϵ_b based on Rosser and Schoenfeld's 1962 approach, to precision d – see Section 3.2.

PsiSmallXEpsilon[b, m, d] ⟶ r

b is a positive real number such that $x = e^b$ is the point at which $\psi(x)$ is estimated,
m is a natural number being the parameter of the estimation,
d is a natural number representing the precision for internal computations,
r is a floating-point number representing the value of ϵ_b.

■ RedhefferMatrix (rem)

This function gives the divisibility matrix of Redheffer – see Section 10.4.

RedhefferMatrix[n] ⟶ LL

n is a natural number,
LL is a list of n lists, being the square divisibility matrix. It can be printed in standard matrix form using the function MatrixForm.

■ RightAbundantQ (raq)

This function determines whether n is a right abundant number up to multiples $a \cdot n$ with $1 \leq a \leq b$ – see Chapter 9.

RightAbundantQ[n, b, d] ⟶ P

n is a natural number,
b is a natural number,

b is a natural number,
P is True if $G(n) \geq G(an)$ for $1 < a \leq b$, False otherwise.

■ **RobinsInequalityQ (riq)**

This function checks the inequality $\sigma(n)/n < e^{\gamma} \log\log n$, to precision d – see Chapter 5.

RobinsInequalityQ[n,d] ⟶ P

n is a natural number,
d is a natural number being the precision for internal computations,
P is True if the inequality is satisfied, otherwise False.

■ **S**

This function computes the argument of $\zeta(\frac{1}{2} + it)$ as in Section 10.9.

S[t] ⟶ r

t is a positive real number,
d is a natural number being the precision for internal computations,
r is a floating-point number representing the value of $S(t) = N(t) - (\vartheta(t)/\pi + 1)$.

■ **SetA, SetB, SetC, SetD**

These symbols evaluate to the sets defined in Chapter 8. For example

SetB $\longrightarrow \{2, 3, 5, 6, 10, 30\}$

■ **SigmaOverN (son)**

This function evaluates $\sigma(n)/n$ as an exact rational.

SigmaOverN[n] ⟶ q

n is a natural number,
q is a rational number being the value of $\sigma(n)/n$.

■ **SuperAbundantQ (saq)**

This function determines whether n is a superabundant integer – see Chapter 6.

SuperAbundantQ[n] ⟶ c

n is a natural number,
P is True if n is superabundant, False otherwise.

■ **UnitaryDivisorSigma (uds)**

This function evaluates the sum the unitary divisors of n – see Section 7.12.

UnitaryDivisorSigma[n] ⟶ m

n is a natural number,
m is a natural number being the sum of the unitary divisors of n.

■ VerifyThetaX (vth)

This function verifies the equation $\theta(x) < x$ for $a \leq x \leq b$ to precision d – see Section 3.3

VerifyThetaX[a, b, d] ⟶ bad

a is a real number being the initial point of the x values to be tested,
b is a real number being the end point of the domain of x values to be tested,
d is a natural number being the precision for internal computations,
bad is a list of integers representing points in the domain where the inequality $\theta(x) < x$ fails at precision $\lfloor d/2 \rfloor$.

■ Xi

This function evaluates the Riemann $\xi(s)$ function of Section 10.9 at complex numbers s.

Xi[s, d] ⟶ c

s is a real or complex number,
d is a natural number being the precision for internal computations,
c is a floating-point complex number.

■ ZetaZeroCount (zzc)

This function returns a count of the zeta zeros up to height T up to height 10^6 – see Section 2.2.

ZetaZeroCount[T, d] ⟶ n

T is a positive real number not greater than 10^6,
d is a natural number being the precision for internal computations,
n is a non-negative integer, representing the number of zeros of $\zeta(s)$ on the critical line, above the x-axis, with imaginary part satisfying $0 \leq \gamma \leq T$, to precision d.

■ ZetaZeroHeight (zzh)

This function computes the ordinate of the nth highest positive imaginary part of a zeta zero, to precision d – see Section 2.2.

ZetaZeroHeight[n, d] ⟶ r

n is a positive integer being the rank of the zero with $n \leq 10^7$,
d is a natural number being the precision for internal computations,
r is a floating-point number representing the y-coordinate of the nth zero.

References

[1] A. Akbary, Z. Friggstad and R. Jurecevic, *Explicit upper bounds for* $\prod_{p \le p_{\omega(n)}} p/(p-1)$, Cont. Discrete Math. **2** (2007), 153–160.

[2] A. Akbary and Z. Friggstad, *Superabundant numbers and the Riemann hypothesis*, Amer. Math. Monthly **116** (2009), 273–275.

[3] L. Alaoglu and P. Erdős, *On highly composite and similar numbers*, Trans. Amer. Math. Soc. **56** (1944), 448–469.

[4] L. Alfors, *Complex Analysis*, 2nd edn, McGraw-Hill, 1966.

[5] T. M. Apostol, *Modular Functions and Dirichlet Series in Number Theory*, Springer, 1976.

[6] T. M. Apostol, *Introduction to Analytic Number Theory*, 2nd edn, Springer, 1990.

[7] R. J. Backlund, *Über die nullstellen der Riemannschen Zetafunktion*, Acta Math. **41** (1918), 345–375.

[8] R. Balasubramanian, S. Kanemitsu and M. Yoshimoto, *Euler products, Farey series and the Riemann hypothesis II*, Publ. Math. Debrecen **69** (2006), 1–16.

[9] W. W. Barrett, R. W. Forcade and A. D. Pollington, *On the spectral radius of a (0,1) matrix related to Mertens' function*, Linear Algebra Appl. **107** (1988), 151–159.

[10] B. C. Berndt, *Ramanujan's Notebooks*, Parts I–V, Springer, 1985–1998.

[11] M. V. Berry and J. P. Keating, *The Riemann zeros and eigenvalue asymptotics*, SIAM Rev. **41** (1999), 236–266.

[12] A. R. Booker, *Turing and the Riemann hypothesis*, Notices Amer. Math. Soc. **53** (2006), 1208–1211.

[13] P. Borwein, S. Choi, B. Rooney and A. Weirathmueller, *The Riemann Hypothesis: A Resource for the Afficionado and Virtuoso Alike*, CMS Books in Mathematics, Springer, 2008.

[14] R. P. Brent, *On the zeros of the Riemann zeta function in the critical strip*, Math. Comp. **33** (1979), 1361–1372.

[15] R. P. Brent, J. van de Lune, H. J. J. te Riele and D. T. Winter, *On the zeros of the Riemann zeta function in the critical strip II*, Math. Comp. **39** (1982), 681–688.

[16] K. Briggs, *Abundant numbers and the Riemann hypothesis*, J. Exp. Math. **15** (2006), 251–256.

[17] K. A. Broughan, *Structure of sectors of zeros of entire flows*, Topology Proc. **27** (2002), 1–16.

[18] K. A. Broughan, *Vanishing of the integral of the Hurwitz zeta function*, Bull. Austral. Math. Soc. **65** (2002), 121–127.

[19] K. A. Broughan, *Holomorphic flows on simply connected subsets have no limit cycles*, Meccanica **38** (2003), 699–709.

[20] K. A. Broughan and A. R. Barnett, *The holomorphic flow of the Riemann zeta function*, Math. Comp. **73** (2004), 987–1004.

[21] K. A. Broughan, *Holomorphic flow of the Riemann xi function*, Nonlinearity **18** (2005), 1269–1294.

[22] K. A. Broughan and A. R. Barnett, *Linear law for the logarithms of the Riemann periods at simple critical zeros*, Math. Comp. **75** (2006), 891–902.

[23] K. A. Broughan and A. R. Barnett, *Corrigendum to "The holomorphic flow of the Riemann zeta function"*, Math. Comp. **76** (2007), 2249–2250.

[24] K. A. Broughan, *Extension of the Riemann ξ-function's logarithmic derivative positivity region to near the critical strip*, Can. Math. Bull. **52** (2009), 186–194.

[25] K. A. Broughan and A. R. Barnett, *Gram lines and the average of the real part of the Riemann zeta function*, Math. Comp. **81** (2012) 1669–1679.

[26] K. A. Broughan, Website for phase portraits of the zeros of $\zeta(s)$ and $\xi(s)$. www.math.waikato.ac.nz/~kab.

[27] K. A. Broughan, J.-M. De Koninck, I. Katai and F. Luca, *On integers for which the sum of divisors is the square of the squarefree core*, J. Integer Seq. **15** (2012), art. 12.7.5.

[28] K. A. Broughan and D. Delbourgo, *On the ratio of the sum of divisors and Euler's totient function I*, J. Integer Seq. **16** (2013), art. 13.8.8 (17pp.).

[29] K. A. Broughan, K. Ford and F. Luca, *On square values of the product of the Euler totient function and the sum of divisors function*, Colloq. Math. **130** (2013), 127–137.

[30] K. A. Broughan and Q. Zhou, *On the ratio of the sum of divisors and Euler's totient function II*, J. Integer Seq. **17** (2014), art. 14.9.2 (22pp.).

[31] K. A. Broughan and T. S. Trudgian, *Robin's inequality for 11-free integers*, Integers **15** (2015), #A12 (5pp.).

[32] K. A. Broughan, *Equivalents of the Riemann Hypothesis II: Analytic Equivalences*, Cambridge University Press, 2017.

[33] J. Büthe, J. Franke, A. Jost and T. Kleinjung, *A practical analytic method for calculating $\pi(x)$*, Preprint.

[34] G. Caveney, J.-L. Nicolas and J. Sondow, *Robin's theorem, primes, and a new elementary reformulation of the Riemann hypothesis*, Integers **11** (2011), #A33 (10pp.).

[35] G. Caveney, J.-L. Nicolas and J. Sondow, *On SA, CA and GA numbers*, Ramanujan J. **29** (2012), 359–384.

[36] Y. Cheng, *An explicit upper bound for the Riemann zeta-function near the line $\sigma = 1$*, Rocky Mountain J. Math. **29** (1999), 115–140.

[37] Y. Cheng and S. W. Graham, *Explicit estimates for the Riemann zeta function*, Rocky Mountain J. Math. **34** (2004), 1261–1280.

[38] Y. Choie, N. Lichiardopol, P. Moree and P. Solé, *On Robin's criterion for the Riemann hypothesis*, J. Théor. Nombres Bordeaux **19** (2007), 357–372.

[39] S. Chowla, *On abundant numbers*, J. Indian Math. Soc. (2) **1** (1934), 41–44.

[40] J. B. Conrey, *More than two fifths of the zeros of the Riemann zeta function are on the critical line*, J. reine angew. Math. **399** (1989), 1–16.

[41] J. B. Conrey, *L-functions and random matrices*, in *Mathematics Unlimited – 2001 and Beyond*, pp. 331–352, Springer, 2001.

[42] H. Davenport, *Über numeri abundantes*, Sitzungsber. Preuss. Akad. Wiss. Berlin **27** (1933), 830–837.

[43] H. Davenport, *Multiplicative Number Theory*, 3rd edn, Springer, 2000.

[44] M. Davis, Y. Matijasevič and J. Robinson, *Hilbert's tenth problem: diophantine equations: positive aspects of a negative solution*, in *Mathematical Developments*

Arising from Hilbert Problems, Proc. Symp. Pure Math., XXVIII, De Kalb, IL, 1974, pp. 323–378. American Mathematical Society, 1976.

[45] A. Derbal, *Grandes valeurs de la fonction* $\sigma(n)/\sigma^*(n)$, C. R. Math. Acad. Sci. Paris **364** (2008), 125–128.

[46] L. E. Dickson, *History of the Theory of Numbers*, vol. I, Dover, 2005.

[47] R. D. Dixon and L. Schoenfeld, *The size of the Riemann zeta-function at places symmetric with respect to the point* $1/2$, Duke Math. J. **33** (1966), 291–292.

[48] P. Dusart, *Inégalités explicites pour* $\psi(X)$, $\theta(X)$, $\pi(X)$ *et les nombres premiers*, C. R. Math. Acad. Sci. Soc. R. Can. **21** (1999), 53–59.

[49] P. Dusart, *The kth prime is greater than* $k(\log k + \log\log k - 1)$ *for* $k \geq 2$, Math. Comp. **68** (1999), 411–415.

[50] P. Dusart, *Estimates of some functions over primes without RH*, Preprint, arXiv:1002.0442v1, 2010.

[51] H. M. Edwards, *Riemann's Zeta Function*, Academic Press, 1974; reprinted by Dover, 2001.

[52] P. Erdős, *On the density of abundant numbers*, J. London Math. Soc. **9** (1934), 278–282.

[53] P. Erdős, *On the normal number of prime factors of* $p - 1$ *and some related problems concerning Euler's* ϕ*-function*, Quart. J. Math. **6** (1935), 205–213.

[54] P. Erdős, *Some remarks on Euler's* φ *function*, Acta Arith. **4** (1958), 10–19.

[55] P. Erdős and P. Turan, *On some problems of a statistical group theory I*, Z. Wahrscheinlichkeitstheorie verw. Geb. **4** (1965), 175–186.

[56] P. Erdős and J.-L. Nicolas, *Repartition des nombres superabondants*, Bull. Soc. Math. France **103** (1975), 65–90.

[57] L. Faber and H. Kadiri, *New bounds for* $\psi(x)$, Math. Comp. **84** (2015), 1339–1357.

[58] K. Ford, *The distribution of totients*, Ramanujan J. **2** (1998), 67–151 (Paul Erdős memorial issue).

[59] K. Ford, *Zero-free regions for the Riemann zeta function*, in *Number Theory for the Millennium II*, Urbana, IL, 2000, pp. 25–56, A. K. Peters, 2002.

[60] K. Ford, *Vinogradov's integral and bounds for the Riemann zeta function*, Proc. London Math. Soc. (3) **85** (2002), 565–633.

[61] K. Ford, F. Luca and C. Pomerance, *Common values of the arithmetic functions* ϕ *and* σ, Bull. London Math. Soc. **42** (2010), 478–488.

[62] K. Ford and P. Pollack, *On common values of* $\phi(n)$ *and* $\sigma(n)$ *I*, Acta Math. Hungar. **133** (2011), 251–271.

[63] K. Ford and P. Pollack, *On common values of* $\phi(n)$ *and* $\sigma(n)$ *II*, Algebra Number Theory **6** (2012), 1669–1696.

[64] J. Franel and E. Landau, *Les suites de Farey et le probléme des nombres premiers*, Göttinger Nachr. (1924), 198–206.

[65] J. A. Gallian, *Contemporary Abstract Algebra*, 5th edn, Houghton Mifflin, 2002.

[66] R. Garunkstis, *On a positivity property of the Riemann* ξ*-function*, Lithuanian Math. J. **43** (2002), 140–145.

[67] B. R. Gillespie, *Extending Redheffer's matrix to arbitary arithmetic functions*, Bachelor with Honors Thesis, Pennsylvania State University, 2011.

[68] X. Gourdon, *The* 10^{13} *first zeros of the Riemann zeta function, and zeros computation at very large height*, Preprint, 2004.

[69] T. H. Grönwall, *Some asymptotic expressions in the theory of numbers*, Trans. Amer. Math. Soc. **14** (1913), 113–122.

[70] E. Grosswald, *Oscillation theorems of arithmetical functions*, Trans. Amer. Math. Soc. **126** (1967), 1–28.

[71] J. L. Hafner, *New omega theorems for two classical lattice point problems*, Invent. Math. **63** (1981), 181–186.

[72] R. R. Hall and W. K. Hayman, *Hyperbolic distance and distinct zeros of the Riemann zeta-function in small regions*, J. reine angew. Math. **526** (2000), 35–59.

[73] G. H. Hardy, *The average order of the arithmetical functions $P(x)$ and $\Delta(x)$*, Proc. London Math. Soc. **15** (1916), 192–213.

[74] G. H. Hardy and J. E. Littlewood *The zeros of Riemann's zeta function on the real line*, Math. Z. **10** (1921), 283–317.

[75] G. H. Hardy and J. M. Wright, *An Introduction to the Theory of Numbers*, 6th edn, Oxford University Press, 2008.

[76] D. R. Heath-Brown, *Simple zeros of the Riemann zeta function on the critical line*, Bull. London Math. Soc. **11** (1979), 17–18.

[77] D. R. Heath-Brown, *Zero-free regions for Dirichlet L-functions, and the least prime in an arithmetic progression*, Proc. London Math. Soc. **64** (1992), 265–338.

[78] D. A. Hejhal and A. M. Odlyzko, *Alan Turing and the Riemann zeta function*, in *Alan Turing – His Work and Impact*, eds S. B. Cooper and J. van Leeuwen, Elsevier, 2012.

[79] G. A. Hiary and A. M. Odlyzko, *The zeta function on the critical line: numerical evidence for moments and random matrix theory models*, Math. Comp. **81** (2012), 1723–1752.

[80] A. E. Ingham, *The Distribution of Prime Numbers*, Cambridge University Press, 1932.

[81] A. Ivić, *Two inequalities for the sum of divisors functions*, Univ. Novom Sadu, Zb. Rad. Prirod.-Mat. Fak. **7** (1977), 17–22.

[82] A. Ivić, *Riemann Zeta Function: Theory and Applications*, Dover, 2003.

[83] W.-J. Jang and S-H. Kwon, *A note on Kadiri's explicit zero free region for the Riemann zeta function*, J. Korean Math. Soc. **51** (2014), 1291–1304.

[84] H. Kadiri, *Une région explicite sans zéros pour la fonction ζ de Riemann*, Acta Arith. **117** (2005), 303–339.

[85] S. Kanemitsu and M. Yoshimoto, *Farey series and the Riemann hypothesis*, Acta Arith. **75** (1996), 351–374.

[86] S. Kanemitsu and M. Yoshimoto, *Farey series and the Riemann hypothesis III*, Ramanujan J. **1** (1997), 363–378.

[87] S. Kanemitsu and M. Yoshimoto, *Euler products, Farey series and the Riemann hypothesis*, Pub. Math. Debrecen **56** (2000), 431–449.

[88] A. A. Karatsuba and S. M. Voronin, *Riemann zeta function* (transl. from Russian by N. Koblitz), De Gruyter, 1994.

[89] N. M. Katz and P. Sarnak, *Zeroes of zeta functions and symmetry*, Bull. Amer. Math. Soc. **36** (1999), 1–26.

[90] N. M. Katz and P. Sarnak, *Random matrices, Frobenius eigenvalues, and monodromy*, AMS Colloquium Publications, 45. American Mathematical Society, 1999.

[91] J. P. Keating and N. C. Snaith, *Random matrix theory and $\zeta(\frac{1}{2} + it)$*, Comm. Math. Phys. **214** (2000), 57–89.

[92] D. E. Knuth, *The Art of Computer Programming I: Fundamental Algorithms*, 3rd edn, Addison-Wesley, 1997.

[93] H. von Koch, *Sur la distribution des nombres premiers*, Acta Math. **24** (1901), 159–182.

[94] H. von Koch, *Über die Riemannsche Primzahlfunction*, Math. Ann. **55** (1902), 440–464.

[95] J.-M. De Koninck and F. Luca, *Analytic Number Theory: Exploring the Anatomy of Integers*, American Mathematical Society, 2012.

[96] N. M. Korobov, *Estimates of trigonometric sums and their applications*, Uspehi Mat. Nauk **13** (1958), 185–192.

[97] T. Kotnik, *The prime counting function and its analytic approximations*, Adv. Comp. Math. **29** (2008), 55–70.

[98] J. C. Lagarias, *On a positivity property of the Riemann ξ-function*, Acta Arith. **99** (1999) 217–213.

[99] J. C. Lagarias, *Errata: On a positivity property of the Riemann ξ-function*, Acta Arith. **99** (1999) 765.

[100] J. C. Lagarias, *An elementary problem equivalent to the Riemann hypothesis*, Amer. Math. Monthly **109** (2002), 534–543.

[101] J. C. Lagarias, *Problem 10949*, Amer. Math. Monthly **109** (2002), 569.

[102] J. C. Lagarias and W. Janous, *A generous bound for divisor sums: problem 10949*, Amer. Math. Monthly **111** (2004), 264–265.

[103] E. Landau, *Über die Maximalordnung der Permutationen gegebenen Grades*, Arch. Math. Phys. Ser. 3 **5** (1903).

[104] E. Landau, *Über einen satz von Tschebyschef*, Math. Ann. **61** (1905), 527–550.

[105] E. Landau, *Handbuch der lehre von der Verteilung der Primzahlen*, 2nd edn, vols 1 and 2, Chelsea, 1953.

[106] S. Lang, *Introduction to Transcendental Numbers*, Addison-Wesley, 1966.

[107] N. Levinson and H. L. Montgomery, *Zeros of the derivatives of the Riemann zeta-function*, Acta Math. **133** (1974), 49–65.

[108] J. E. Littlewood, *Quelques conséquences de l'hypothése que la fonction $\zeta(s)$ n'a pas de zéros dans le demi-plan $\Re s > \frac{1}{2}$*, C. R. Acad. Sci. Paris Sér. I Math. **158** (1912), 263–266.

[109] J. E. Littlewood, *Sur la distribution des nombres premiers*, C. R. Acad. Sci. Paris Sér. I Math. **158** (1914), 1869–1872.

[110] E. R. Lorch, *Spectral Theory*, Oxford University Press, 1962.

[111] J. van de Lune, H. J. J. te Riele and D. T. Winter, *On the zeros of the Riemann zeta function in the critical strip IV*, Math. Comp. **46** (1986), 667–681.

[112] J.-P. Massias, J.-L. Nicolas and G. Robin, *Evaluation asymptotique de l'ordre maximum d'un element du groupe symétrique*, Acta Arith. **50** (1988), 221–242.

[113] J.-P. Massias, J.-L. Nicolas and G. Robin, *Effective bounds for the maximal order of an element in the symmetric group*, Math. Comp. **53** (1989), 665–678.

[114] J.-P. Massias and G. Robin, *Bornes effectives pour certaines fonctions concernant les nombres premiers*, J. Théor. Nombres Bordeaux **8** (1996), 215–242.

[115] F. Mertens, *Über einize asymptotische Gesetse der Zahlentheorie*, J. reine angew. Math. **77** (1874), 46–62.

[116] M. Mikolás, *Farey series and their connection with the prime number problem I*, Acta Univ. Szeged. Sect. Sci. Math. **13** (1949), 93–117.

[117] M. Mikolás, *Farey series and their connection with the prime number problem II*, Acta Sci. Math. Szeged **14** (1951), 5–21.

[118] W. Miller, *The maximum order of an element of a finite symmetric group*, Amer. Math. Monthly **94** (1987), 497–506.

[119] H. L. Montgomery, *The pair correlation of zeros of the zeta function*, in *Analytic Number Theory*, Proc. Symp. Pure Math. XXIV, pp. 181–193, American Mathematical Society, 1973.

[120] F. Morain, *Tables sur la fonction $g(n)$*, Département de Mathématiques, Université de Limoges, 1988.

[121] M. J. Mossinghoff and T. S. Trudgian, *Non-negative trigonometric polynomials and a zero-free region for the Riemann zeta-function*, Preprint, arXiv:1410.3926v1, 2014.

[122] S. Nazardonyavi and S. Yakubovich, *Superabundant numbers, their subsequences and the Riemann hypothesis*, Preprint, arXiv:1211.2147v3, 2013.

[123] S. Nazardonyavi and S. Yakubovich, *Extremely abundant numbers and the Riemann hypothesis*, J. Integer Seq. **17** (2014), art. 14.2.8 (23pp.).

[124] M. B. Nathanson, *Elementary Methods in Number Theory*, Springer, 2000.

[125] J.-L. Nicolas, *Ordre maximal d'un element du groupe des permutations et 'highly composite numbers'*, Bull. Soc. Math. France **97** (1969), 129–191.

[126] J.-L. Nicolas, *Petites valeurs de la fonction d'Euler*, J. Number Theory **17** (1983), 375–388.

[127] J.-L. Nicolas, *Small values of the Euler function and the Riemann hypothesis*, Acta Arith. **155** (2012), 311–321.

[128] J.-L. Nicolas, *Ramanujan, Robin, highly composite numbers, and the Riemann hypothesis*, Preprint, arXiv:1211.6944v4, 2013.

[129] A. M. Odlyzko and H. J. J. te Riele, *Disproof of the Mertens' conjecture*, J. reine angew. Math. **357** (1985), 138–160.

[130] A. M. Odlyzko, *On the distribution of spacings between zeros of the zeta function*, Math. Comp. **48** (1987), 273–308.

[131] S. J. Patterson, *An Introduction to the Theory of the Riemann Zeta-Function*, Cambridge Studies in Advanced Mathematics 14, Cambridge University Press, 1988.

[132] D. J. Platt, *Computing degree 1 L-functions rigorously*, Ph.D. Thesis, University of Bristol, 2011.

[133] D. J. Platt, *Computing $\pi(x)$ analytically*, Math. Comp. **84** (2015), 1521–1535.

[134] D. J. Platt and S. A. Trudgian, *An improved explicit bound on $|\zeta(\frac{1}{2} + it)|$*, J. Number Theory **147** (2015), 842–851.

[135] D. J. Platt and S. A. Trudgian, *On the first sign change of $\theta(x) - x$*, Math. Comp. **85** (2016), 1539–1547.

[136] S. Ramanujan, *Highly composite numbers (Part 1)*, Proc. London Math. Soc. **14** (1915), 347–409.

[137] S. Ramanujan, annotated by J.-L. Nicolas and G. Robin, *Highly composite numbers (Part 2)*, Ramanujan J. **1** (1997), 119–153.

[138] R. Redheffer, *Eine explizit lösbare Optimierungsaufgabe*, in *Numerische Methoden bei Optimierungsaufgaben*, Band 3 (Tagung, Math. Forschungsinst., Oberwolfach, 1976), Int. Ser. Numer. Math., Vol. 36, pp. 213–216, Birkhäuser, 1977.

[139] H.-E. Richert, *Zur Abschätzung der Riemannschen Zetafunktion in der Nähe der Vertikalen $\sigma = 1$*, Math. Ann. **169** (1967), 97–101.

[140] B. Riemann, *Gesammelte Werke*, Teubner, 1893; reprinted by Dover, 1953.

[141] G. Robin, *Estimation de la fonction de Tchebychef θ sur le k-ième nombre premier et grandes valeurs de la fonction $\omega(n)$ nombre de diviseurs premiers de n*, Acta Arith. **42** (1983), 367–389.

[142] G. Robin, *Grandes valeurs de la fonction sommes des diviseurs et hypotheses de Riemann*, J. Math. Pures Appl. **63** (1984), 187–213.

[143] J. B. Rosser, *The n-th prime is greater than $n \log n$*, Proc. London Math. Soc. **45** (1938), 21–44.

[144] J. B. Rosser, *Explicit bounds for some functions of prime numbers*, Amer. J. Math. **63** (1941), 211–232.

[145] J. B. Rosser and L. Schoenfeld, *Approximate formulas for some functions of prime numbers*, Illinois J. Math. **6** (1962), 64–94.

[146] J. B. Rosser and L. Schoenfeld, *Sharper bounds for the Chebyshev functions $\theta(x)$ and $\psi(x)$*, Math. Comp. **29** (1975), 243–269.

[147] Z. Rudnick and P. Sarnak, *Zeros of principal L-functions and random matrix theory. A celebration of John F. Nash, Jr.*, Duke Math. J. **81** (1996), 269–322.

[148] K. Sabbagh, *The Riemann Hypothesis*, Farrar, Straus and Giroux, 2003.

[149] B. Sagan, *The Symmetric Group: Representations, Combinatorial Algorithms, and Symmetric Functions*, Springer, 2001.

[150] J. Sándor, D. S. Mitrinović and B. Crstici, *Handbook of Number Theory I*, Springer, 2006.

[151] M. du Sautoy, *The Music of the Primes*, HarperCollins, 2003.

[152] E. Schmidt *Über die Anzahl der Primzahlen unter gegebener Grenze*, Math. Ann. **57** (1903), 195–204.

[153] L. Schoenfeld, *Sharper bounds for the Chebyshev functions $\vartheta(x)$ and $\psi(x)$ II*, Math. Comput. **30** (1976), 337–360.

[154] L. Schoenfeld, *Corrigendum: Sharper bounds for the Chebyshev functions $\vartheta(x)$ and $\psi(x)$ II*, Math. Comput. 30 (1976), 900.

[155] D. Schumayer and D. A. W. Hutchinson, *Physics of the Riemann hypothesis*, Rev. Mod. Phys. **83** (2011), 307–330.

[156] S. M. Shah, *An inequality for the arithmetic function $g(x)$*, J. Indian Math. Soc. **3** (1939), 316–318.

[157] S. Skews, *On the difference $\pi(x) - \mathrm{li}(x)$*, J. London Math. Soc. **8** (1933), 277–283.

[158] N. J. A. Sloan, *Online Encyclopedia of Integer Sequences* (OEIS), https://oeis.org.

[159] P. Solé and M. Planat, *The Robin inequality for 7-free integers*, Integers **11** (2011), #A65 (8pp.).

[160] K. Soundararajan, *Moments of the Riemann zeta function*, Annals Math. **170** (2009), 981–993.

[161] A. Speiser, *Geometrisches zur Riemannschen Zetafunktion*, Math. Ann. **110** (1934), 514–521.

[162] R. Spira, *An inequality for the Riemann zeta function*, Duke Math. J. **32** (1965), 247–250.

[163] R. Spira, *Zeros of $\zeta'(s)$ and the Riemann hypothesis*, Illinois J. Math. **17** (1973), 147–152.

[164] A. I. Suriajaya, *On the zeros of the k-th derivative of the Riemann zeta function under the Riemann hypothesis*, Funct. Approx. Comment. Math. **53** (2015), 69–95.

[165] S. M. Szalay, *On the maximal order in S_n and S_n^**, Acta Arith. **37** (1980), 321–331.

[166] G. Tenenbaum, *Introduction to Analytic and Probabilistic Number Theory*, Cambridge University Press, 1995.

[167] E. C. Titchmarsh and D. R. Heath-Brown, *The Theory of the Riemann Zeta-Function*, 2nd edn, Oxford University Press, 1986.

[168] T. S. Trudgian, *Improvements to Turing's method*, Math. Comp. **80** (2011), 2259–2279.

[169] T. S. Trudgian, *An improved upper bound for the argument of the Riemann zeta-function on the critical line II*, J. Number Theory **134** (2014), 280–292.

[170] T. S. Trudgian, *The sum of the unitary divisor function*, Publ. Inst. Math. (Belgrad) (N.S.) **97** (2015), 175–180.

[171] T. S. Trudgian, *A short extension of two of Spira's results*, J. Math. Inequalities **9** (2015), 795–798.

[172] T. S. Trudgian, *Improvements to Turing's method II*, Rocky Mountain J. Math. **46** (2016), 325–332.

[173] T. S. Trudgian, *Updating the error term in the prime number theorem*, Ramanujan J. **39** (2016), 225–234.

[174] A. M. Turing, *A method for the calculation of the zeta-function*, Proc. London Math. Soc. **48** (1943), 180–197.

[175] A. M. Turing, *Some calculations of the Riemann zeta-function*, Proc. London Math. Soc. **3** (1953), 99–117.

[176] I. M. Vinogradov, *A new estimate for* $\zeta(1 + it)$, Izv. Akad. Nauk SSSR, Ser. Mat. **22** (1958), 161–164.

[177] A. Walfisz, *Weylsche Exponentialsummen in der neueren Zahlentheorie*, Mathematische Forschungsberichte, XV, VEB Deutscher Verlag der Wissenschaften, 1963.

[178] S. Wedeniwski, *Section 1.1 History*, ZetaGrad website, www.zetagrid.net/zeta/math/zeta.result.100billion.zeros.html.

[179] A. Weil, *Sur les "formules explicites" de la théorie des nombres premiers*, Comm. Sém. Math. Univ. Lund (1952), 252–265.

[180] H. S. Wilf, *The Redheffer matrix of a partially ordered set*, Electron. J. Combin. **11** (2004/2006), Research Paper 10 (5pp.).

[181] M. Yoshimoto, *Farey series and the Riemann hypothesis II*, Acta Math. Hungar. **78** (1998), 287–304.

[182] D. Zagier, *The first 50 million prime numbers*, Math. Intelligencer (1977), 7–19.

[183] S. Zegowitz, *On the positive region of* $\pi(x) - \mathrm{li}(x)$. Master's Thesis, University of Manchester, 2010.

Index